CAMBRIDGE LIBRARY COLLECTION

Books of enduring scholarly value

Mathematical Sciences

From its pre-historic roots in simple counting to the algorithms powering modern desktop computers, from the genius of Archimedes to the genius of Einstein, advances in mathematical understanding and numerical techniques have been directly responsible for creating the modern world as we know it. This series will provide a library of the most influential publications and writers on mathematics in its broadest sense. As such, it will show not only the deep roots from which modern science and technology have grown, but also the astonishing breadth of application of mathematical techniques in the humanities and social sciences, and in everyday life.

Statics

Sir Horace Lamb (1849–1934) the British mathematician, wrote a number of influential works in classical physics. A pupil of Stokes and Clerk Maxwell, he taught for ten years as the first professor of mathematics at the University of Adelaide before returning to Britain to take up the post of professor of physics at the Victoria University of Manchester (where he had first studied mathematics at Owens College). As a teacher and writer his stated aim was clarity: 'somehow to make these dry bones live'. His Statics was first published in 1912, and the third edition, offered here, in 1928. It was intended as a textbook for students with some knowledge of mechanics, and deals mainly with two-dimensional problems: examples are provided at the end of each section.

Cambridge University Press has long been a pioneer in the reissuing of out-of-print titles from its own backlist, producing digital reprints of books that are still sought after by scholars and students but could not be reprinted economically using traditional technology. The Cambridge Library Collection extends this activity to a wider range of books which are still of importance to researchers and professionals, either for the source material they contain, or as landmarks in the history of their academic discipline.

Drawing from the world-renowned collections in the Cambridge University Library, and guided by the advice of experts in each subject area, Cambridge University Press is using state-of-the-art scanning machines in its own Printing House to capture the content of each book selected for inclusion. The files are processed to give a consistently clear, crisp image, and the books finished to the high quality standard for which the Press is recognised around the world. The latest print-on-demand technology ensures that the books will remain available indefinitely, and that orders for single or multiple copies can quickly be supplied.

The Cambridge Library Collection will bring back to life books of enduring scholarly value across a wide range of disciplines in the humanities and social sciences and in science and technology.

Statics

Including Hydrostatics and the Elements of the Theory of Elasticity

HORACE LAMB

CAMBRIDGE UNIVERSITY PRESS

Cambridge New York Melbourne Madrid Cape Town Singapore São Paolo Delhi

Published in the United States of America by Cambridge University Press, New York

www.cambridge.org
Information on this title: www.cambridge.org/9781108005319

This edition first published 1928
This digitally printed version 2009

ISBN 978-1-108-00531-9

STATICS

INCLUDING HYDROSTATICS AND THE ELEMENTS OF THE THEORY OF ELASTICITY

CAMBRIDGE
UNIVERSITY PRESS
LONDON : Fetter Lane

NEW YORK
The Macmillan Co.
BOMBAY, CALCUTTA and MADRAS
Macmillan and Co., Ltd.
TORONTO
The Macmillan Co. of Canada, Ltd.
TOKYO
Maruzen-Kabushiki-Kaisha

STATICS

INCLUDING HYDROSTATICS AND THE ELEMENTS OF THE THEORY OF ELASTICITY

BY

HORACE LAMB, M.A., LL.D., Sc.D., F.R.S.

HONORARY FELLOW OF TRINITY COLLEGE, CAMBRIDGE

LATELY PROFESSOR OF MATHEMATICS IN THE VICTORIA UNIVERSITY OF MANCHESTER

THIRD EDITION

Cambridge:

at the University Press

1928

First Edition 1912
Reprinted 1916, 1921
Second Edition 1924
Third Edition 1928

PRINTED IN GREAT BRITAIN

PREFACE TO THE THIRD EDITION

THIS book had its origin in a course of lectures given for a number of years in the Manchester University, but has received various additions and developments. It is intended for students who have already some knowledge of elementary Mechanics, and who have arrived at the stage at which they may usefully begin to apply the methods of the Calculus. It deals mainly with two-dimensional problems, but occasionally, where the extension to three dimensions is easy, theorems are stated and proved in their more general form.

The present volume differs from many academical manuals in the prominence given to geometrical methods, and in particular to those of Graphical Statics. These methods, especially in relation to the theory of frames, have imported a new interest into a subject which was in danger of becoming fossilized. I have not attempted, however, to enter into details which are best learned from technical treatises, or in engineering practice.

It seemed natural and convenient to treat of Hydrostatics, to a similar degree of development, and I have also, for reasons stated at the beginning of Chap. XV, included the rudiments of the theory of Elasticity. A number of important problems are here discussed by quite elementary methods.

The chapter on Mass-Systems has an interest in the present subject, but I have also had in view, here and elsewhere, the requirements of the companion volume on Dynamics.

I am indebted for some valuable suggestions to Prof. A. Föppl's excellent *Vorlesungen über technische Mechanik*. I have also derived some interesting references from the *Encyclopädie der mathematischen Wissenschaften*.

The examples for practice have been selected (or devised) with some care, and it is hoped that most of them will serve as genuine illustrations of statical principles rather than as exercises in

Algebra or Trigonometry. Problems of a mainly mathematical character have been excluded, unless there appeared to be some special interest or elegance in the results.

The present edition has been carefully revised; a few sections have been re-written, and some rather serious oversights have been corrected. I shall be much obliged to any readers who will call my attention to such errors or omissions as have still escaped detection.

H. L.

CAMBRIDGE, 1928

CONTENTS

INTRODUCTION.

THEORY OF VECTORS.

CHAPTER I.

STATICS OF A PARTICLE.

CHAPTER II.

PLANE KINEMATICS OF A RIGID BODY.

CHAPTER III.

PLANE STATICS.

CHAPTER IV.

GRAPHICAL STATICS.

CHAPTER V.

THEORY OF FRAMES.

CHAPTER VI.

WORK AND ENERGY.

CHAPTER VII.

ANALYTICAL STATICS.

CHAPTER VIII.

THEORY OF MASS-SYSTEMS.

CHAPTER IX.

FLEXIBLE CHAINS.

CHAPTER X.

LAWS OF FLUID PRESSURE.

CHAPTER XI.

EQUILIBRIUM OF FLOATING BODIES.

CHAPTER XII.

GENERAL CONDITIONS OF EQUILIBRIUM OF A FLUID.

CHAPTER XIII.

EQUILIBRIUM OF GASEOUS FLUIDS.

CHAPTER XIV.

CAPILLARITY.

CHAPTER XV.

STRAINS AND STRESSES.

CHAPTER XVI.

EXTENSION OF BARS.

CHAPTER XVII.

FLEXURE AND TORSION OF BARS.

CHAPTER XVIII.

STRESSES IN CYLINDRICAL AND SPHERICAL SHELLS.

INTRODUCTION

THEORY OF VECTORS

1. Definition of a Vector.

The statement and the proof of many theorems in Mechanics are so much simplified by the terminology of the Theory of Vectors that it is worth while to begin with a brief account of the more elementary notions of this subject. The student who is already conversant with it may at once pass on.

The quantities with which we deal in mathematical physics may be classified into 'vectors' and 'scalars,' according as they do or do not involve the idea of direction.

A quantity which is completely specified by a numerical symbol, positive or negative, and has no intrinsic reference to direction in space, is called a 'scalar,' since it is defined by its position on the proper scale of measurement. Thus such quantities as mass, length, time, energy, hydrostatic pressure-intensity, belong to this category.

A 'vector' quantity, on the other hand, involves essentially the idea of direction as well as magnitude. To take a simple geometrical example, the position of a point B relative to another point A is specified by means of a straight line drawn from A to B. It may equally well be specified by any equal and parallel straight line drawn in the same sense from (say) C to D, since the position of D relative to C is the same as that of B relative to A. A straight line regarded in this way as having a definite magnitude and direction, but no definite location in space, is called a 'vector'*. Occasionally,

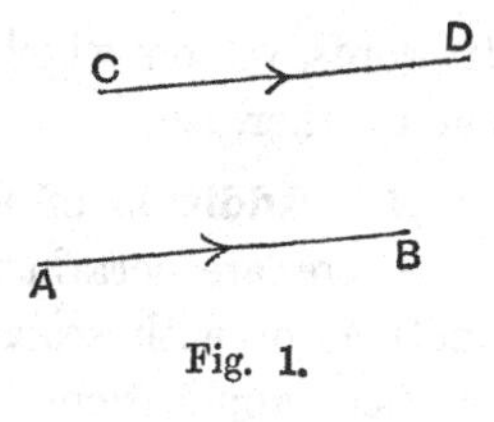

Fig. 1.

* Or 'carrier,' since (in the above instance) it indicates the operation by which a point is transferred from A to B. The terms 'vector' and 'scalar' are due to Sir W. R. Hamilton (1853).

when we wish to lay stress on the latter particular, it is called a 'free' vector. For example, if a rigid body be moved from one position to another without rotation, so that the lines joining the initial and final positions of the various points of the body are all equal and parallel, the displacement of the body as a whole is completely specified by a free vector, which may be any one of these lines.

As regards notation, a vector may be specified by means of the letters denoting the terminal points of a representative line, written in the proper order*. It is sometimes convenient, however, to denote vectors by single symbols. For this purpose what is called 'clarendon' type (**A**, **a**, ...) is often employed, whilst scalar quantities are denoted as usual by italic symbols.

For reasons which have already been indicated, two vectors **P** and **Q** which, like AB and CD in Fig. 1, have the same magnitude and direction, are regarded as equal, or rather identical, and the equation

$$\mathbf{P} = \mathbf{Q}$$

is used to express this complete identity. We have here the definition of the sign '=' as used in the present connection; and it is to be particularly noticed that there can be no question of equality between vectors whose directions are different. Since straight lines which are equal and parallel to the same straight line are equal and parallel to one another, it follows that if

$$\mathbf{P} = \mathbf{R} \text{ and } \mathbf{Q} = \mathbf{R},$$

then

$$\mathbf{P} = \mathbf{Q}.$$

In words, vectors which are equal to the same vector are equal to one another.

2. Addition of Vectors.

There are certain modes of combination of vectors with one another, or with scalars, which have important geometrical and physical applications. As regards combinations of two or more vectors, the only kind which we need consider at present is that suggested by composition of displacements of pure translation of a

* Thus AB is to be distinguished from BA. In this book we shall use Roman letters when denoting a vector in this way, the italics *AB* or *BA* being used when the *length* only of the line is referred to. In manuscript work a bar may be drawn over two letters which are meant to denote a vector.

rigid body. Thus if such a body receives in succession two translations represented by AB and BC, the final result is equivalent to a translation represented by AC. It is therefore natural to speak of AC as in a sense the 'geometric sum,' or simply the 'sum,' of the vectors AB and BC, and to write

$$AB + BC = AC.$$

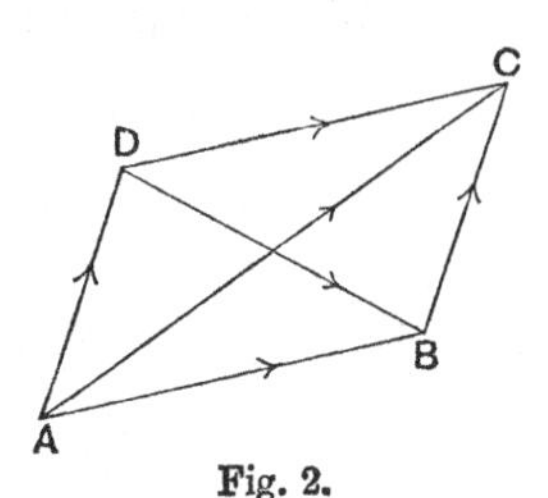

Fig. 2.

Hence to construct the sum of any two vectors $\mathbf{P}$, $\mathbf{Q}$, we draw a line AB to represent $\mathbf{P}$, and then BC to represent $\mathbf{Q}$; the sum $\mathbf{P} + \mathbf{Q}$ is then represented by AC. This definition of vector addition is of course conventional and arbitrary, and it remains to be seen whether the process is subject to the same rules as those which govern ordinary algebraical addition.

If we complete the parallelogram *ABCD*, as in Fig. 2, we have in virtue of our conventions,

$$DC = AB = \mathbf{P}, \quad AD = BC = \mathbf{Q},$$

and therefore
$$\mathbf{Q} + \mathbf{P} = AD + DC = AC,$$

or
$$\mathbf{Q} + \mathbf{P} = \mathbf{P} + \mathbf{Q}. \qquad \text{(1)}$$

This is the 'commutative law' of addition; it is not self-evident, but depends, as we see, on the Euclidean theory of parallels.

When we wish to indicate that a particular vector which occurs in a formula arises as the sum of two vectors $\mathbf{P}$ and $\mathbf{Q}$, we enclose the sum in brackets, as $(\mathbf{P} + \mathbf{Q})$. There is accordingly a distinction of meaning in the first instance between, say, $(\mathbf{P} + \mathbf{Q}) + \mathbf{R}$ and $\mathbf{P} + (\mathbf{Q} + \mathbf{R})$. Thus if (see Fig. 3) we make

$$AB = \mathbf{P}, \quad BC = \mathbf{Q}, \quad CD = \mathbf{R},$$

we have

$$(\mathbf{P} + \mathbf{Q}) + \mathbf{R} = AC + CD, \quad \mathbf{P} + (\mathbf{Q} + \mathbf{R}) = AB + BD,$$

but since each of these results is equal to AD, we have

$$(\mathbf{P} + \mathbf{Q}) + \mathbf{R} = \mathbf{P} + (\mathbf{Q} + \mathbf{R}). \qquad \text{(2)}$$

This is known as the 'associative law' of addition. It easily follows from this and from the commutative law that three or more vectors may be added in any order without affecting the result.

For this reason the brackets, which are in strictness necessary to define the succession of the operations, are in practice often omitted, either side of (2), for instance, being denoted by

$$\mathbf{P}+\mathbf{Q}+\mathbf{R}.$$

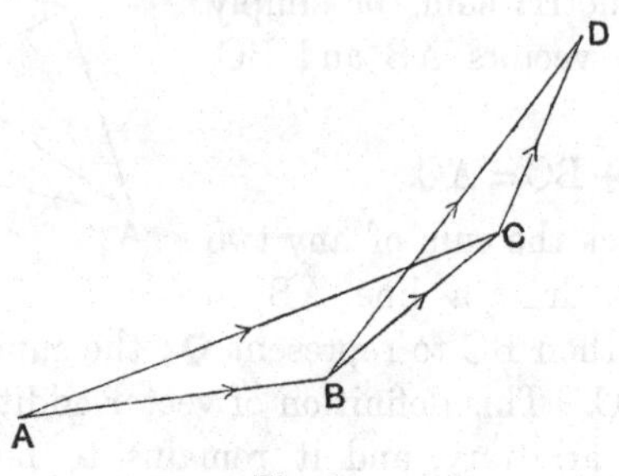

Fig. 3.

It is to be noticed that the points A, B, C, D need not be in the same plane, and consequently that the vectors $\mathbf{P}$, $\mathbf{Q}$, $\mathbf{R}$ may have any directions whatever in space.

The symbol '$-$' prefixed to a vector is used to indicate that its direction is reversed, thus

$$\text{BA} = -\text{AB}. \qquad\qquad (3)$$

It is also usual to write for shortness

$$\mathbf{P}-\mathbf{Q} \text{ in place of } \mathbf{P}+(-\mathbf{Q}).$$

Thus in Fig. 2 we have

$$\mathbf{P}-\mathbf{Q} = \text{AB}-\text{BC} = \text{AB}+\text{CB} = \text{DA}+\text{AB} = \text{DB}.$$

The difference of two vectors has a simple interpretation in the theory of displacements. Thus if $\mathbf{P}$, $\mathbf{Q}$ denote the absolute displacements (of pure translation) of two bodies, the vector $\mathbf{P}-\mathbf{Q}$ represents the displacement of the first body *relative* to the second.

A vector whose terminal points coincide is denoted by the symbol '0,' and it is plain that all such evanescent vectors may be regarded as equivalent. Thus in Fig. 2 we have

$$\text{AA} = 0, \quad \text{AB}+\text{BA} = 0, \quad \text{AB}+\text{BC}+\text{CA} = 0. \qquad (4)$$

Moreover

$$\text{AB}+\text{BB} = \text{AB},$$

or

$$\mathbf{P}+0 = \mathbf{P}. \qquad\qquad (5)$$

Hence, also,

$$(\mathbf{P}-\mathbf{Q})+\mathbf{Q}=\mathbf{P}+(-\mathbf{Q}+\mathbf{Q})=\mathbf{P}+0=\mathbf{P}. \quad \text{......(6)}$$

This will be recognized as the fundamental property of the sign '$-$' in formal algebra.

3. Multiplication by Scalars.

Finally, we have to consider the multiplication of a vector $\mathbf{P}$ by a scalar m. We define $m\mathbf{P}$ to mean a vector whose length is to that of $\mathbf{P}$ in the ratio denoted by the absolute value of m, and whose direction is that of $\mathbf{P}$, or the reverse, according as m is positive or negative. It follows that

$$\text{if } \mathbf{P}=\mathbf{Q}, \text{ then } m\mathbf{P}=m\mathbf{Q}. \quad \text{..................(1)}$$

It only remains to examine whether the distributive law

$$m(\mathbf{P}+\mathbf{Q})=m\mathbf{P}+m\mathbf{Q}, \quad \text{..................(2)}$$

which is fundamental in ordinary algebra, holds on the above definition. The proof depends on the properties of similar triangles. If we make

$$\mathrm{OA}=\mathbf{P}, \quad \mathrm{OA}'=m\mathbf{P}, \quad \mathrm{AB}=\mathbf{Q}, \quad \mathrm{A'B'}=m\mathbf{Q},$$

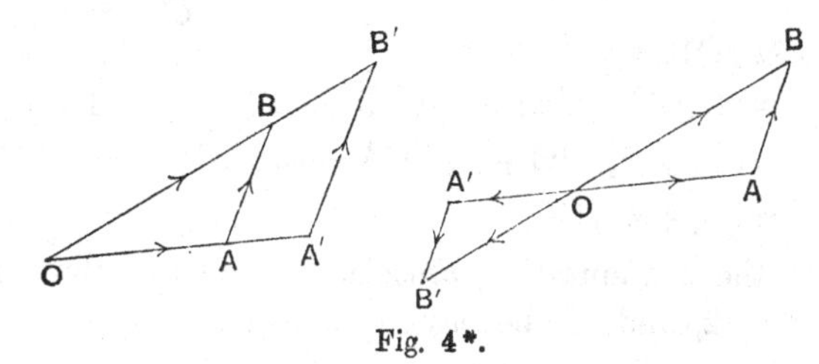

Fig. 4*.

then in the triangles OAB, $OA'B'$ we have

$$OA' : A'B' :: OA : AB,$$

whilst the angles $OA'B'$, OAB are equal. It follows that the points O, B, B' are collinear, and that

$$OB' : OB :: OA' : OA.$$

Hence

$$m\mathbf{P}+m\mathbf{Q}=\mathrm{OA}'+\mathrm{A'B'}=\mathrm{OB}'=m\,.\,\mathrm{OB}=m(\mathbf{P}+\mathbf{Q}),$$

and the theorem (2) is established.

* The two diagrams relate to the cases where m is positive and negative, respectively.

4. Geometrical Applications.

We have been at some pains to shew that although the literal symbols **P**, **Q**, ... no longer denote mere magnitudes, and although the signs '=,' '+,' '−,' '0' have received meanings different from, or rather more general* than, those which they bear in ordinary quantitative algebra, yet they are subject to precisely the same laws of operation as in that science. The conclusions which follow from the application of these laws will therefore possess the same validity. The theory of vectors furnishes us in this way with a convenient shorthand by which many interesting theorems of Geometry can be obtained in a concise manner. We shall see later that some of these theorems have important applications in Mechanics.

For example, if C be a point in a straight line AB such that

$$m_1 . \mathrm{CA} + m_2 . \mathrm{CB} = 0, \quad \text{.........(1)}$$

and O any point whatever, we have

$$m_1 . \mathrm{OA} + m_2 . \mathrm{OB} = (m_1 + m_2)\,\mathrm{OC}. \quad \text{...(2)}$$

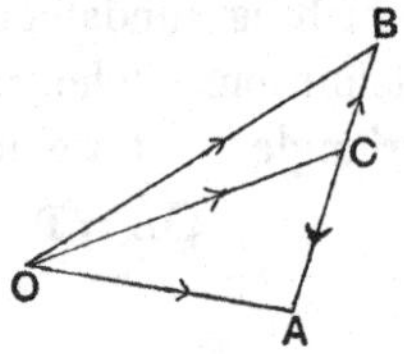

For

$$\begin{aligned} m_1 . \mathrm{OA} + m_2 . \mathrm{OB} &= m_1(\mathrm{OC} + \mathrm{CA}) + m_2(\mathrm{OC} + \mathrm{CB}) \\ &= (m_1 + m_2)\,\mathrm{OC} + (m_1 . \mathrm{CA} + m_2 . \mathrm{CB}) \\ &= (m_1 + m_2)\,\mathrm{OC}, \end{aligned}$$

Fig. 5.

in virtue of the commutative, associative, and distributive laws proved in Art. 2, and of the assumption (1).

In the particular case where $m_1 = m_2$, C is the middle point of AB, and the theorem becomes

$$\mathrm{OA} + \mathrm{OB} = 2.\mathrm{OC}. \quad \text{.....................(3)}$$

This may be interpreted as expressing that the diagonal through O of the parallelogram constructed with OA, OB as adjacent sides has the same direction as OC and double the length; in other words, the diagonals of a parallelogram bisect one another.

It is to be noticed that if m_1, m_2 have opposite signs C will lie in the prolongation of AB, beyond A or beyond B according as

* The processes of ordinary algebra have their representation in the addition &c. of vectors *in the same line* (or of a system of *parallel* vectors).

m_1 or m_2 is the greater in absolute magnitude. The theorem fails when $m_1 + m_2 = 0$, since C is then at infinity; but in this case we have obviously

$$m_1.\mathrm{OA} + m_2.\mathrm{OB} = m_1(\mathrm{OA} - \mathrm{OB}) = m_1.\mathrm{BA}. \quad \text{......(4)}$$

The formula (2) has many applications. Thus if AA', BB', CC' be the median lines of a triangle ABC, and if in AA' we take G so that $\mathrm{AG} = 2.\mathrm{GA}'$, we have, by (3),

$$\mathrm{BB}' = \tfrac{1}{2}(\mathrm{BC} + \mathrm{BA}),$$

and, by (2),

$$\mathrm{BC} + \mathrm{BA} = 2.\mathrm{BA}' + \mathrm{BA} = 3.\mathrm{BG}.$$

Hence $$\mathrm{BG} = \tfrac{2}{3}\mathrm{BB}'. \quad \text{..................(5)}$$

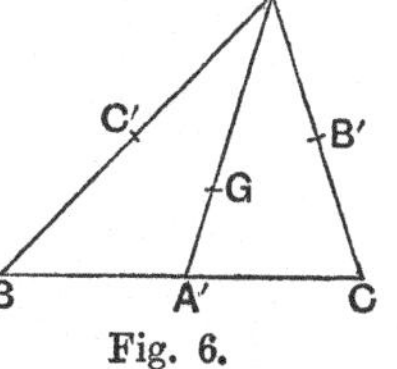

Fig. 6.

This, being a vector equation, implies that G lies in BB', and is a point of trisection on this line. In other words, we have proved that the three median lines of a triangle intersect in one point, which is a point of trisection on each.

It is also easily proved that

$$\mathrm{GA} + \mathrm{GB} + \mathrm{GC} = 0, \quad \text{.............................(6)}$$

and that if O be any point whatever (not necessarily in the same plane with A, B, C),

$$\mathrm{OG} = \tfrac{1}{3}(\mathrm{OA} + \mathrm{OB} + \mathrm{OC}). \quad \text{..........................(7)}$$

The point G which possesses these properties is called the 'mean centre' of A, B, C.

In a subsequent chapter these relations will be greatly extended.

5. Parallel Projection of Vectors.

The particular kind of projection here contemplated is by means of systems of parallel lines or planes. Taking first the case of two dimensions, where all the points and lines considered lie in one plane, the 'projection' of a point A on a given straight line OX is defined as the point A' in which a straight line drawn through A in some *prescribed* direction meets OX. Again if AB represent any vector, and A', B' be the projections of the points A, B, the vector A′B′ is called the projection of AB.

A particular case of great importance is that of 'orthogonal' projection, where the projecting lines are perpendicular to OX.

The most important theorem in the present connection is that the projection of the sum of two or more vectors is equal to the sum of the projections of the several vectors. Thus, if AB, BC be drawn to represent any two vectors, and A', B', C' be the projections of A, B, C, respectively, we have obviously

$$A'B' + B'C' = A'C'.$$

Now A′C′ is the projection of the vector AC, which is the geometric sum of AB and BC.

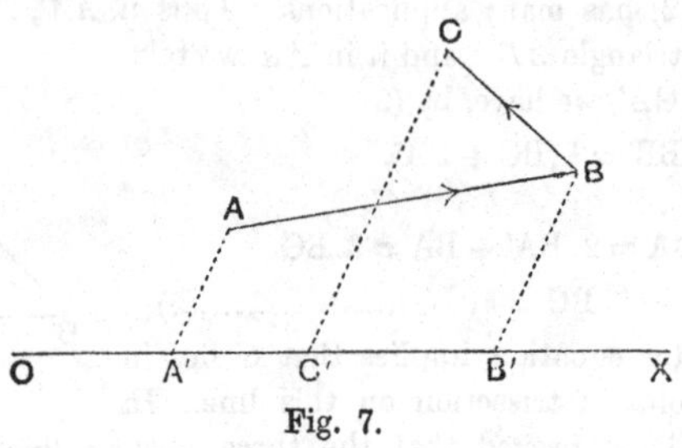

Fig. 7.

These projections of vectors on OX may evidently be specified by a series of scalar quantities, provided we fix on one direction along OX, say that from O to X, as the standard or positive direction. Thus if we specify the projection of AB by a, we mean that the length $A'B'$ is equal to the absolute value of a, and that the direction from A' to B' agrees with, or is opposed to, that from O to X, according as a is positive or negative. On this convention, the algebra of vectors in OX becomes identical in all respects with ordinary algebra.

In the particular case of orthogonal projection, the projection of a vector $\mathbf{P}$ is $P \cos \theta$, where P denotes the absolute value of $\mathbf{P}$, without regard to sign, and θ is the angle which the direction

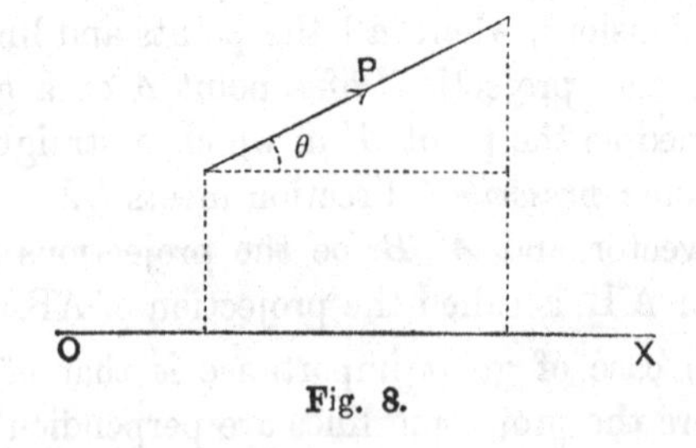

Fig. 8.

of $\mathbf{P}$ makes with the direction OX. This hardly needs proof, since the general definition of a cosine in Trigonometry is essentially that it is the projection of a unit vector on the initial line.

We here come in contact with the principles of Analytical Geometry. If we take two fixed lines of reference Ox, Oy, and project any point A on each of these by a line drawn parallel to the other, the projections of the vector OA are simply the ordinary Cartesian coordinates of A relative to the axes Ox, Oy. The vector OA, it may be added, is sometimes called the 'position vector' of A relative to the fixed origin O.

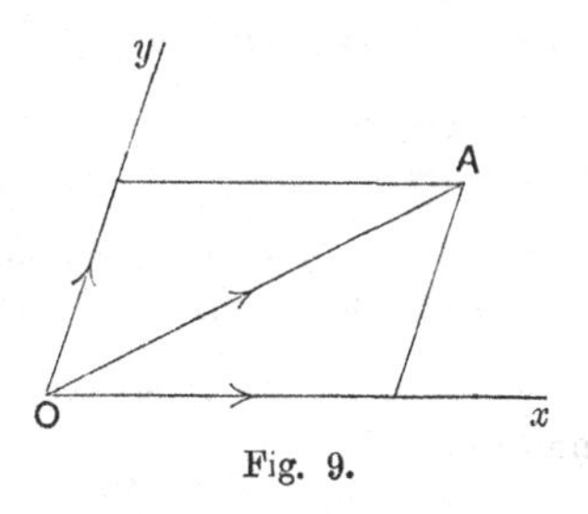

Fig. 9.

The preceding conventions are easily extended to the case of three dimensions. The only modification is that we project by a system of parallel *planes*. The points A', B', C', ... in which the planes of the system which pass through A, B, C, ... meet OX are called the projections of A, B, C, ..., respectively; the vector A′B′ is the projection of AB, and so on. Again, if we project on each of a system of three fixed axes Ox, Oy, Oz by planes parallel to the other two, the projections of a position vector OA are identical with the Cartesian coordinates of A.

EXAMPLES. I.

1. Illustrate geometrically the formulæ

$$\mathbf{A} = \tfrac{1}{2}(\mathbf{A}+\mathbf{B}) + \tfrac{1}{2}(\mathbf{A}-\mathbf{B}),$$
$$\mathbf{B} = \tfrac{1}{2}(\mathbf{A}+\mathbf{B}) - \tfrac{1}{2}(\mathbf{A}-\mathbf{B}).$$

2. Find a point O in the plane of a quadrilateral $ABCD$ such that

$$\mathrm{OA} + \mathrm{OB} + \mathrm{OC} + \mathrm{OD} = 0.$$

3. If O, O' be the middle points of any two straight lines AB, $A'B'$, prove that

$$\mathrm{AA'} + \mathrm{BB'} = 2 . \mathrm{OO'}.$$

4. If AB, $A'B'$ be any two parallel straight lines, the line joining the middle points of AA', BB' is parallel to AB and $A'B'$, and equal to

$$\tfrac{1}{2}(AB + A'B').$$

What is the corresponding result for the line joining the middle points of AB', $A'B$?

5. If A, B, C, D be any four points, prove that

$$\mathrm{AB} + \mathrm{AD} + \mathrm{CB} + \mathrm{CD} = 4.\mathrm{PQ},$$

where P, Q are the middle points of AC and BD, respectively.

6. The middle points of the sides of any quadrilateral (plane or skew) are corners of a parallelogram.

7. $ABCD$ is a parallelogram, and H, K are the middle points of AB, CD. Prove that if DH, BK be drawn, they trisect the diagonal AC.

8. If G be the mean centre of A, B, C, and G' that of A', B', C', prove that

$$\mathrm{AA'} + \mathrm{BB'} + \mathrm{CC'} = 3.\mathrm{GG'}.$$

9. If points P, Q, R be taken in the sides of a triangle ABC such that

$$\mathrm{BP} = m.\mathrm{BC}, \quad \mathrm{CQ} = m.\mathrm{CA}, \quad \mathrm{AR} = m.\mathrm{AB},$$

the mean centre of P, Q, R will coincide with that of A, B, C.

10. If points P, Q, R, S be taken in the sides AB, BC, CD, DA of a parallelogram, so that

$$\mathrm{AP} = m.\mathrm{AB}, \quad \mathrm{BQ} = n.\mathrm{BC}, \quad \mathrm{CR} = m.\mathrm{CD}, \quad \mathrm{DS} = n.\mathrm{DA},$$

then $PQRS$ will be a parallelogram having the same centre as $ABCD$.

11. If I be the centre of the circle inscribed in the triangle ABC, prove that

$$a.\mathrm{IA} + b.\mathrm{IB} + c.\mathrm{IC} = 0,$$

where a, b, c denote the lengths of the sides.

What is the corresponding statement when I is the centre of an escribed circle?

12. If OA, OB, OC be concurrent edges of a parallepiped, and

$$\mathrm{OA} = \mathbf{P}, \quad \mathrm{OB} = \mathbf{Q}, \quad \mathrm{OC} = \mathbf{R},$$

interpret the vectors

$$\mathbf{P} + \mathbf{Q} + \mathbf{R}, \quad \mathbf{Q} + \mathbf{R} - \mathbf{P}, \quad \mathbf{R} + \mathbf{P} - \mathbf{Q}, \quad \mathbf{P} + \mathbf{Q} - \mathbf{R}.$$

13. Prove that the four diagonals of a parallelepiped meet in a point and bisect one another.

14. If OA, OB, OC be three concurrent edges of a parallelepiped, prove that the point G where the line joining O to the opposite corner D meets the plane ABC is the mean centre of A, B, C. Also that

$$\mathrm{OG} = \tfrac{1}{3}\mathrm{OD}.$$

15. Prove that the three straight lines which join the middle points of opposite edges of a tetrahedron all meet, and bisect one another.

16. If $$a.\text{OP} + b.\text{OQ} + c.\text{OR} = 0$$
and $$a + b + c = 0,$$
the three points P, Q, R are in a straight line.

17. If $$a.\text{OP} + b.\text{OQ} + c.\text{OR} + d.\text{OS} = 0$$
and $$a + b + c + d = 0,$$
the four points P, Q, R, S are in one plane.

18. If the position vector of a point P with respect to a fixed point O be $\mathbf{A} + \mathbf{B}t$, where t is variable, prove that the locus of P is a straight line.

19. If the position vector be $\mathbf{A} + \mathbf{B}t + \mathbf{C}t^2$, the locus of P is a parabola.

20. If it be $\mathbf{A}t + \mathbf{B}/t$, the locus is a hyperbola.

CHAPTER I

STATICS OF A PARTICLE

6. Preliminary Notions.

When we speak of a body as a 'particle' we merely mean to indicate that for the time being we are not concerned with its actual dimensions, so that its position may be adequately represented by a mathematical point. There is no implication that the dimensions are infinitely small, or even that they are small compared with ordinary standards. All that is in general essential is that they should be small compared with the other linear magnitudes which enter into the particular problem. In physical astronomy, for instance, even such vast bodies as the Earth, the other planets, and the Sun can for many purposes be treated as material particles, their actual dimensions being negligible compared with their mutual distances.

A 'force' acting on a particle is conceived as an effort, of the nature of a push or a pull, having a certain direction and a certain magnitude. It may therefore, for mathematical purposes, be sufficiently represented by a straight line AB drawn in the direction in question, of length proportional (on some convenient scale) to the magnitude of the force. In other words, a force is mathematically of the nature of a vector. The force is to be regarded of course as acting in a line through the point which represents the particle; but in auxiliary diagrams it is convenient to treat it as a 'free' vector (Art. 1).

In many statical problems we are concerned mainly with the *ratios* of the various forces to one another, so that the question of the unit of measurement does not arise. For practical purposes a gravitational system of measurement is often adopted; thus we speak of a force of one pound, meaning a force equal to that which a mass of one pound at rest exerts on its supports. It is true that this force is not exactly the same in all latitudes, but the degree of vagueness thus introduced is slight, and seldom important, in view of other unavoidable sources of error. If we wish to be more precise it is necessary to specify the place at which the measurements are supposed to be made.

7. Composition of Forces.

The fundamental postulate of this part of the subject is that two forces acting simultaneously on a particle may be replaced by a single force, or 'resultant,' derived from them by the law of vector addition (Art. 2). In the notation already explained, two forces $\mathbf{P}$, $\mathbf{Q}$ have a resultant $\mathbf{P}+\mathbf{Q}$. This is of course a physical assumption, whose validity must rest ultimately on experience. As shewn in books on Dynamics, it is implied in Newton's Second Law of Motion.

To construct graphically the resultant of two given forces $\mathbf{P}$, $\mathbf{Q}$, we have only to draw vectors AB, BC* to represent them; the resultant $\mathbf{P}+\mathbf{Q}$ is then represented by AC. This is equivalent (see Fig. 2, p. 3) to using the familiar 'parallelogram of forces,' but requires the drawing of fewer lines.

The process of composition can be extended, step by step, to the case of any number of forces. Thus a system of forces $\mathbf{P}, \mathbf{Q}, \ldots, \mathbf{W}$, acting on a particle can be replaced by a single resultant $\mathbf{P}+\mathbf{Q}+\ldots+\mathbf{W}$. This resultant may be found graphically by a 'force-polygon'; viz. if we make

$$AB = \mathbf{P}, \quad BC = \mathbf{Q}, \ldots, \quad HK = \mathbf{W},$$

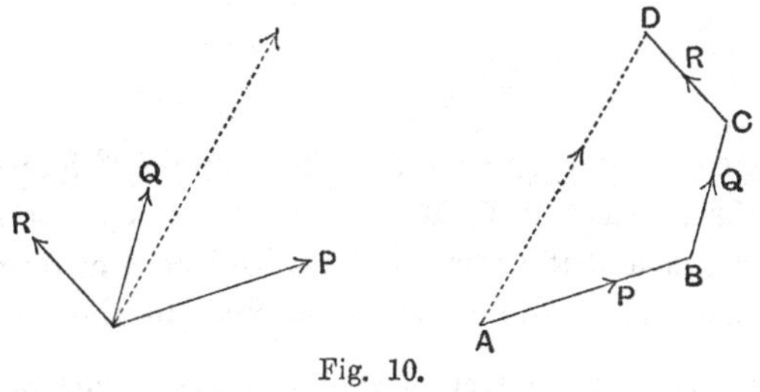

Fig. 10.

* For the convention as to the use of Roman capitals see p. 2.

it is represented by the 'closing line,' as it is called, i.e. the line AK joining the first and last points of the open polygon of lines thus drawn*. It is known (Art. 2) that the order in which the forces are taken will not affect the final result. It may be noticed, also, that the given forces may have any directions whatever in space, and consequently that the force-polygon is not necessarily a plane figure.

It follows from the physical assumption that any result of vector addition has an immediate interpretation in the composition of forces acting on a particle. For instance, three forces represented by OA, OB, OC have a resultant represented by 3.OG, where G is the mean centre of the points A, B, C (Art. 4).

As a particular case of the polygon of forces, the first and last points of the polygon may coincide, and the resultant is then represented by a zero vector. The forces are then said to be 'in equilibrium,' i.e. the particle could remain permanently at rest under their joint action. This is the proposition known as the 'polygon of forces'; viz. if a system of forces acting on a particle be represented in magnitude and direction by the sides of a *closed* polygon taken in order, they are in equilibrium. The simplest case is that of two equal and opposite forces, represented by (say) AB and BA. The next is that of the 'triangle of forces,' which asserts that three forces represented by AB, BC, CA are in equilibrium†.

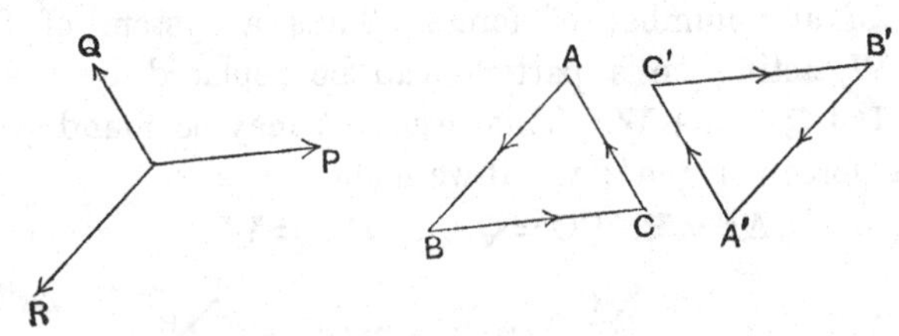

Fig. 11.

A sort of converse proposition to the triangle of forces is often useful. If three forces **P**, **Q**, **R** acting on a particle are known to be in equilibrium, and if a triangle be constructed having its sides respectively parallel to these forces, the sides of the triangle will

* For simplicity, the diagram is limited to the case of three forces.

† The proposition is ascribed to S. Stevinus of Bruges (1586).

be to one another in the ratios of the magnitudes of the corresponding forces. For if we draw AB, BC to represent **P**, **Q**, respectively, the vector CA will represent **R**, since otherwise we should not have $\mathbf{P}+\mathbf{Q}+\mathbf{R}=0$. And any triangle $A'B'C'$ whose sides are parallel to those of ABC is equiangular to ABC, and therefore similar to it*. The statement will of course also be valid if the sides of $A'B'C'$ are drawn respectively *perpendicular* to those of ABC, since this is merely equivalent to a subsequent rotation of $A'B'C'$ through 90°.

Since the sides of any triangle are proportional to the sines of the opposite angles, it appears on inspection of Fig. 11, that if three forces are in equilibrium, each force is proportional to the sine of the angle between the directions of the other two. This is known as Lamy's Theorem†.

8. Analytical Method.

Just as two forces can be combined into a single force, or resultant, so a given force may be 'resolved' into 'components' acting in any two assigned directions Ox, Oy in the same plane with it. The process is simply that of projection of vectors explained in Art. 5. Thus a force P can be uniquely resolved into two components X, Y along two assigned directions in the same plane with it, by a parallelogram construction. The value of the component force in either of the standard directions will of course depend also on what the other standard direction is. If, as is usually most convenient, the two assigned directions are at right angles, we have

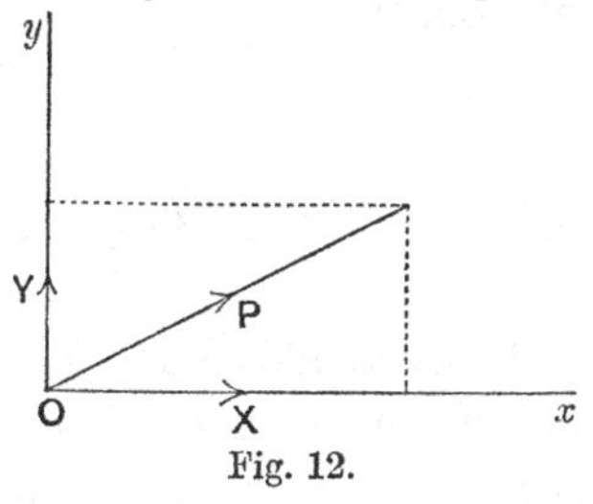

Fig. 12.

$$X = P\cos\theta, \quad Y = P\sin\theta, \quad \text{.................(1)}$$

where θ denotes the angle which the direction of P makes with Ox. Hence

$$P^2 = X^2 + Y^2, \quad \tan\theta = \frac{Y}{X}, \quad \text{..............(2)}$$

* This converse proposition cannot be extended to the case where there are more than *three* forces, the shape of a polygon of more than three sides being indeterminate when only the angles are given.

† B. Lamy, *Traité de mécanique*, 1679.

which determine P, θ when X, Y are given. It should be noticed that in these formulæ P denotes the absolute magnitude of the force, whilst X, Y are the scalar quantities (positive or negative) which are its projections.

We have seen (Art. 5) that the sum of the projections of two or more vectors on any assigned direction is equal to the projection of the geometric sum of the several vectors. Hence the sum of the components of any system of forces acting on a particle, in any assigned direction, is equal to the corresponding component of the resultant. Thus if the components of a plane system of forces $P_1, P_2, \ldots$ in the two directions Ox, Oy be

$$X_1, Y_1; \; X_2, Y_2; \ldots;$$

respectively, the components of their resultant will be

$$X_1 + X_2 + \ldots, \quad Y_1 + Y_2 + \ldots.$$

These may be more concisely denoted by $\Sigma(X)$, $\Sigma(Y)$. Hence if R be the resultant, and ϕ the angle which its direction makes with Ox, we have

$$R\cos\phi = \Sigma(X), \quad R\sin\phi = \Sigma(Y), \; \ldots\ldots\ldots\ldots(3)$$

whence $$R^2 = \{\Sigma(X)\}^2 + \{\Sigma(Y)\}^2, \quad \tan\phi = \frac{\Sigma(Y)}{\Sigma(X)}. \; \ldots\ldots\ldots(4)$$

In particular, if the given forces be in equilibrium, the sum of their components in any assigned direction must vanish, so that

$$\Sigma(X) = 0, \quad \Sigma(Y) = 0. \ldots\ldots\ldots\ldots\ldots\ldots\ldots\ldots(5)$$

Conversely, it is evident from (4) that if these are satisfied R will vanish. The conditions (5) are therefore sufficient as well as necessary conditions of equilibrium of a particle subject to a two-dimensional system of forces.

Since the conditions are *two* in number, it appears that the problem of ascertaining the possible positions of equilibrium of a particle subject to forces in one plane, which are known functions of its position in that plane, is in general a determinate one. For we have two equations to determine the two coordinates (x, y) of the particle.

9. Equilibrium under Constraint. Friction.

In many problems the particle considered is subject to some geometrical condition, or constraint; e.g. it may be attached to an inextensible string, or constrained to lie on a given material curve. In such cases the tension, or pull, exerted by the string in the direction of its length, or the pressure exerted by the curve, are usually in the first instance unknown forces which have to be allowed for in addition to the known forces such as gravity.

By a 'smooth' curve is meant one which can only exert a pressure in the direction of the normal. The notion of a perfectly smooth curve or surface, though often met with in illustrative examples, is seldom realized even approximately in practice. In actual cases the pressure may be oblique to the surface, and is then conveniently resolved into a normal component, called the 'normal pressure,' and a tangential component called the 'friction.'

The usual empirical law of friction* is that equilibrium can subsist only so long as the amount of the friction F requisite to satisfy the mathematical conditions does not bear more than a certain ratio μ to the normal pressure R. This ratio μ is called the 'coefficient of friction.' If the resultant pressure S of the curve on the particle make an angle θ with the direction of the normal pressure R, we have, resolving,

$$R = S\cos\theta, \quad F = S\sin\theta, \qquad (1)$$

and the condition $F \not> \mu R$ is therefore equivalent to

$$\tan\theta \not> \mu. \qquad (2)$$

In other words, the inclination of the resultant pressure to the normal cannot exceed a certain value λ, determined by the equation

$$\tan\lambda = \mu. \qquad (3)$$

This angle λ is called the 'angle of friction.'

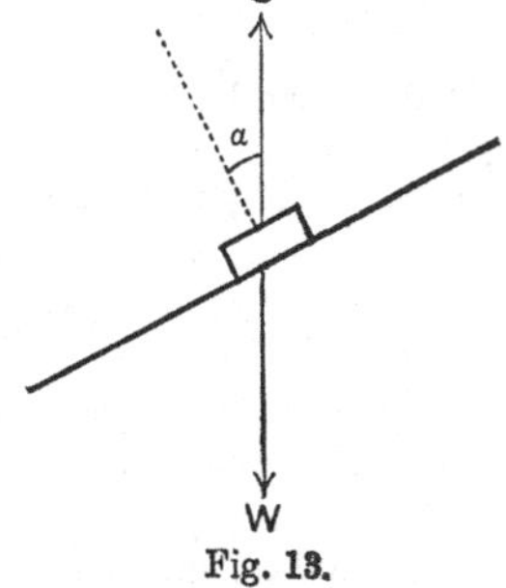

Fig. 13.

In the case of a particle resting on an inclined plane, under no forces except gravity and the reaction of the plane, this reaction must balance the weight, and therefore be vertical. Hence equilibrium

* Due to C. A. Coulomb (1821).

is possible only if the inclination of the plane to the horizontal does not exceed λ. For this reason λ is sometimes called the 'angle of repose.'

It should be said that the above law of friction can only claim to be a rough representation of the facts, and that the values of μ found by experiment for surfaces of given materials may vary appreciably with different specimens. For the friction of wood on wood μ may range from about $\frac{1}{5}$ to $\frac{1}{2}$; for metal on metal it may be about $\frac{1}{5}$ or $\frac{1}{6}$. For lubricated surfaces it has much smaller values, such as $\frac{1}{20}$ or $\frac{1}{100}$.

Ex. A body is in equilibrium on a plane of inclination α under its own weight W, a force P applied in a vertical plane through a line of greatest slope, and the pressure S of the plane. It is required to find the relations between these forces.

The question resolves itself into the construction of a triangle of forces HKL, such that the vector HK (say) shall represent W, KL the force P, and LH the reaction S. The first-mentioned side HK is to be regarded as given.

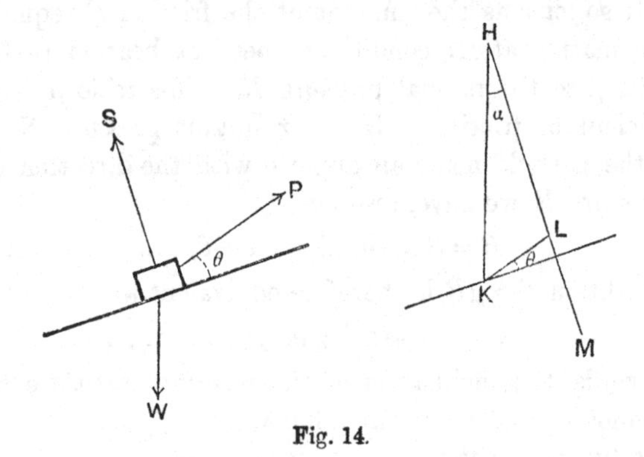

Fig. 14.

If the plane be smooth, we have only to draw HM normal to it, as in Fig. 14; then *any* point on this line will be a possible position of L. If θ be the angle which P makes with the plane, we have

$$\frac{P}{W}=\frac{KL}{HK}=\frac{\sin\alpha}{\sin(\frac{1}{2}\pi+\theta)}=\frac{\sin\alpha}{\cos\theta}. \quad \text{.....................(4)}$$

If the plane be rough, we draw two lines HM_1, HM_2 making equal angles λ with HM on opposite sides. (See Fig. 15, which corresponds to the case of $\alpha > \lambda$.) Then any point within the angle M_1HM_2 is a possible position of L. If the angle θ which P makes with the plane be given, we draw KL_1L_2 in the required direction, meeting HM_1, HM_2 in L_1, L_2, respectively. Then KL_1.

KL$_2$ represent the extreme admissible values of P. Denoting these by P_1, P_2, respectively, we have

$$\frac{P_1}{W}=\frac{KL_1}{HK}=\frac{\sin(a-\lambda)}{\sin(\frac{1}{2}\pi+\lambda+\theta)}=\frac{\sin(a-\lambda)}{\cos(\theta+\lambda)}, \qquad (5)$$

$$\frac{P_2}{W}=\frac{KL_2}{HK}=\frac{\sin(a+\lambda)}{\sin(\frac{1}{2}\pi-\lambda+\theta)}=\frac{\sin(a+\lambda)}{\cos(\theta-\lambda)}. \qquad (6)$$

If a force be applied, in the given direction, less than P_1, equilibrium is impossible and the body will slide down the plane. If the force be greater than P_2, the body will slide upwards.

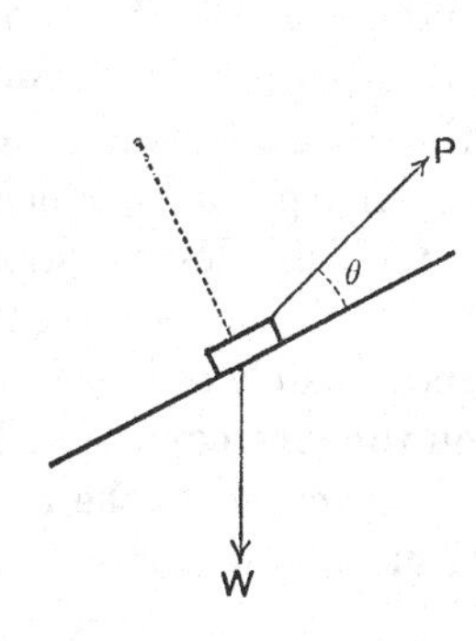

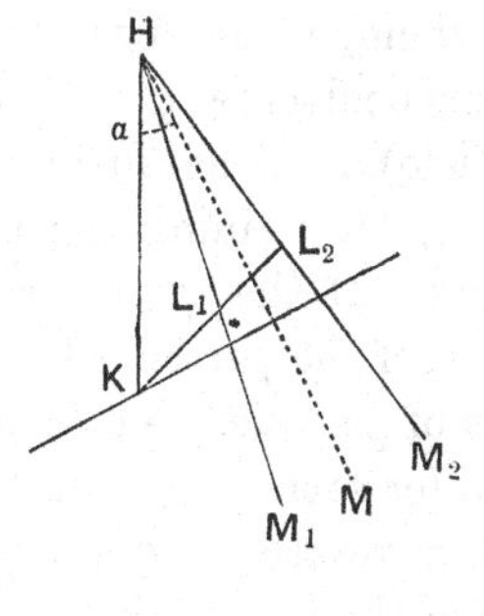

Fig. 15.

It appears, further, that if θ be varied P_1 is least when KL_1 is perpendicular to HM_1. The force then makes an angle $-\lambda$ with the plane, and its magnitude is

$$P_1 = W\sin(a-\lambda). \qquad (7)$$

This is the least force which will support the body. Similarly, P_2 is least when it makes an angle λ with the plane, and its magnitude is then

$$P_2 = W\sin(a+\lambda). \qquad (8)$$

This is the limit below which the applied force must not fall if it is to drag the body up the plane.

The case where $a<\lambda$ can be treated in the same manner.

10. Equilibrium of a System of Particles.

We assume that the mutual forces, whatever their nature, between the pairs of particles are subject to Newton's Law of the equality of Action and Reaction; i.e. that the force exerted by a particle A on a particle B, and the force exerted by B on A, are equal and opposite in the straight line AB. In many statical problems these forces are due to the tension of a string, or the tension or thrust of a rod, which is supposed to be itself free from extraneous force except for the reactions at A and B.

To find the conditions of equilibrium of the system, we have to formulate the conditions of equilibrium of each particle separately, and combine the results, taking account, of course, of the internal forces, or mutual actions, referred to.

The problem of ascertaining the possible configurations of equilibrium of a system of n particles subject to extraneous forces which are known functions of the positions of the particles, as well as to internal forces which are given functions of the distances between them, is in general a determinate one. Thus, in the two-dimensional case, the $2n$ conditions of equilibrium (two for each particle) are equal in number to the $2n$ Cartesian (or other) coordinates, determining the positions of the particles, which are to be found. If the system be subject to frictionless constraints, e.g. if some of the particles be constrained to lie on smooth curves, or if pairs of particles be connected by inextensible strings or light rods, then for each geometrical condition thus introduced we have an unknown reaction, e.g. the pressure of the curve, or the tension or thrust of the rod, so that the number of equations is still equal to that of the unknown quantities.

When friction is taken into account, however, cases of indeterminateness may arise; see Ex. 2 below.

Ex. 1. Two weights P, Q are suspended from a fixed point O by strings OA, OB, and are kept apart by a light rod AB; to find the thrust (T) in the latter.

If G be the point of the rod vertically beneath O, OAG will be a triangle of forces for the particle A, and we have

$$\frac{P}{T}=\frac{OG}{AG}, \text{ and similarly } \frac{Q}{T}=\frac{OG}{GB}. \quad \text{...(1)}$$

Hence $$P.AG=Q.GB, \quad \text{.........(2)}$$

which determines the position of the point G on the rod, and thence the position of equilibrium. The value of T is then given by either of the equations (1). From these we may derive the more symmetrical formula

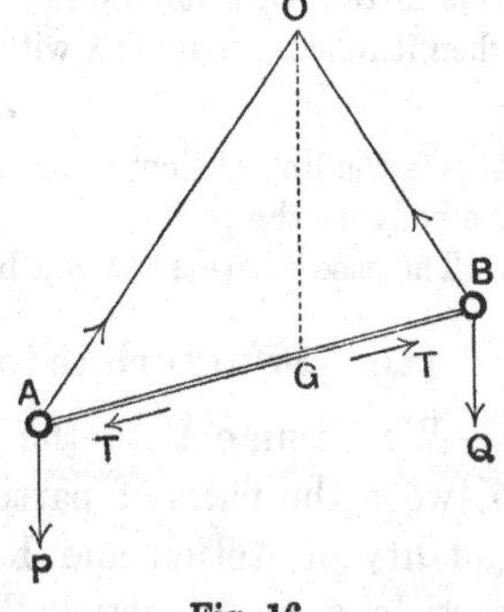

Fig. 16.

$$T=\frac{PQ}{P+Q}\cdot\frac{AB}{OG}. \quad \text{..............................(3)}$$

Ex. 2. Two rings A, B, of weights P, Q, connected by a string, can slide on two rods in the same vertical plane, whose inclinations to the horizontal are α, β, respectively; it is required to find the inclination θ of the string to the horizontal in the position of equilibrium.

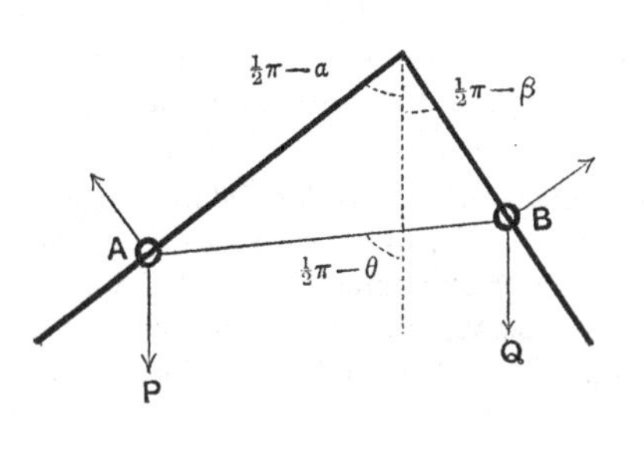

Fig. 17.

Resolving the forces on each ring in the direction of the corresponding rod, we have, if there is no friction,

$$P \sin \alpha = T \cos(\alpha - \theta), \quad Q \sin \beta = T \cos(\beta + \theta). \quad \text{...... .....(4)}$$

These equations determine the tension T of the string and the angle θ. Eliminating T we find

$$\tan \theta = \frac{P \cot \beta - Q \cot \alpha}{P + Q}. \quad \text{........................(5)}$$

The string will be horizontal if $P \tan \alpha = Q \tan \beta$.

The problem also admits of a simple graphical solution. We draw HK, KL to represent the weights P, Q, and HM, LM parallel to the normals at A and B. A triangle of forces for the ring A, constructed on HK, must have its third vertex in HM; and a triangle for the ring B constructed on KL, must have its third vertex in LM. Since the sides of these triangles which are opposite to H and K respectively represent the tension of the string, the vertices in question must coincide at M. Hence KM gives the direction of the string. The formula (5) can now be deduced from the figure without difficulty.

This graphical method also gives a clear view of the relations when friction is taken into account. We will suppose for definiteness that the inclination of each rod to the horizontal exceeds the corresponding angle of friction. We draw through H two lines HM_1, HM_2 making with HM, on opposite sides, angles equal to the angle (λ) of friction at A; and similarly we draw through L lines LN_1, LN_2 making with LM angles equal to the angle (λ') of friction at B. Then *any* point R within the quadrilateral ($M_1N_2M_2N_1$ in the figure) formed by these lines is a possible position of the third vertex of the triangles of forces, and KR a

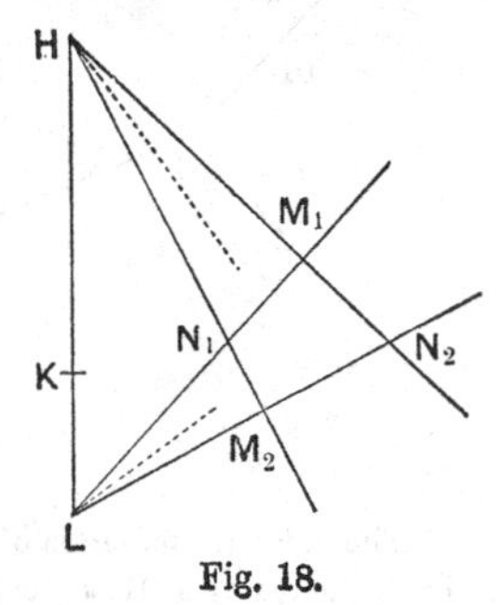

Fig. 18.

possible direction of the string. The extreme directions of the string are given by the lines joining K, M_1 and K, M_2 in the figure.

It will be noticed not only that the direction of the string is indeterminate, but that even when this is given (within the limits for which equilibrium is possible) the tension of the string and the reactions of the rod are indeterminate.

The reason for this indeterminateness is that the data are insufficient. To obtain a definite result we should need to take account of the elasticity of the string. A real string is more or less extensible; and if we know its actual as well as its natural length, and the law of its elasticity, the tension becomes determinate in amount. The values of the reactions then follow from the respective triangles of forces*.

We shall have, later, various other instances of problems which are 'statically indeterminate,' i.e. they cannot be completely solved by the principles of pure Statics alone.

11. The Funicular Polygon.

This problem is interesting in itself, and will serve as an introduction to important graphical methods which will occupy us later (Chap. IV).

A number of particles attached to various points of a string are acted on by given extraneous forces. We will suppose that these forces, and the string, are all in the same plane, although this is not strictly necessary.

We distinguish the several particles by the numerals 1, 2, ..., and denote the corresponding extraneous forces by P_1, P_2, ... The tension in the string joining the mth and nth particles may then be denoted by T_{mn}. Each particle is in equilibrium under

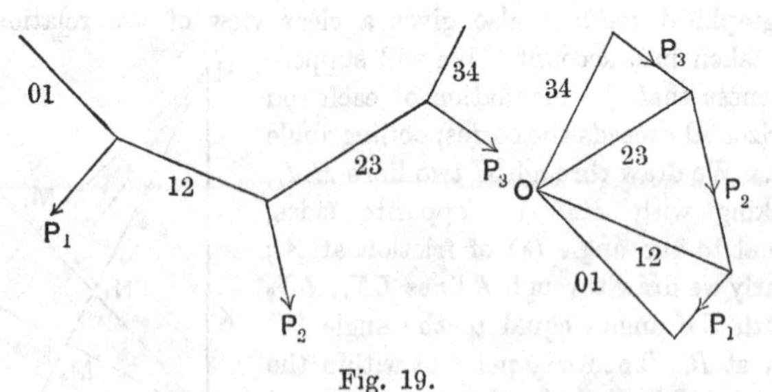

Fig. 19.

* The indeterminateness of problems involving friction seems to have been first fully elucidated by J. H. Jellett, *Theory of Friction*, Dublin, 1872.

three forces, viz. the extraneous force acting on it, and the tensions of the two adjacent portions of the string. The relation between these forces can in each case be exhibited by a triangle of forces; and if the triangles corresponding to successive particles be drawn to the same scale, they can be fitted together into a single 'force-diagram' as it is called, two consecutive triangles having one side in common, viz. that which indicates the tension in the portion of string connecting the corresponding particles. This diagram is seen to consist of the polygon of the extraneous forces, constructed as in Art. 7, together with a series of straight lines connecting the vertices with a point O. These latter lines represent the tensions in the several sides of the funicular.

A special, but very important, case arises when the forces P_1, P_2, ... are all parallel. For instance, they may be the weights of a system of particles attached at various points of a string whose ends are fixed, but which otherwise hangs freely. The polygon of the extraneous forces then consists of segments of the same vertical

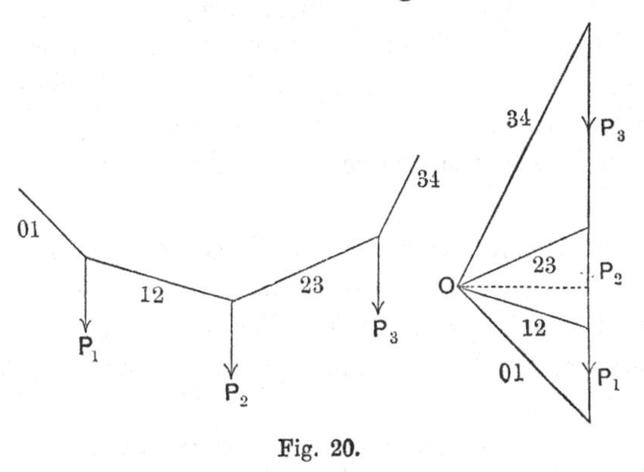

Fig. 20.

line. It is to be noticed that the tensions have now the same horizontal component, represented by the dotted line in the figure.

12. The Parabolic Funicular.

If, further, the weights be all equal, and at equal *horizontal* intervals, the vertices of the funicular will all lie on a parabola whose axis is vertical. This has an interesting application in the theory of suspension bridges.

To prove the statement, let $A, B, C, D, E, \ldots$ be successive vertices; and let AH, DK be drawn vertically to meet BC produced. If, in the auxiliary force-diagram, the distance of O

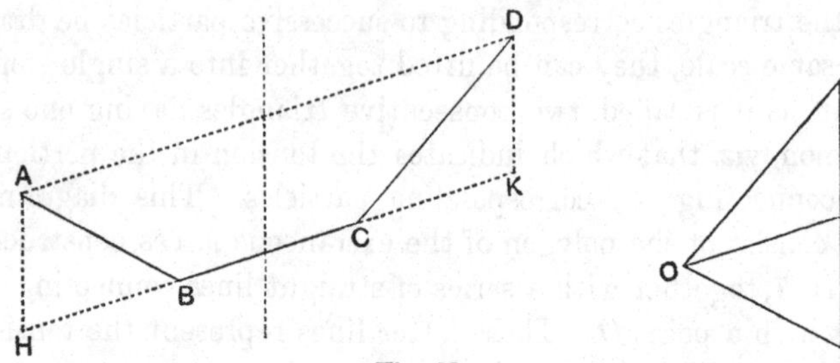

Fig. 21.

from the vertical line be taken equal to the constant horizontal interval between the lines of action of the weights, the lines which represent the tensions will be equal as well as parallel to the corresponding sides of the funicular. It follows, since the weights are equal, that $AH = DK$, so that AD and BC are parallel. They are moreover bisected by the same vertical straight line. Hence the four points A, B, C, D lie on a parabola whose axis is vertical. Similarly for B, C, D, E, and so on. But since a parabola is completely determined by the direction of its axis and by *three* points upon it*, it follows that the successive parabolas $ABCD, BCDE, \ldots$ must coincide.

In the application to suspension bridges the portions of string are represented by the links of the chain, and the weights of the particles by the tensions of the equidistant vertical rods, each of which is supposed to bear an equal portion of the weight of the roadway.

If in a funicular polygon the extraneous forces are taken to be infinitely small, and their points of application infinitely close, we pass to the case of a chain subject to a continuous distribution of force. This question will be considered independently in Chap. IX, but two results may already be anticipated: (1) if the extraneous forces be due to gravity, the horizontal tension is uniform; (2) if the weight of any portion of a chain varies as its projection on the horizontal, the chain will hang in the form of a parabola whose axis is vertical.

* Its equation being of the form

$$y = Ax^2 + Bx + C$$

which contains three arbitrary constants.

EXAMPLES. II.

1. A light ring can slide freely on a string 15 feet long whose ends are attached to two points of a fixed rod 10 feet apart. In what direction must a force be applied to the ring in order that the latter may be in equilibrium at a distance of 6 feet from one end of the string?

[The inclination of the force to the rod is 76° 45′, about.]

2. Two unequal weights are attached to the ends of a string which passes over two smooth pegs (at different levels), and a third weight is attached to the part of the string between the pegs. Find, by simple geometrical construction, the position of equilibrium; and ascertain in what cases the equilibrium is impossible.

Also find the pressures on the pegs.

3. A ring placed on a table has attached to it three strings which pass through small holes A, B, C in the table and carry given weights P, Q, R hanging vertically. Give a geometrical construction for finding the position of equilibrium of the ring.

4. A weight of 200 lbs. hangs by two ropes inclined at angles of 60° to the horizontal. If a horizontal pull of 50 lbs. be applied to the weight, find the changes produced in the tensions of the strings.

5. A weight of 50 lbs. is suspended by two equal ropes 5 feet long from two points of a horizontal bar 4 feet apart; find the tension of the ropes (1) graphically, and (2) by calculation. [27·3 lbs.]

Also find the tensions when the bar is tilted so as to make an angle of 20° with the horizontal. [47·0 lbs., 4·26 lbs.]

6. A weight of 15 lbs. is supported on a smooth plane whose inclination to the horizontal is 25° by a string which passes over a smooth pulley and carries a weight of 10 lbs. hanging vertically. Find the angle which the string makes with the plane. [50° 40′.]

7. The poles N and S of a magnet respectively repel and attract a magnetic pole at any point P with forces proportional to $1/NP^2$ and $1/SP^2$, respectively. Prove that the resultant force at P cuts the line NS produced in a point Q such that

$$NQ : SQ :: NP^3 : SP^3.$$

8. A string of length l is fastened to two points A, B at the same level, at a distance a apart. A ring of weight W can slide on the string, and a horizontal force X is applied to it such that it is in equilibrium vertically beneath B. Prove that $X = (a/l) . W$, and that the tension of the string is $W(l^2 + a^2)/2l^2$.

9. A tense string passes through a number of fixed rings $A, B, C, D, \dots$ at the corners of an equilateral polygon. Prove that the pressures on $B, C, D, \dots$ are proportional to the curvatures of the circles ABC, BCD, $CDE, \dots$, respectively.

10. Two rings whose weights are P, Q can slide on a smooth vertical circular hoop, and are connected by a string of length l which passes over a peg vertically above the centre of the hoop. Prove that in the position of equilibrium the distances r, r' of the rings from the peg are given by

$$\frac{r}{Q}=\frac{r'}{P}=\frac{l}{P+Q}.$$

11. Two equal weights W are suspended from a point O by unequal strings OA, OB, and are kept asunder by a light rod AB. Prove that if the angle AOB be a right angle, the thrust in the rod will be equal to W.

12. Two weights P, Q are attached to the ends of a string which passes over a smooth circular cylinder whose axis is horizontal. Find the condition of equilibrium when P hangs vertically whilst Q rests on the cylinder.

13. If in the preceding problem both weights rest on the cylinder, prove that the inclination (θ) to the horizontal of the line joining them is given by

$$\tan\theta=\frac{P-Q}{P+Q}\tan\alpha,$$

if 2α be the angle which this line subtends at the nearest point of the axis of the cylinder.

14. Prove that the same result applies to the case of two rings which can slide along the circumference of a smooth circular hoop in a vertical plane, and are connected by a straight string whose length is less than the diameter of the hoop.

15. A weight W can slide on the circumference of a smooth vertical hoop of radius a, and is attached to a string which passes over a smooth peg at a height c vertically above the centre, and carries a weight P hanging vertically. Give a geometrical construction for finding the positions of equilibrium, if any, other than those in which W is at the highest or lowest point of the hoop. If $c > a$, prove that such positions are possible only if the ratio P/W lies between $1-a/c$ and $1+a/c$. What is the corresponding condition if $c < a$?

16. Examine graphically the condition of equilibrium of a particle on a rough inclined plane whose inclination is less than the angle of friction.

Find the directions and magnitudes of the least forces which will drag the particle up and down the plane, respectively.

17. Find (1) by a diagram drawn to scale, and (2) by calculation, the least horizontal force which will push a weight of 50 lbs. up an incline of 2 (vertical) in 5 (horizontal), having given that the friction is such that the weight could just rest on the plane by itself if the gradient were 3 in 5. [65·8 lbs.]

18. Two equal rings can slide along a rough horizontal rod, and are connected by a string which carries a weight W at its centre. Prove that the greatest possible distance between the rings is $l \sin \theta$, where θ is determined by

$$\tan \theta = \left(1 + \frac{2w}{W}\right) \mu,$$

l being the length of the string and w the weight of each ring.

19. A weight is to be conveyed from the bottom to the top of an inclined plane (a); prove that a smaller force will be required to drag it along the plane than to lift it, provided the coefficient of friction be less than

$$\tan \left(\tfrac{1}{4}\pi - \tfrac{1}{2}a\right).$$

20. Two rings of equal weight, connected by a string, can slide on two fixed rough rods which are in the same vertical plane and are inclined at equal angles a in opposite ways to the horizontal. Prove that the extreme angle θ which the string can make with the horizontal is given by

$$\tan \theta = \frac{\mu}{\sin^2 a - \mu^2 \cos^2 a},$$

where μ is the coefficient of friction.

21. Two rings connected by a string can slide on two rods in the same vertical plane, as in Art. 10, Ex. 2; discuss graphically the case where the inclination of each rod to the horizontal is less than the corresponding angle of friction.

22. Prove that if in a funicular polygon the weights of the particles be all given, and the inclinations of any two of the sides, the inclinations of the remaining strings, and the tensions of all, can be determined.

23. An endless string is maintained in the shape of a given parallelogram by four forces applied at the corners. The forces at two opposite corners being given in magnitude and direction, find by a graphical construction, as simple as possible, the magnitudes and directions of the forces at the remaining corners.

24. A number of weights W_1, W_2,... hang from various points of a light string whose ends are fixed. If a, β be the inclinations of the extreme portions of the string to the horizontal, prove that the horizontal pull on the points of attachment is

$$\frac{\Sigma(W)}{\tan a + \tan \beta}.$$

25. Prove that in the parabolic funicular (Art. 12) the horizontal tension bears to any one of the weights the ratio $l/2h$, where l is the latus-rectum of the parabola, and h is the horizontal projection of any side of the funicular.

26. Prove from statical principles that in the parabolic funicular the sides are tangents to an equal parabola, at their middle points.

27. Prove that if in a funicular polygon the weights are equal, the tangents of the angles which successive portions of the string make with the horizontal are in arithmetic progression.

28. If a string be loaded with equal particles W at equal horizontal intervals a, and if y_n be the (vertical) ordinate of the nth weight, prove that

$$y_{n+1} - 2y_n + y_{n-1} = \frac{Wa}{T_0},$$

where T_0 is the horizontal tension. Hence shew that

$$y_n = \frac{n^2 Wa}{2T_0} + An + B,$$

where A, B are constants; and deduce the equation of the parabola on which all the particles lie.

CHAPTER II

PLANE KINEMATICS OF A RIGID BODY

13. Degrees of Freedom.

The purely geometrical theory of displacements and motions, apart from any consideration of the forces which are in operation, is called 'Kinematics' ($\kappa\acute{\iota}\nu\eta\mu\alpha =$ a movement). The present chapter treats of some geometrical propositions relating to the two-dimensional displacements of a body of invariable form. This will be sufficiently typified by a rigid lamina, or plate, moveable in its own plane. For some purposes it is convenient to regard the lamina as indefinitely extended.

The position of such a lamina is completely determinate when we know the positions of any two points A, B of it. Since the four coordinates (Cartesian or other) of these points are connected by a relation which expresses that AB is a known length, we see that virtually *three* independent elements are necessary and sufficient to specify the position of the lamina. These may be chosen in various ways, e.g. they may be the Cartesian coordinates of the point A, and the angle which AB makes with some fixed direction; but the number of independent measured data is always the same, viz. three. These three independent elements, in whatever way they are defined, are called, in a generalized sense, the 'coordinates' of the body. Hence also the lamina, when unrestricted, is said to possess three 'degrees of freedom.'

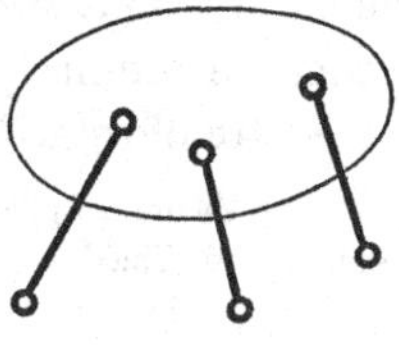

Fig. 22.

As an important practical consequence of this principle, a plane rigid frame or structure of any kind will in general be securely fixed by three links connecting three points of it to three fixed points in the plane. We say 'in general' in order to allow for a case of exception to be referred to presently (Art. 15). Similarly, the position of a lamina moveable in its own plane is in general determinate if three studs on the lamina bear against three fixed curves.

14. Centre of Rotation.

We proceed to shew that any displacement of a lamina in its own plane is equivalent to a rotation about some finite or infinitely distant point*.

For suppose that in consequence of the displacement a point of the lamina which was originally at P is brought to Q, whilst the point which was at Q is brought to R, and let I be the centre of the circle PQR. Since PQ and QR are merely different positions of the same line in the lamina, they are equal, and the triangles PIQ, QIR are congruent. It appears therefore that the displacement is equivalent to a rotation about the point I, through an angle equal to PIQ.

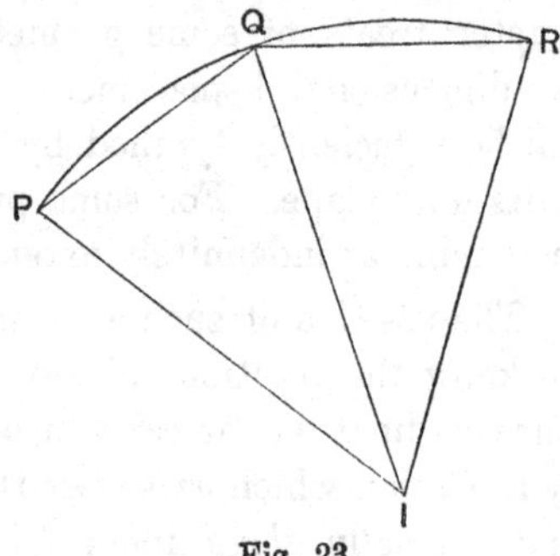

Fig. 23.

As a special case, the three points P, Q, R in the figure may be in a straight line, so that I is at infinity. The displacement is then equivalent to a pure translation, every point in the lamina being moved parallel to PQ through a space equal to PQ.

15. Instantaneous Centre.

In Mechanics we are more particularly concerned with the theory of *infinitesimal* displacements. If the two positions of the lamina be infinitely near to one another, the limiting position of the centre of rotation is called the 'instantaneous centre†.'

* The theorem appears to have been first stated explicitly (for finite displacements) by M. Chasles (1830), although the corresponding proposition in spherical geometry was known to Euler (1776).

† The existence of an instantaneous centre (centrum spontaneum rotationis) was known to J. Bernoulli (1742).

If P, P' be consecutive positions of any point of the lamina, and $\delta\theta$ the corresponding angle of rotation, the centre (I) of rotation is on the line bisecting PP' at right angles, and the angle PIP' is equal to $\delta\theta$. Hence, ultimately, the infinitesimal displacement of any point P is at right angles to the line joining it to the instantaneous centre I, and equal to $IP.\delta\theta$.

Hence if we know the directions of displacement of any *two* points, the position of the instantaneous centre is at once determined. Thus if the ends A, B of a bar be constrained to move on given curves, the instantaneous centre, in any position, is at the intersection of the normals to these curves at A and B.

An important case is where the two curves are circles. This is exemplified in the problem of 'three-bar motion.' Let $ABCD$ be a plane quadrilateral of jointed rods. If, AB being held fixed, the quadrilateral, which has now one degree of freedom, be slightly deformed, the displacement of the point D will be at right angles to AD, and the displacement of C will be at right angles to BC. The instantaneous centre of the bar CD will therefore be at the intersection of the straight lines AD, BC.

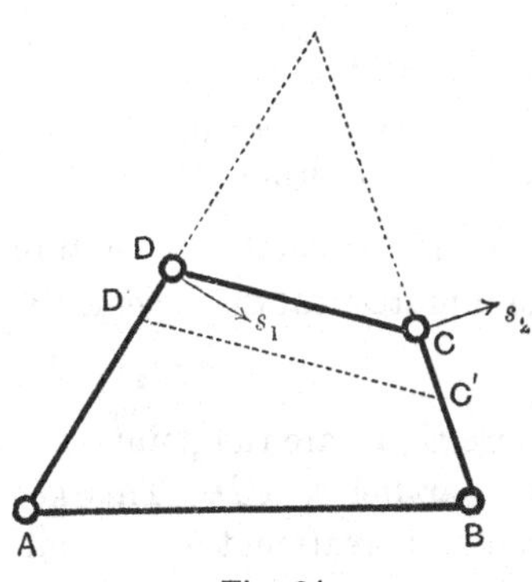

Fig. 24.

Ex. An arm OQ revolves about one extremity O; a bar QP is hinged to it at Q; and P is constrained to move in a straight line through O. See Fig. 25. (The arrangement is that of the crank and connecting rod of a steam-engine; it is a particular case of a three-bar mechanism, one bar, viz. that which guides the motion of P, being infinitely long.)

The instantaneous centre I is at the intersection of OQ produced with the perpendicular to the fixed straight line at P. Hence if $OP=x$, and θ denote the angle POQ, we have

$$-\delta x : OQ.\delta\theta = IP : IQ. \qquad \text{.........................(1)}$$

Let PQ, produced if necessary, meet the perpendicular to the line of motion of P at O in the point R. Then

$$-\delta x = OQ.\delta\theta \times \frac{IP}{IQ} = OQ.\delta\theta \times \frac{OR}{OQ} = OR.\delta\theta. \qquad \text{..............(2)}$$

Hence if the angular velocity of the crank be constant, the velocity of the point *P* of the connecting rod, and consequently of the piston to which it is attached, varies as *OR*.

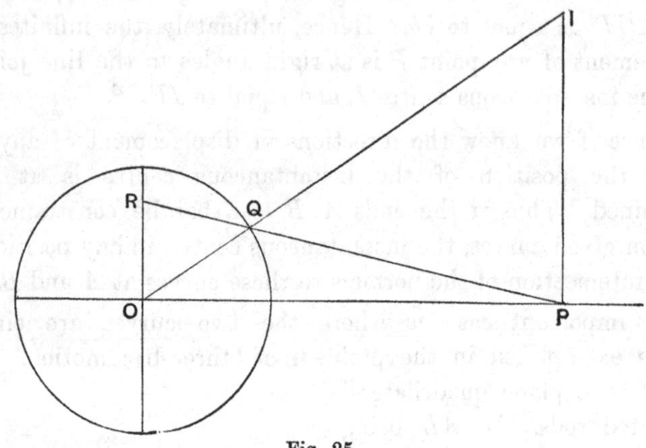

Fig. 25.

It will be observed that it is not essential that the line of motion of *P* should pass through *O*.

In the general case of three-bar motion (Fig. 24), if s_1, s_2 be the infinitesimal displacements of *D* and *C*, respectively, we have

$$s_1 : s_2 = ID : IC = DD' : CC', \quad \text{..............(3)}$$

where *C'*, *D'* are the points in which *BC*, *AD* are met by a straight line parallel to *CD*. This geometrical relation is useful in the graphical treatment of 'virtual velocities' (Chap. VI).

An important consequence of the preceding principles is that if a plane structure is to be firmly secured by means of three links, as in Fig. 22, their directions must not be concurrent or parallel. If we imagine one link to be removed, the structure acquires one degree of freedom, the instantaneous centre being at the intersection of the lines of the remaining links. If this centre be in a line with the points to which the removed link was attached, an infinitesimal rotation about it does not affect the distance between these points, to the first order of small quantities, and can therefore take place even if the link be restored. If the links are parallel and equal, even finite displacements are possible.

As a second illustration, take the case where a curve in the lamina rolls, without slipping, on a fixed curve. The instantaneous centre in any position is then at the point of contact.

Suppose, in the figure, that it is the *lower* curve which is fixed. Let A be the point of contact, and let equal infinitely small arcs AP, AP' ($= \delta s$) be measured off on the same side along the two curves. Let the normals at P, P' meet the common normal at A in the points O and O'. Then ultimately we have

$$OA = R, \quad O'A = R',$$

where R, R' are the radii of curvature of the two curves at A. After an infinitely small displacement, $P'O'$ will come into the same straight line with OP, the two curves being then in contact at P. Hence the angle ($\delta\theta$) through which the lamina has turned, being equal to the acute angle between OP and $P'O'$, is equal to the sum of the angles at O and O', so that

$$\delta\theta = \frac{\delta s}{R} + \frac{\delta s}{R'}, \quad \ldots\ldots\ldots\ldots(4)$$

ultimately. Since the distance PP' is ultimately of the second order in δs, the limiting position of the centre of rotation (I) must coincide with A, for if it were at a finite distance from this point, the displacement of P', being equal to $IP'.\delta\theta$, would, by (4), be of the first order in δs.

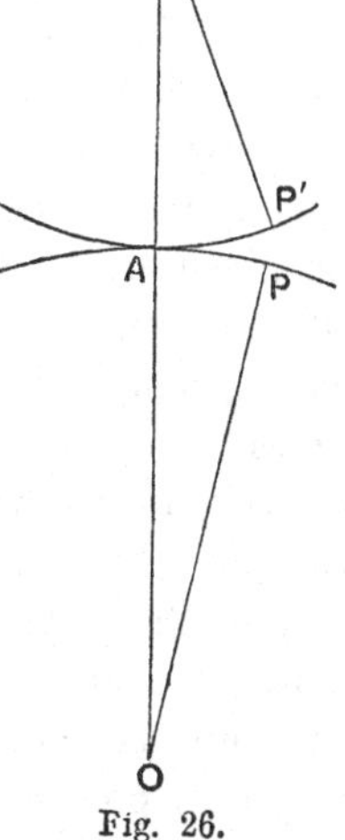

Fig. 26.

It follows that when a curve rolls on a fixed curve, the normals to the paths of all points connected with the moving curve pass through the point of contact. This principle is useful in the geometry of cycloidal and other curves*.

16. Pole-Curves†.

Conversely, we can shew that any continuous motion whatever of the lamina in its own plane can be regarded as produced in the

* It was so employed by R. Descartes (1638).

† This Art. can be postponed, as it is mainly of geometrical interest.

manner just described, viz. by the rolling of a certain curve fixed in the lamina on a certain curve fixed in the plane*.

In each position which the lamina assumes there is a certain position of the instantaneous centre. This point will therefore have a certain locus in the lamina, and a certain locus in space. The two curves thus defined are the curves referred to in the preceding enunciation. They are variously called 'pole-curves' or 'centrodes.'

Consider in the first place any series of positions 1, 2, 3, ... through which the lamina passes in succession; and let $I_{12}, I_{23}, I_{34}, \ldots$ be the centres of the rotations by which the lamina could be brought from position 1 to position 2, from position 2 to position 3, and so on, respectively. Further, let $I_{12}, I'_{23}, I'_{34}, \ldots$ be the points *of the lamina* which would become the successive centres of rotation, as they are situated in position 1. It is plain that the given series of positions 1, 2, 3, ... will be assumed in succession by the moving lamina if we imagine this to rotate about I_{12} until I'_{23} comes into coincidence with I_{23}, then about I_{23} until I'_{34} comes into coincidence with I_{34}, and so on. In other words, the lamina will pass through the actual series of positions 1, 2, 3, ... if we imagine the polygon $I_{12}, I'_{23}, I'_{34}, \ldots$, supposed fixed in the lamina, to roll on the polygon $I_{12}, I_{23}, I_{34}, \ldots$, which is fixed in space. The intermediate positions assumed by the lamina, in this imaginary process, will of course be different from those assumed in the actual motion; and the path of any point P of the figure will consist of a succession of circular arcs, described with the points $I_{12}, I_{23}, I_{34}, \ldots$ as centres, instead of a curve of continuous curvature. It is evident, however, that by taking the positions 1, 2, 3, ... sufficiently close to one another, the path of any point P can be made to deviate from the true path as little as we please. At the same time the polygons tend to become identical with the pole-curves as above defined†.

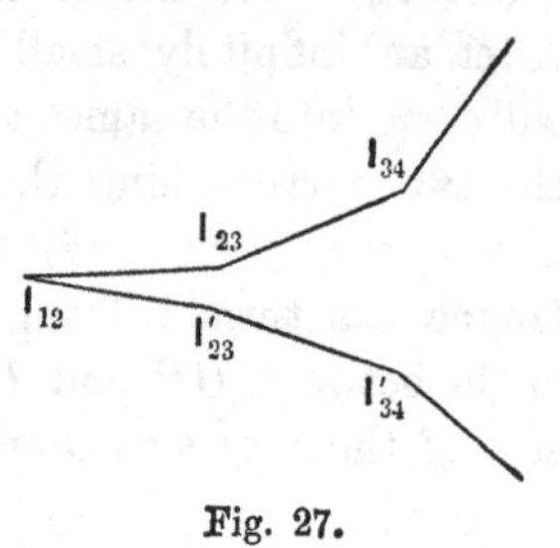

Fig. 27.

* M. Chasles (1830).

† This proof is generally accepted as sufficient. In the author's *Infinitesimal Calculus* an attempt is made to give it a more rigorous form.

The actual determination of the pole-curves is in general difficult, even in the case of three-bar motion. There are, however, a few special cases in which simple results can be obtained.

Ex. 1. Let $ABCD$ be a 'crossed parallelogram*' formed of jointed bars, the alternate bars being equal in length, viz. $AB=DC$, $AD=BC$. If the bar AD be held fixed, the instantaneous centre for the bar BC is at the point I where the bars AB and DC cross, and it is plain from the symmetries of the figure that the sums

$$AI+ID \quad \text{and} \quad BI+IC$$

are constant, being equal to AB or CD. Hence the locus of I relative to AD is an ellipse with A, D as foci, whilst that of I relative to BC is an equal ellipse with B, C as foci. The motion of BC relative to AD is therefore represented by the rolling of an ellipse on an equal ellipse.

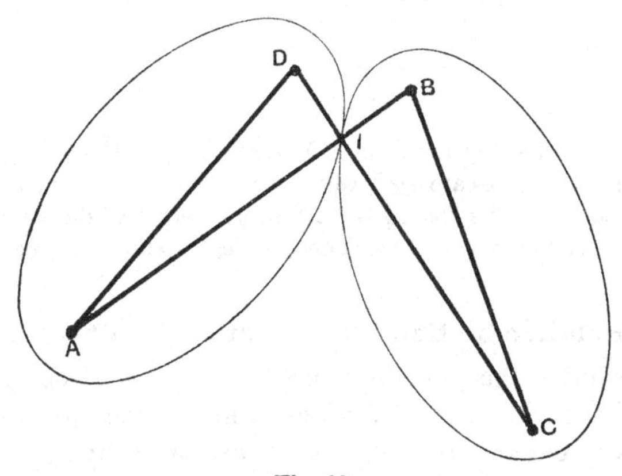

Fig. 28.

On the other hand, if AB be fixed, it may be shewn that the relative motion of CD will be represented by the rolling of a hyperbola with C, D as foci on an equal hyperbola having A, B as foci.

Ex 2. A straight line AB moves with its ends on two fixed straight lines OX, OY. This, like the preceding, may be regarded as a particular case of 'three-bar motion,' the links which constrain the points A, B being now infinitely long.

The instantaneous centre I (Fig. 29) is at the intersection of the perpendiculars to OX, OY at the points A, B, respectively. These points therefore lie on the circle described on OI as diameter; and since in this circle the chord AB, of constant length, subtends a constant angle AOB at the circumference, the diameter is constant. Hence the space-locus of I is a circle with centre O.

* The figure would become a parallelogram if the bars AD, DC were rotated through 180° about AC.

Again, since the angle AIB is constant, the locus of I relative to AB is a circle, and it is evident that the diameter of this circle is equal to the constant value of OI. Hence in the motion of the bar AB a circle rolls on the inside of a fixed circle of twice its size. It is known that in this case the

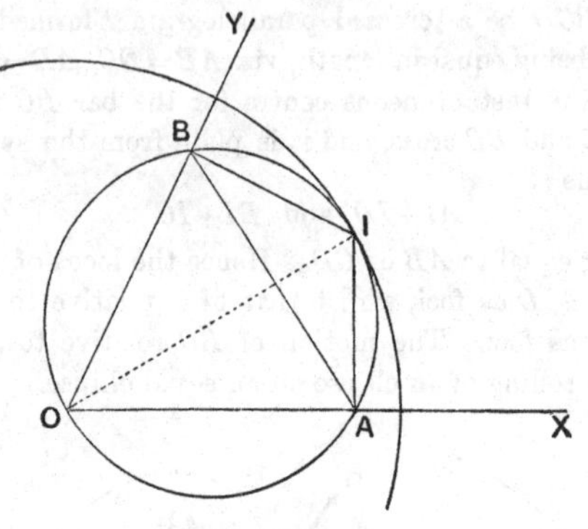

Fig. 29.

hypotrochoid described by any point P fixed relatively to the rolling circle is an ellipse whose semiaxes are $a \pm k$, where a is the radius of the fixed circle, and k the distance of the tracing point from the centre of the rolling circle. If P lie on the circumference of the rolling circle, the ellipse degenerates into a straight line.

17. Relative Motion. Theorem of the Three Centres.

In explaining the preceding theory it has been convenient to speak of the displacements of a lamina in a plane space which is regarded as fixed, but the same ideas are obviously applicable to the *relative* motion of two plane figures α, β. It is also evident that the instantaneous centre for the motion of α relative to β coincides with that of β relative to α.

In this connection we have the following simple theorem. If three figures α, β, γ be in motion in the same plane, then in any configuration the instantaneous centres for the pairs $\beta\gamma$, $\gamma\alpha$, and $\alpha\beta$ lie in a straight line.

For simplicity, suppose γ to be fixed; and let Q, P be the instantaneous centres for the motions of α and β, respectively, relative to γ. Further, let $\delta\theta$, $\delta\phi$ be the infinitesimal rotations, relative to γ, by which α, β pass, respectively, to their consecutive positions. Then if in the line PQ we take a point R such that

$$\mathrm{QR}.\delta\theta = \mathrm{PR}.\delta\phi, \quad \text{....................(1)}$$

this point will have the same displacement whether it be considered as a point of α or a point of β. It will therefore be the instantaneous centre for the relative motion of α and β. If we denote by $\delta\omega_1$, $\delta\omega_2$, $\delta\omega_3$ the absolute rotations of α, β, γ, we have $\delta\theta = \delta\omega_1 - \delta\omega_3$, $\delta\phi = \delta\omega_2 - \delta\omega_3$, and the relation (1) takes the symmetrical form

$$\text{QR}.\delta\omega_1 + \text{RP}.\delta\omega_2 + \text{PQ}.\delta\omega_3 = 0, \quad \ldots\ldots\ldots\ldots(2)$$

on usual conventions as to sign.

EXAMPLES. III.

1. A lamina is rotated about two points A, B in succession, through angles α, β, respectively. Determine the centre C of the equivalent rotation $\alpha + \beta$.

Examine the case where α, β have opposite signs.

2. Prove that if a lamina be rotated in succession through angles $2A$, $2B$, $2C$ about the vertices of a triangle ABC it will be restored to its original position, provided the sense of the rotations be the opposite to that indicated by the order of the letters A, B, C.

3. Prove that in the motion of a lamina in its own plane the directions of motion of all points on a circle through the instantaneous centre pass through a point.

4. Prove that the directions of motion of all points on a given straight line are tangents to a parabola.

5. Prove that in three-bar motion (Fig. 24), the simultaneous small angles of rotation of the bars AD, BC are to one another as

$$ID.BC : IC.AD.$$

6. A loose frame in the form of a quadrilateral $ABCD$ formed of rods jointed at their extremities is in motion in its own plane, in any manner. Prove that at any instant the instantaneous centres of the four rods form a quadrilateral whose sides pass through A, B, C, D, respectively.

If $AB = CD$, $AD = BC$, prove that the four sides of this quadrilateral are divided in the same ratio by the points A, B, C, D.

7. A plane frame consists of nine bars forming the sides of an irregular hexagon $ABCDEF$ together with the three diagonals AD, BE, CF, and the bars are smoothly jointed together at the vertices. If one bar (say AB) be removed, the frame becomes deformable; indicate on a diagram the positions of the instantaneous centres of AD relative to BE, BE relative to CF, and CF relative to AD.

What becomes of the result when the hexagon is regular?

8. A point of a lamina moves in a straight line with the constant velocity u, whilst the lamina rotates with the constant angular velocity ω; determine the pole-curves.

9. A locomotive is moving with the velocity u, whilst the driving wheel (of radius a) is skidding so that the velocity of its lowest point is v, backwards. Find the pole-curves for the wheel.

10. A point of a lamina describes a circle with constant angular velocity ω, and the lamina rotates with the constant angular velocity ω', determine the pole-curves.

11. A plane figure moves so that two straight lines in it pass each through a fixed point; determine the two pole-curves, and prove that every straight line in the figure envelopes a circle.

Also find the pole-curves when each of two straight lines in the figure touches a fixed circle.

12. A bar moves so as always to pass through a fixed point, whilst a point on it describes a fixed straight line. Prove that the motion is equivalent to the rolling of a curve of the type $r = a \sec^2 \theta$ on a parabola.

13. A bar moves so as always to touch a fixed circle, whilst a point on it describes a straight line through the centre of the circle. Prove that the motion is equivalent to the rolling of a parabola on the curve $r = a \sec^2 \theta$.

14. A bar moves so as always to touch the circumference of a fixed circular disk, whilst one point of it moves along a fixed tangent to the disk. Prove that the pole-curves are equal parabolas.

15. A bar moves so as always to pass through a small fixed ring at O, whilst a point Q on it is made to describe a circle through O. Find the instantaneous centre, and determine the two pole-curves.

CHAPTER III

PLANE STATICS

18. Fundamental Postulates.

The ideal 'rigid body' of theory is such that the distance between any two points of it is invariable. Actual solids do not of course quite fulfil this definition, since they become slightly strained by the application of force, but the changes in their dimensions are for many purposes negligible.

The complete specification of a force involves a statement of its line of action, its magnitude (and sense), and its point of application, but for a reason to be given immediately, the latter element is in pure Statics unessential.

The Statics of a body treated as rigid is based in fact on the following physical assumptions:

1°. A force may be supposed to be applied indifferently at any point of the body in the line of action. This is sometimes expressed by saying that a force is of the nature of a 'localized' vector; i.e. it is regarded as resident in a certain line, and is in this respect to be distinguished from the 'free' vectors treated of in the Introduction, but there is no necessary reference to any particular point of the line. A force which is known to lie in a given plane is therefore for our purposes sufficiently specified by *three* elements, e.g. by the two constants which on the principles of Analytical Geometry determine its line of action, and by its magnitude.

The principle here formulated is known as that of the 'Transmissibility of Force.'

2°. Two forces in intersecting lines may be replaced by a single force (called their 'resultant') which is their geometric or vector sum, and acts through the intersection. In particular, two equal but opposite forces in the same line cancel one another, i.e. they produce no effect.

It follows that any proposition relating to the addition of vectors has an application to the theory of composition of *concurrent* forces. For instance, the Triangle of Forces (Art. 7) holds with this limitation.

By means of these two principles any given system of forces may be replaced by another system in an endless variety of ways; and the various systems thus obtained are said to be 'equivalent.' If a system can be reduced to two equal and opposite finite forces in the same straight line, it is said to be 'in equilibrium.'

It will be understood that the preceding statements, as applied to a real solid, are valid only as regards the equilibrium of the body *as a whole*. Two systems of forces which are statically equivalent may have very different effects as regards the internal forces, or stresses, which they evoke.

3°. We assume, as always, that the mutual actions between two different bodies, or between two parts of the same body, when internal forces are considered, are exactly equal and opposite.

The force (or system of forces) which a body A exerts on a body B, and the opposite force (or system of forces) which B exerts on A, may be distinguished by the terms 'action' and 'reaction' respectively; but the word 'reaction' is often used loosely for either of these forces (or systems). It must of course be borne in mind that when we are considering the equilibrium of A we are concerned with the reaction of B upon it, but not at all with the action of A upon B. Similarly for the equilibrium of B. The student will often find it convenient, for clearness, to draw diagrams indicating the forces which act upon A and B separately (see Art. 23, Ex. 4).

The consideration of internal forces in general need not occupy us at present, but we so often have occasion to speak of the longitudinal stress in a bar, or a string, subject to external force at two points only, that it is worth while to notice what is involved.

Let AB be such a bar, subject to external forces at A and B; for equilibrium these must be equal and opposite in the line AB. If S be any cross-section, the matter which lies immediately to the right of S exerts on the matter immediately to the left of S

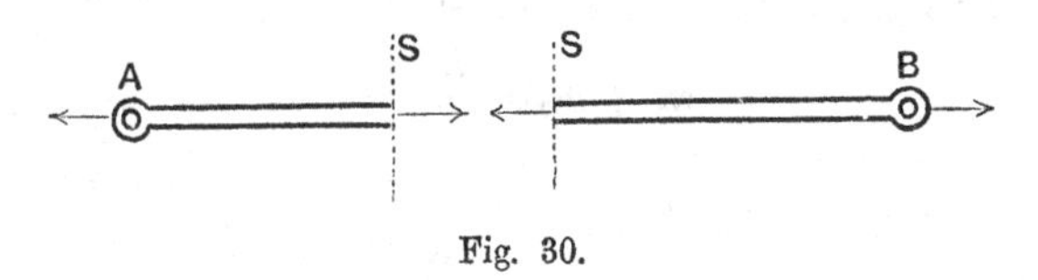

Fig. 30.

a system of forces whose resultant must be equal and opposite to the external force at A, as we see by considering the equilibrium of the portion of the bar between A and S. Similarly, the matter immediately to the left of S exerts on the matter immediately to the right a force equal and opposite to the external force at B. This combination of equal and opposite forces across the section S is called the stress at S; its amount may be specified by the magnitude of either force, reckoned positive when it is a *tension* (as in the figure), and negative when a *thrust*.

19. Concurrent and Parallel Forces.

The following construction for the resultant of two intersecting forces depends on the theorem of Art. 4.

Let O be the point of concurrence, and let any transversal line be drawn meeting the lines of action of the given forces P, Q, and

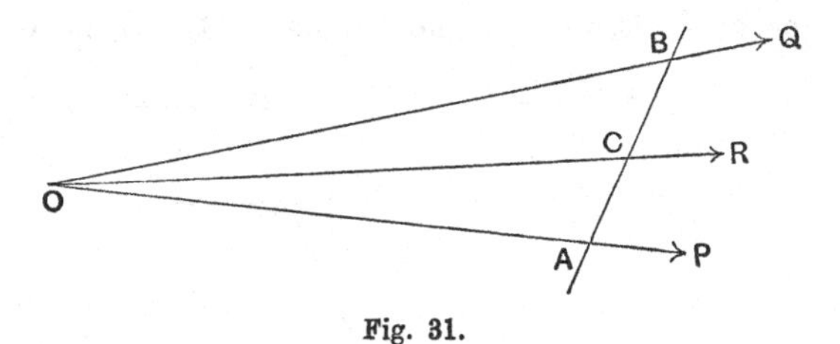

Fig. 31.

of their resultant R, in the points A, B, C, respectively. By the theorem referred to, we have

$$m_1.\mathrm{OA} + m_2.\mathrm{OB} = (m_1 + m_2).\mathrm{OC}, \quad \text{...........(1)}$$

provided

$$m_1.\mathrm{AC} = m_2.\mathrm{CB}. \quad \text{.....................(2)}$$

The first two terms in (1) are identified with the vectors representing the forces P, Q, provided

$$m_1 = \frac{P}{OA}, \quad m_2 = \frac{Q}{OB}; \quad \text{.....................(3)}$$

and since the right-hand member must then represent R, we have

$$m_1 + m_2 = \frac{R}{OC}. \quad \text{.......................(4)}$$

Hence we have the relations

$$\frac{P}{OA} + \frac{Q}{OB} = \frac{R}{OC}, \quad \text{.......................(5)}$$

and $$P.\mathrm{AC} : Q.\mathrm{CB} = OA : OB. \quad \text{..................(6)}$$

The latter relation determines the position of C on the transversal, and the value of R is then given by (5). The proof includes all cases, provided P, Q, R be reckoned positive or negative according as they respectively act from or towards O.

The rule for compounding two *parallel* forces follows as a limiting case if we imagine the point O to recede to infinity. The lines OA, OB, OC are then ultimately in a ratio of equality, and we have

$$P + Q = R, \quad \text{.........................(7)}$$

subject to a proper convention as to signs, whilst the position of C on the transversal AB is determined by

$$P.\mathrm{AC} = Q.\mathrm{CB}. \quad \text{.......................(8)}$$

These results have full generality, if the usual conventions as to the signs of P, Q, and of the vectors AC, CB, be observed.

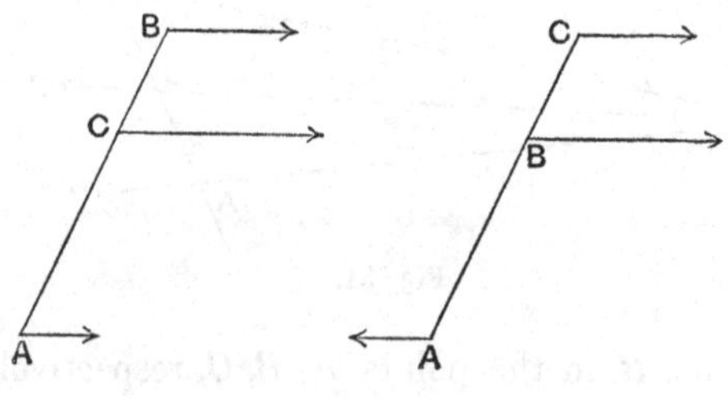

Fig. 32.

The point C divides AB internally or externally according as P, Q have the same or opposite senses, and is in any case nearer to the numerically greater force, with which the resultant agrees in direction.

The rule fails to give an intelligible result when the parallel forces have opposite signs, and are equal in absolute magnitude. As this condition is approached, the point C recedes to infinity, whilst the resultant becomes infinitely small. A combination of two numerically equal, parallel, but oppositely directed forces constitutes in fact an irreducible entity in Statics, which cannot be replaced by anything simpler. It is called a 'couple'*

20. Theorem of Moments.

From this point onwards we contemplate mainly a system of forces whose lines of action are all in one plane. The discussion of such systems is much simplified by the conception of the moment of a force about a point†.

The 'moment' of a force P about a point O is defined as the product of the force into the perpendicular OM drawn to its line of action from O, this perpendicular being reckoned positive or negative according as it lies to the left or right of the direction of P. If we mark off a segment AB on the line of action, so as to represent the force completely, the moment is measured by twice the area of the triangle OAB, this area being reckoned positive or negative according as it lies to the left or right of AB.

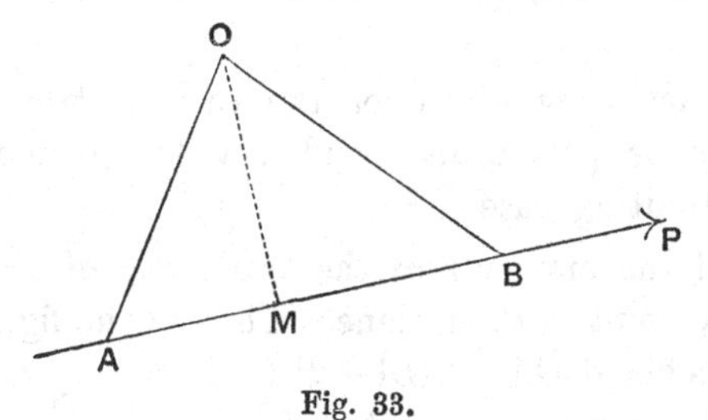

Fig. 33.

The usefulness of the conception depends on Varignon's theorem‡:

The sum of the moments of two forces in the same plane,

* The name, and the subsequent theory, are due to L. Poinsot, *Éléments de statique*, 1806.

† More strictly, about an *axis* through the point normal to the plane.

‡ P. Varignon, *Nouvelle mécanique*, 1725.

about any point of the plane, is equal to the moment of their resultant.

Let AB, AC represent the two forces, AD their resultant. If O be the point about which moments are to be taken, we have to prove that the sum of the triangles OAB, OAC is equal to the triangle OAD, regard being had to sign, according to the convention above stated. Since the side OA is common to these triangles, this requires that the sum of the perpendiculars from B and C on OA should be equal to the perpendicular from D on OA, these perpendiculars being reckoned positive or negative according as they lie to the right or left of the line drawn from

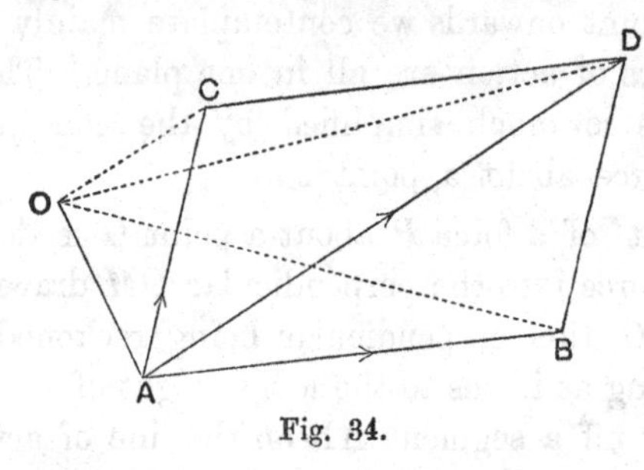

Fig. 34.

A to O. But this is merely equivalent to the statement that the sum of the orthogonal projections, on a direction perpendicular to OA, of the vectors AB, AC is equal to the projection of their sum AD.

The proof has been given for the case of two intersecting forces; it may be proved independently for parallel forces, or inferred as a limiting case.

The sum of the moments of the two forces of a *couple* is the same about any point in their plane. Thus in the figure, the sum of the moments about O is $P.OA - P.OB$, or $P.AB$. This product of either force into the perpendicular distance between their lines of action, taken with the proper sign, is called the 'moment of the couple.' We shall see presently (Art. 24) that a couple in a given plane is sufficiently specified by its moment.

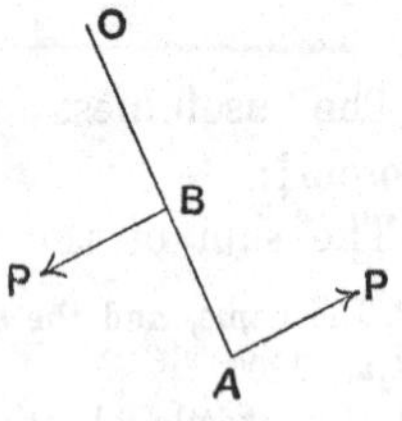

Fig. 35

21. Reduction of a Plane System of Forces.

The postulates of Art. 18 enable us in general to reduce any plane system of forces to a single resultant. For any two forces of the system may be replaced by a single force, unless they are equal and opposite in parallel lines; and by repetitions of this process the reduction in question can in general be effected.

As exceptional cases, the system may reduce to a couple, or again to two equal and opposite forces in the same line, in which case we have equilibrium.

Moreover, it is evident that at no stage in the process of reduction is any change made in the sum of the projections of the forces on any line, or in the sum of their moments about any point. It follows that as regards magnitude and direction the single resultant is the geometric sum of the given forces, and that its line of action is also determinate, i.e. it is independent of the order in which the forces have been combined.

If the geometric sum of the given forces vanishes, the system reduces to a couple whose moment is determinate, although the magnitude and the lines of action of the two residual forces may vary with the particular mode of reduction adopted.

If the geometric sum is zero, and the sum of the moments of the forces about any point also vanishes, the system is in equilibrium.

A plane system of forces may also (in general) be replaced by a single force acting through *any* assigned point O and a couple. The force is the geometric sum of the given forces, and the moment of the couple is equal to the sum of the moments of the given forces about O. For let R be the single resultant to which the system reduces, and at O introduce two equal and contrary forces R_1, R_2, of which R_1 is parallel as well as equal to R. The original system is therefore equivalent to R_1, together with the couple constituted by R and R_2, whose moment is the moment of R about O. This proposition has an interesting application in the

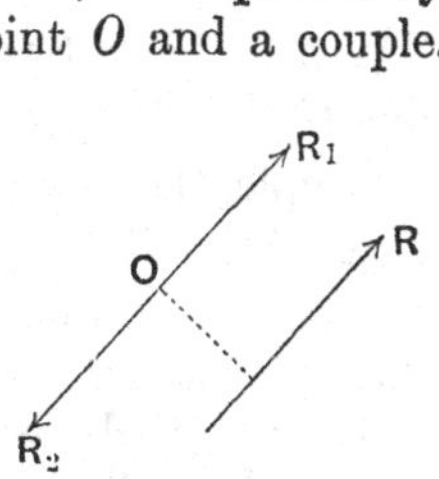

Fig. 36.

theory of the distribution of shearing stress and bending moment in a horizontal beam (Art. 27).

Again, a force, and consequently any plane system of forces, can be resolved into three components, as they may be called, acting in three assigned lines, provided these be not all parallel, or concurrent.

Thus if the three lines form a triangle ABC, and if the given force F meet BC in H, then F can be resolved into two components acting in HA, BC, respectively, and the force in HA can be resolved into two components acting in BC, CA, respectively. A simple graphical construction is shewn in Fig. 37, where the dotted line in the second diagram is parallel to AH, and the three components are denoted by P, Q, R. As an example, in the case

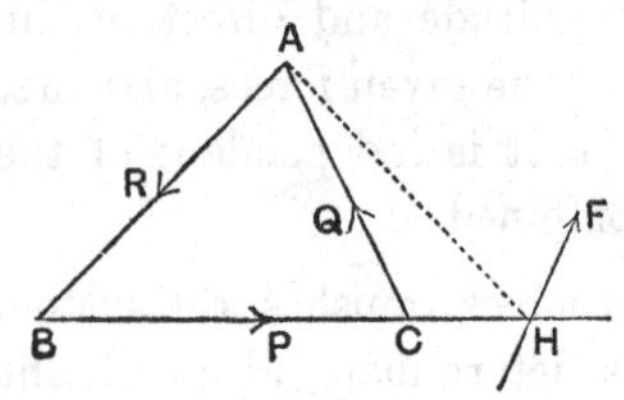

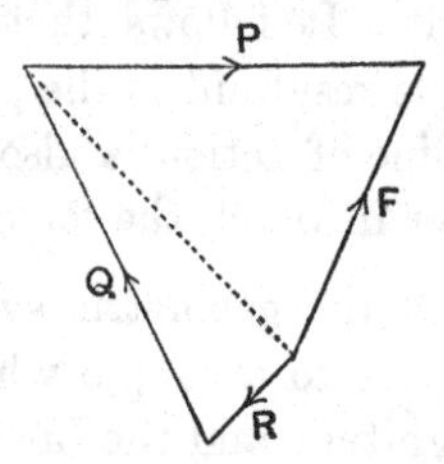

Fig. 37.

of a lamina fixed by three links (Fig. 22, p. 29), any system of forces acting on the lamina in its own plane is balanced by three determinate tensions (or thrusts) exerted by the three links, provided these have not the special configuration referred to in Art. 15, i.e. the links must not be concurrent or parallel.

It may be noticed that when F is infinitely small, and the point H at infinity, as in the case of a plane system which reduces to a couple, the components P, Q, R are proportional to BC, CA, AB (cf. Art. 24).

In 'trilinear' coordinates, with ABC as fundamental triangle, the equation of the line of action of the resultant of three forces P, Q, R acting along BC, CA, AB, respectively, is found from the consideration that the sum of the moments of the three forces about any point (α, β, γ) of the line must vanish, viz. it is

$$P\alpha + Q\beta + R\gamma = 0. \quad \text{.................................(1)}$$

If $P : Q : R = a : b : c$, this becomes

$$a\alpha + b\beta + c\gamma = 0, \quad \text{.................................(2)}$$

which is the equation of the 'line at infinity.'

22. Parallel Forces. Centre of Gravity.

The particular case where all the forces of the system under consideration are parallel is of importance on account of its application to gravity.

If parallel forces acting in opposite senses are distinguished by opposite signs, the geometric sum, which gives the magnitude and sense of the resultant, becomes identical with the sum in the ordinary algebraic meaning of the word. The line of action of the resultant is found from the consideration that the moment of the resultant about any axis is equal to the sum of the moments of the given forces. If the geometric sum vanishes, the system either reduces to a couple, or is in equilibrium.

Suppose we have a system of parallel forces P_1, P_2, P_3, ..., acting through assigned points A_1, A_2, A_3, ..., respectively. Then P_1, P_2 are equivalent to a force $P_1 + P_2$ acting through a certain point B in A_1A_2; this force $(P_1 + P_2)$ together with P_3 is equivalent to a force $P_1 + P_2 + P_3$ acting through a certain point C in BA_3; and so on. Proceeding in this way we finally obtain (in general) a resultant equal to the algebraic sum of the given forces acting through a certain point G. Since the positions of the successive points $B, C, \ldots G$ depend only on the *ratios* of the given forces, and not on their absolute magnitudes or directions, it is plain that the same point G will be arrived at if the forces be altered in any the same ratio, or if they act in any other (mutually parallel) directions through the assigned points $A_1, A_2, \ldots$. The point G thus found is called the 'centre' of the given system of parallel forces.

This proposition contains the theory of the 'centre of gravity' as ordinarily understood. For, suppose we have a system of particles whose mutual distances are small compared with the dimensions of the Earth. The forces of gravity acting on the several particles constitute a system of forces which are practically parallel, and proportional to the respective masses. If the system be brought into any other position relative to the Earth, without alteration of the relative configuration, this is equivalent to a change of direction of the forces relative to the system, but the forces are still parallel, and their mutual ratios are as before.

Hence there is a certain point (G), fixed relatively to the system, through which the resultant of gravitational action on the system always passes. This resultant is, moreover, equal to the sum of the forces on the several particles.

The general theory of the 'mass-centre' of a material system is reserved till Chap. VIII.

When the geometric sum vanishes, the system of parallel forces P_1, P_2, P_3, ... acting at given points A_1, A_2, A_3, ... has no finite 'centre.' In this case we may however divide the forces into two groups, according to their sense. One group will have a resultant R acting through a definite centre H, and the second will have a parallel, but opposite, resultant R acting through a centre K. If A_1, A_2, A_3, ... are points of a rigid body, and the direction of the forces be given, the body will tend to turn, under the action of the couple thus constituted, until the line HK is in the direction of the forces. This has an application in Magnetism.

When the parallel forces are in one plane, the line of action of the resultant is given by a simple formula. If x_1, x_2, ... be the abscissæ, measured from a fixed point O, of the points in which the lines of action meet a fixed straight line OX, and if $\bar{x}$ denote the abscissa of the point in which the line of action of the resultant meets OX, we have

$$P_1 x_1 + P_2 x_2 + \ldots = (P_1 + P_2 + \ldots)\,\bar{x}, \quad \text{.........(1)}$$

or

$$\bar{x} = \frac{\Sigma(Px)}{\Sigma(P)}. \quad \text{.........(2)}$$

This is obtained by taking moments about O, since the perpendicular distances of the respective lines of action from O are proportional to x_1, x_2, ..., $\bar{x}$. The formula holds in all cases provided proper signs be attributed to the abscissæ x and the forces P. If $\Sigma(P) = 0$, we have $\bar{x} = \infty$, unless also $\Sigma(Px) = 0$.

A plane system of parallel forces can be replaced, in one and only one way, by two forces in any two assigned lines which are parallel to the system. This is seen by taking moments about two points lying one on each of the given lines. Thus a system of loads on a beam supported at two points is statically equivalent to two vertical forces at the supports.

23. Conditions of Equilibrium.

It will be noticed that, in each of the forms to which a plane system of forces was reduced in Art. 21, the specification of the equivalent system involves *three* independent measurements. Thus in the case of the single resultant, two measurements are required to fix the line of action, and a third for the magnitude. It follows that the necessary and sufficient conditions of equilibrium, in whatever form they may be expressed, will be three in number.

Thus the system will be in equilibrium if the sum of the projections of the forces on each of two perpendicular directions is zero, and if (further) the sum of the moments about any one point is zero. For the first two conditions ensure that the system cannot reduce to a single resultant (since the geometric sum of the given forces vanishes), and the remaining condition ensures that it cannot reduce to a couple. It is also obvious that these conditions are necessary.

The three conditions of equilibrium may be obtained in other, but of course equivalent, forms. For example, equilibrium is ensured if the sum of the moments of the given forces is zero about each of three points A, B, C which are not collinear. For the system obviously cannot reduce to a couple, and it cannot reduce to a single force, since this would have to pass through each of the points A, B, C.

In the particular case of *three* forces, which is of frequent occurrence, the conditions may be put in a more convenient form. Excluding the case of parallelism, the resultant of any two of the given forces, acting through the intersection of their lines of action, must balance the remaining force. Hence the necessary and sufficient conditions for equilibrium are that the three forces must be concurrent, and that their geometric sum must vanish. The latter statement is of course equivalent to *two* quantitative conditions.

Ex. 1. Three forces perpendicular to the sides of a triangle ABC at their middle points will be in equilibrium provided they are proportional to the respective sides, and act all inwards, or all outwards.

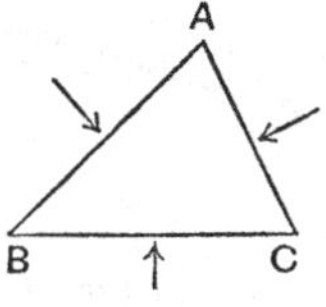

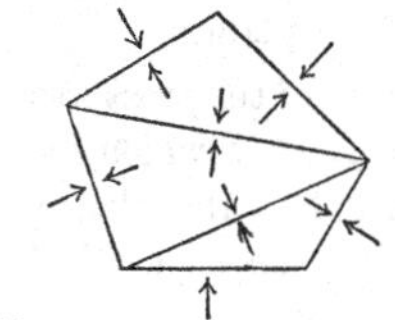

Fig. 38.

For the lines of action meet in a point, viz. at the centre of the circle ABC, and the triangle becomes itself a triangle of forces if turned through 90°.

The result is easily extended to the case of a plane polygon of any number of sides, by drawing diagonals so as to divide the polygon into triangles, and introducing at the middle point of each diagonal a pair of equal and opposite forces, perpendicular to that diagonal, and proportional to its length. The forces can now be divided into sets of three, each of which sets is in equilibrium, by the former case.

Ex. 2. A beam AB is free to turn about the end A as a fixed point, and the end B rests on a smooth inclined plane; it is required to find the reactions at A, B when the beam is loaded in any given manner.

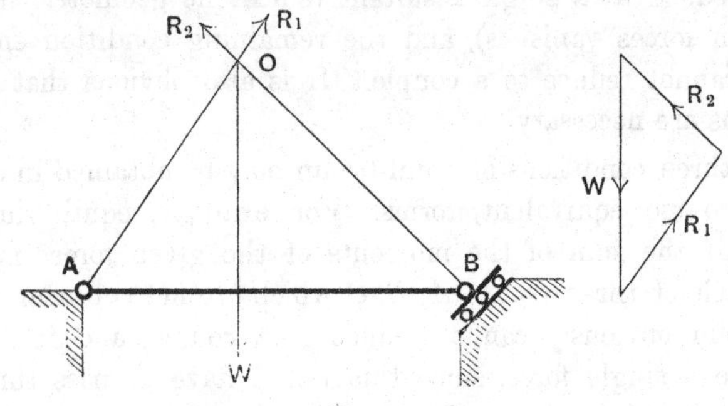

Fig. 39.

There are here *three* unknown elements to be found, viz. the direction and magnitude of the reaction R_1 at A, and the magnitude of the pressure R_2 at B, which is normal to the inclined plane. The problem is therefore determinate. If we assume the loads to be combined into a single resultant W, we have only three forces to deal with, and their lines of action must therefore meet in a point O, which is determined as the intersection of R_2 and W. This fixes the direction of R_1, and the ratios R_1/W and R_2/W can then be found from a triangle of forces.

Ex. 3. A heavy bar AB rests with its ends on two smooth inclined planes which face each other; to find the inclination (θ) to the vertical in the position of equilibrium.

Let G be the centre of gravity, and let $AG=a$, $GB=b$. Let α, β be the inclinations of the planes.

The lines of action of the pressures exerted on the bar at A, B, which are by hypothesis normal to the planes, must meet in a point I on the vertical through G. This gives

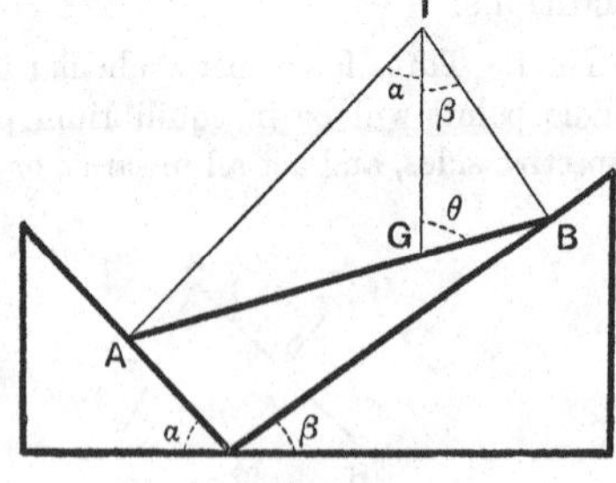

Fig. 40.

$$\frac{IG}{a} = \frac{\sin(\theta - a)}{\sin a}, \quad \frac{IG}{b} = \frac{\sin(\theta + \beta)}{\sin \beta}, \quad \text{.................(1)}$$

and therefore
$$\frac{a \sin(\theta - a)}{\sin a} = \frac{b \sin(\theta + \beta)}{\sin \beta}, \quad \text{.......................(2)}$$

whence we find
$$\cot \theta = \frac{a \cot a - b \cot \beta}{a + b}. \quad \text{..........................(3)}$$

In particular, if the bar is uniform, so that $a = b$, we have

$$\cot \theta = \tfrac{1}{2}(\cot a - \cot \beta). \quad \text{..........................(4)}$$

We are not at present concerned with questions of stability. It will appear later that in the present case the equilibrium is unstable.

The case of a rod suspended by strings attached to its ends is practically identical, the tensions of the strings taking the place of the reactions of the planes.

When two or more moveable bodies A, B, C, ... are concerned, we have to take account of the mutual actions referred to in Art. 18. The conditions to be fulfilled by A and B separately will of course ensure that the conditions of equilibrium of A and B, considered as one system, are satisfied, just as if they were rigidly connected together. In these latter conditions the mutual actions of the two bodies will cancel, and it is therefore often convenient to begin by formulating them at once.

Ex. 4. Two smooth circular cylinders rest in contact in the angle between two inclined planes which face each other.

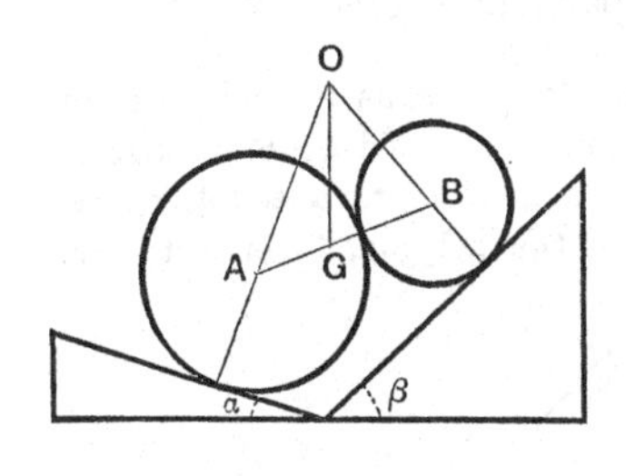

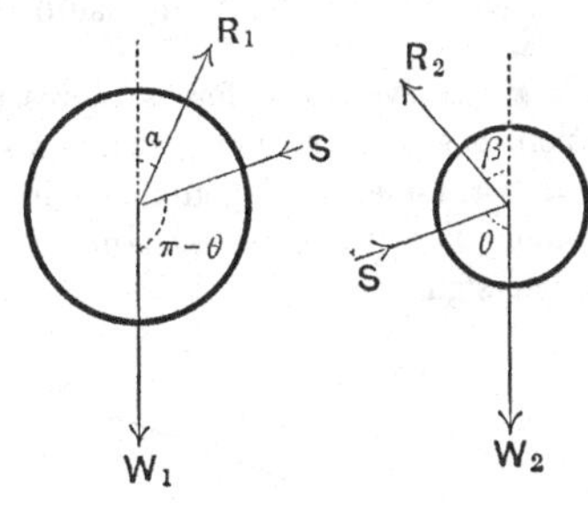

Fig. 41.

The pressures of the planes will act in lines through the axes A, B. If these lines meet in O, the centre of gravity G of the two bodies, considered as one system, must lie in the vertical through O. Hence the inclination θ of the plane AB to the vertical is given by the formula (3) above, provided we make $a = AG$, $b = GB$. If W_1, W_2 be the weights of the spheres, we have

$$AG : GB = W_2 : W_1, \quad \text{..............................(5)}$$

and therefore
$$\cot\theta=\frac{W_2\cot\alpha-W_1\cot\beta}{W_1+W_2}. \quad\text{.......................(6)}$$

To find the pressures R_1, R_2 of the two planes, and the mutual pressure S of the spheres, we must examine the conditions of equilibrium of each sphere separately. Since it is in each case a question of three forces, we find at once

$$R_1=\frac{\sin\theta}{\sin(\theta-\alpha)}.W_1,\quad R_2=\frac{\sin\theta}{\sin(\theta+\beta)}.W_2,\quad S=\frac{\sin\alpha}{\sin(\theta-\alpha)}W_1=\frac{\sin\beta}{\sin(\theta+\beta)}W_2.$$

These determine R_1, R_2 and S, when θ is known from (6).

24. Theory of Couples.

Two couples in the same plane whose *moments* (Art. 20) are equal (both in magnitude and sign) are equivalent. For if we reverse one of the couples we have a system of four forces in equilibrium, since the sum of the moments about any point in the plane is zero.

In the same way we can prove that any number of couples in the same plane can be replaced by a single couple whose moment is the sum of the moments of the given couples, regard being had of course to signs.

Since a couple in a given plane is for the purposes of pure Statics sufficiently defined by its moment, it has been proposed to introduce a name ('torque,' or twisting effort,) which shall be free from the irrelevant suggestion of two particular forces.

Ex. A system of forces represented *completely* (i.e. as to their lines of action, as well as their magnitudes and directions) by the sides $AB, BC, CD, \ldots$ of a *closed* plane polygon taken in order, is equivalent to a couple whose moment is represented by twice the area of the polygon, taken with the proper sign.

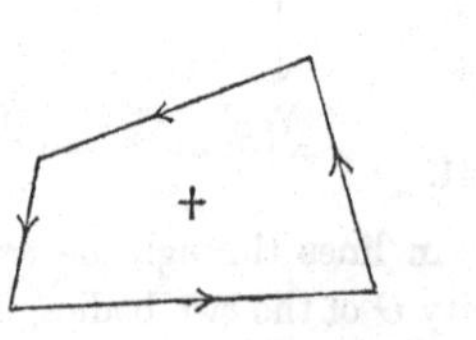

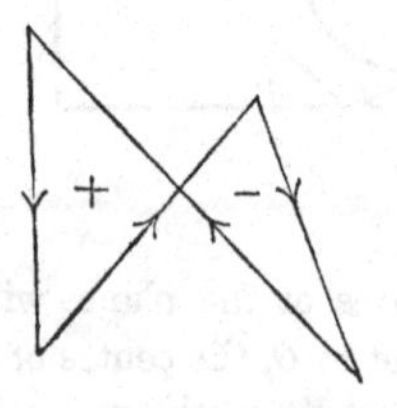

Fig. 42.

For if we take moments about any point in the plane we get a result which is constant, and equal to the area in question. The rule of signs is that the area is to be reckoned positive or negative according as it lies to the left or right of a point describing the boundary in the direction of the forces. If the polygon intersects itself, some portions of the area will be positive and some negative.

25. Determinate and Indeterminate Statical Problems.

The question whether a particular problem is statically 'determinate,' i.e. whether the data are sufficient for a definite solution, is to be decided on the principles of Arts. 13, 23.

For instance, the problem of finding the possible position or positions of equilibrium of a lamina, under forces in its own plane which are known functions of its position, is a determinate one. For the three conditions of equilibrium are just the requisite number to determine the three coordinates by which (Art. 13) the position may be specified. Of course with an arbitrary scheme of forces it may happen that there is no real solution; or, again, there may be more than one solution, but in this case each of the corresponding positions will in general be definite, in the sense that it cannot be departed from, however slightly, without violating the conditions of equilibrium.

Again, in the case of a lamina fixed by three links (Fig. 22), or suspended by three inextensible strings, and subject to any given forces in its plane, the three conditions of equilibrium are just sufficient to determine the three unknown tensions.

Similarly, in the case of a lamina free to turn about a fixed point there are three things to be determined, viz. the angular coordinate which now suffices to specify the position of the lamina, and the direction and magnitude of the reaction at the fixed point.

Questions may however present themselves quite naturally where the principles of pure Statics are inadequate for a definite solution. To take as simple an example as possible, suppose we have a horizontal beam resting on three smooth supports A, B, C, the centre of gravity being at G. One condition of equilibrium, viz. that the sum of the horizontal components of the forces should vanish, is already fulfilled, and the two remaining conditions are insufficient to determine the three unknown pressures P, Q, R. On the other hand it is certain that in an actual case there is a definite pressure at each support. The difficulty disappears when we take account of the fact that a real beam is not absolutely rigid. Let us suppose that one of the supports, say C, can be gradually raised, e.g. by means of a screw, and that to

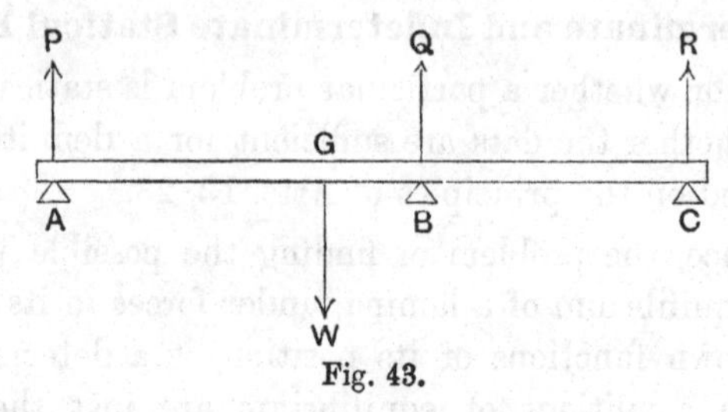

Fig. 43.

begin with it is just clear of the beam, so that $R = 0$. The pressures P and Q in this state are found at once by taking moments, say they are P_1 and Q_1. If C be gently raised, the beam bends slightly, so as to remain in contact for a while at all three points A, B, C. During this stage the pressure R will gradually increase from zero, whilst Q diminishes, and P increases. At length the pressure Q will sink to 0, and the beam will be supported at A and C only. The pressures at these points are then determinate, say they are P_2, R_2. Thus when all three supports are in contact we have a continuous transition of the values of the three pressures from P_1, Q_1, 0 to P_2, 0, R_2. This explains the indeterminateness of the question, when proposed as one of pure Statics. It becomes physically determinate when we know exactly the relative heights of the supports, and the elastic properties of the beam. (See Chap. XVII.)

As a further example, suppose we have a bar AB whose ends are fixed to some rigid structure by means of cylindrical pins

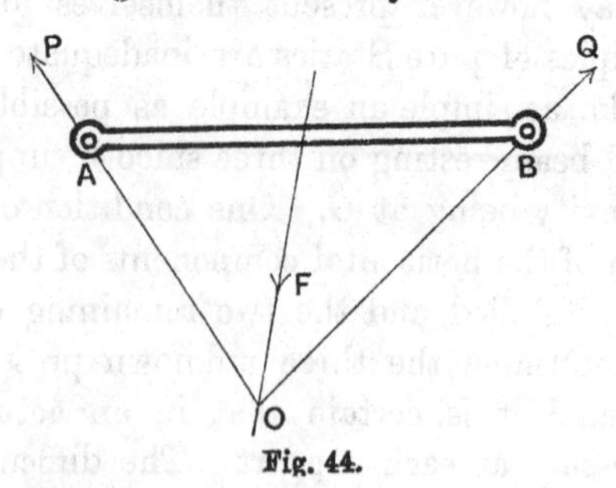

Fig. 44.

fitting into cylindrical 'eyes,' or sockets, and that it is required to find the reactions on the ends when the bar is subject to given forces. Even if we assume perfect smoothness at the contacts, so that the reaction at each end reduces to a single force, there are four unknowns, and the problem is statically indeterminate. If we take any point O whatever on the line of action of the resultant (F) of the external forces, OA and OB will be possible directions of the reactions P, Q of the pins, and the corresponding magnitudes of these reactions can be found from a triangle of forces. To make the question physically determinate it would be necessary to take account in this case also of the elasticity of the bar, and of the precise conditions of the constraint. It should be noticed that the bar may be in a state of stress, and so may exert reactions (equal and opposite) on the two pins, even when there is no extraneous force. Such stresses are introduced in the mere act of fixing the bar if the distance between the centres of the pins be not exactly equal to the natural distance between the centres of the eyes*.

It would be easy to multiply examples, but all are to be elucidated by similar considerations. The question will present itself again in connection with the theory of frames (Chap. v).

26. Equilibrium of Jointed Structures.

This question, which is very important in practice, requires some care on account of the complexity of the actions involved.

We will suppose, for definiteness, that the connection between any two pieces, or members, of the structure, is by means of a cylindrical pin passing through cylindrical holes which it just fits. For simplicity we assume perfect smoothness of the surfaces in contact, so that the pressures exerted on a pin by any piece, and the opposite reactions exerted by the piece on the pin, are in lines meeting the axis of the pin, and so reduce in each case to a force through this axis. The conditions of equilibrium of the structure are to be found by formulating separately the conditions of equilibrium of each member and each pin, and combining the results.

* A similar remark would apply in the former example if the 'supports' are such as to prevent upward as well as downward motion.

Very often, each member has two joints only, and if it be subject to no force except the reactions of the two pins upon it, these reactions must be equal and opposite, in the line joining the centres.

Ex. To take a simple case, suppose we have a bracket formed by two bars AC, BC, of which the former is horizontal while the latter is inclined at an angle α. There are smooth joints at A, B, C, of which the two former are supposed to be carried by some supporting structure; and a given weight W is suspended from the pin at C.

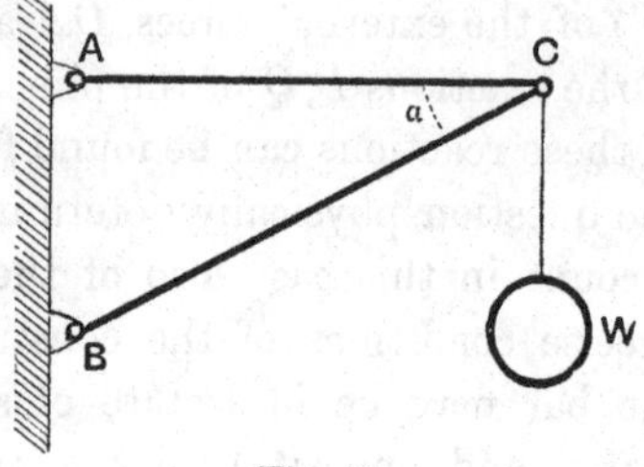

Fig. 45.

If the weight of the bars be neglected, the forces exerted by the bars on the pin C will be in the directions of the lengths, so that ABC is a triangle of forces for the pin C, viz. AB represents the weight W, BC the force exerted by the bar BC, and CA that exerted by the bar AC. Hence we have a tension in AC of amount $W \cot \alpha$, and a thrust in BC of amount $W \operatorname{cosec} \alpha$.

If the bars have weights W_1, W_2, which we will for simplicity suppose to act at the middle points, the most straightforward plan is to introduce symbols for the horizontal and vertical components of the force exerted on each bar by each pin; but some reduction in the amount of notation required can be effected by obvious considerations. Thus the vertical components of the actions of the pins A and C on the bar AC must be each equal to $\frac{1}{2}W_1$, upwards; and if we denote the horizontal component of the action of C by X_1, the horizontal pull of A will be X_1 in the opposite direction, as shewn in

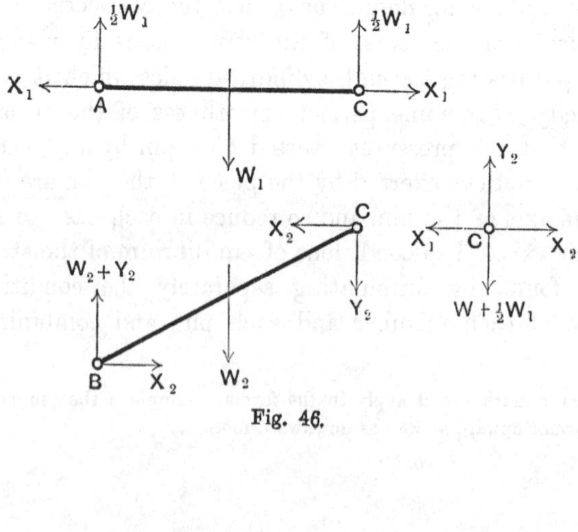

Fig. 46.

Fig. 46, where the forces acting on each bar and on the pin C are indicated separately. It will be noted that we have here applied, informally, the three conditions of equilibrium of the bar AC. Again, if we denote by X_2, Y_2 the horizontal and vertical components of the action of the pin C on the bar BC, as shewn, the components of the action of B will be $W_2 + Y_2$ upwards, and X_2 to the right. The forces on the pin C will consist of the reactions of the bars and the weight W.

As regards the bar BC we have already virtually used two of the conditions of equilibrium. The third is found by taking moments about the centre, viz.

$$2X_2 . a \sin \alpha = (W_2 + 2Y_2) . a \cos \alpha, \quad \text{.....................(1)}$$

if $BC = 2a$. Considering the equilibrium of the pin C, we have

$$X_1 = X_2, \quad Y_2 = W + \tfrac{1}{2} W_1 . \quad \text{...........................(2)}$$

This gives Y_2, and substituting in (1) we find

$$X_1 = X_2 = \{W + \tfrac{1}{2}(W_1 + W_2)\} \cot \alpha. \quad \text{.....................(3)}$$

In this example it was easy to foresee what would be the sense of each component reaction, but it is not really necessary to attend to this. If we assume the positive directions arbitrarily, the signs attached to the symbols in the result will indicate in which sense the respective forces act.

If the centres of gravity are not at the middle points, or if the applied forces are less simple in character, it may become necessary to assume unknown values for the component actions of each pin on each bar. When the structure is at all complicated the method becomes laborious, owing to the number of equations and of unknown quantities involved, and is very liable to error. A graphical method of treating such questions is explained in Chap. V.

27. Shearing Force and Bending Moment in a Beam.

The theorem, proved in Art. 21, that any plane system of forces is equivalent to a single force acting through any assigned point, together with a couple, has an interesting illustration in the theory of the distribution of shearing force and bending moment in a horizontal beam, or other structure, subject to vertical loads.

If we imagine a vertical section S drawn across a beam at any point P of its length, the portions of matter which lie immediately to the right of S exert on those immediately to the left a system of forces which must be in equilibrium with the remaining forces which act on the part of the beam which lies to the left, and must therefore be equivalent to a vertical force F at P, and a couple M. In virtue of the law of equality of action and reaction, the matter which lies to the left of S must exert on that to the right an exactly equal and opposite system of forces, which is therefore reducible to a force $-F$ and a couple $-M$. The force F is called the 'shearing force,' and the couple M the 'bending moment,' at the point P of the beam*. We will assume the positive senses of this force and couple to be those indicated in the annexed figure, where the two portions of the beam are drawn separately.

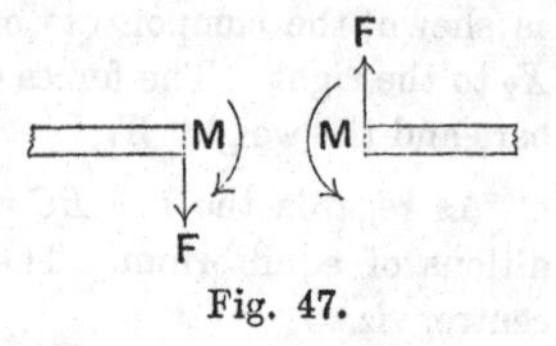

Fig. 47.

If the remaining forces which act on either portion of the beam are known, then resolving vertically we find the value of F, and by taking moments about P we obtain that of M.

Hence if PQ be any portion of the beam which is free from external force, Q lying (say) to the right of P, we have, with an obvious notation,

$$F_P = F_Q, \quad M_Q - M_P = -F.PQ. \quad \text{...........(1)}$$

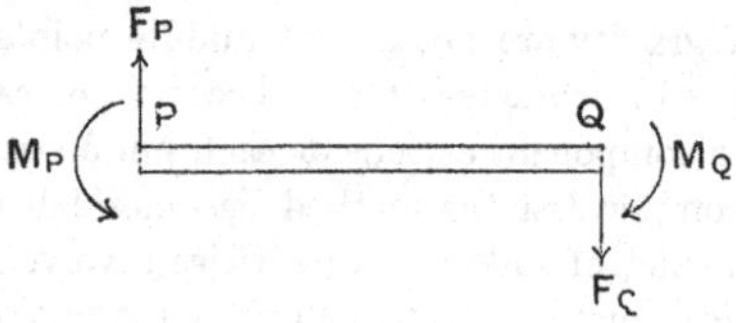

Fig. 48.

Hence, along any portion of the beam which is free from loads and from supporting pressures, the shearing force F is constant, whilst M diminishes as we travel to the right, with a gradient equal to $-F$.

Again, if PQ be a short segment containing an isolated load W, we have

$$\left.\begin{aligned} F_Q - F_P &= -W, \\ M_Q - M_P &= -F_Q.PQ - W.\epsilon.PQ, \end{aligned}\right\} \text{...(2)}$$

where ϵ is some proper fraction. Hence, ultimately, when PQ is regarded as infinitely small, we have

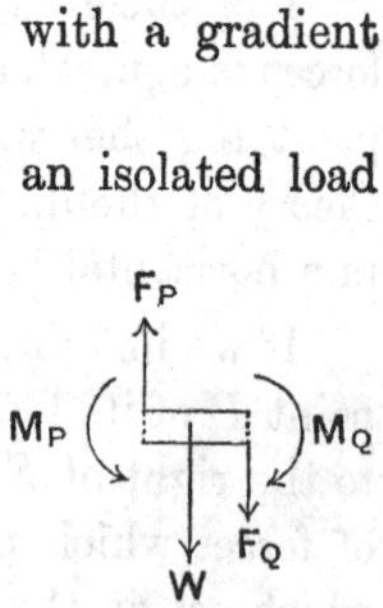

$$M_Q = M_P. \quad \text{...........(3)}$$

Fig. 49.

* It is shewn in Chap. XVII that the state of strain at any point of the length depends mainly on the bending moment there.

Hence F is discontinuous at the point of application of a concentrated load (or at a point of support), diminishing by an amount equal to the downward force as we pass the point in question to the right; whilst M is continuous. Accordingly, if we draw the graphs of F and M, respectively, for any system of concentrated loads and pressures, the former will consist of a series of horizontal lines, and the latter of a continuous chain of sloping lines.

A more purely geometrical way of obtaining the diagrams is explained in Chap. IV.

Ex. 1. Take the case of a single load W at any point C on a beam which is supported at A and B.

If $AC=a$, $AB=l$, the reactions at A, B are $W(l-a)/l$ and Wa/l, respectively. Hence the shearing force F has the value $W(l-a)/l$ from A to C, and the value $-Wa/l$ from C to B. Also, taking moments about C of the forces which act on the segment AC, we find

$$M_C = -Wa(l-a)/l. \qquad (4)$$

Since M vanishes at A and B*, its graph is now easily constructed on any required scale.

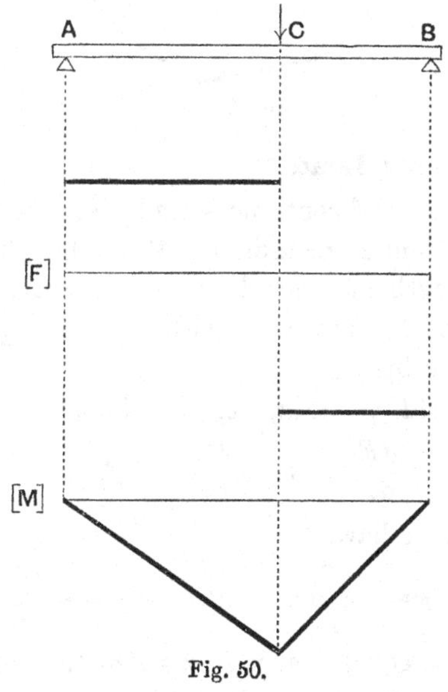

Fig. 50.

The relation between the gradient of the M-graph and the ordinate of the F-graph should be noticed.

* For M vanishes as a matter of course at a free end, and since it is continuous it vanishes at a point immediately to the right of A or immediately to the left of B.

Ex. 2. The annexed Fig. 51 shews a case of several loads. The amounts of the respective loads are indicated by the discontinuities of the F-graph.

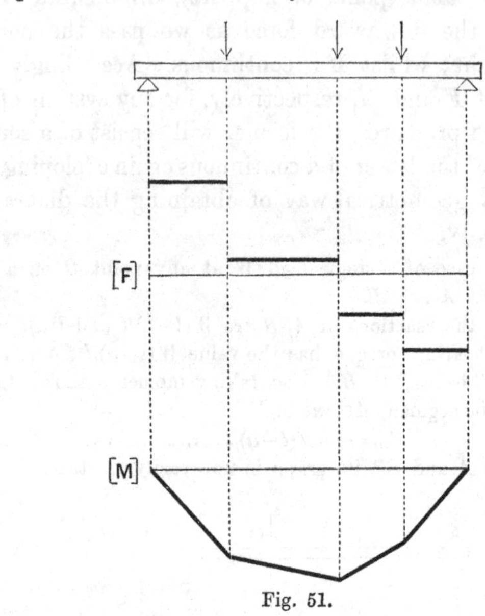

Fig. 51.

28. Continuous Load.

To pass to the case of continuous loads, let x be the horizontal coordinate of any point P, measured to the right. The load on an element of the length may now be denoted by $w\delta x$, where w, the 'load per unit length,' may vary with x. The equations (2) of Art. 27 are replaced by

$$\delta F = -w\delta x, \quad \delta M = -F\delta x,$$

or

$$\frac{dF}{dx} = -w, \quad \frac{dM}{dx} = -F. \quad \text{..................(1)}$$

By integration we have

$$F_Q - F_P = -\int_P^Q w\,dx, \quad M_Q - M_P = -\int_P^Q F\,dx. \quad \text{......(2)}$$

The former of these equations expresses that the difference of the shearing force at any two points is equal to the total intervening load, as is otherwise obvious. The latter shews that the difference of the bending moments is proportional to the area intercepted between the corresponding ordinates in the graph of F.

In the case of a uniform load w is constant, and we have from (1)

$$F = -wx + A, \quad M = \tfrac{1}{2}wx^2 - Ax + B. \quad \text{.........(3)}$$

The arbitrary constants A, B are to be determined by the conditions of the special problem, e.g. the conditions at the ends of the beam. The graph of F is a straight line; that of M is a parabola whose axis is vertical and concavity upwards.

Ex. 1. A uniform heavy beam supported at two points close to the ends.

We take the coordinates of the supports to be $x=0$, $x=l$, and apply the formulæ to the intervening space. Immediately to the left of the first support we have $F=0$, $M=0$. Since at this point there is an upward pressure $\frac{1}{2}wl$, we have, immediately to the right, $F=\frac{1}{2}wl$, $M=0$, in virtue of Art. 27 (1). Hence, in (3), $A=\frac{1}{2}wl$, $B=0$. The formulæ are therefore

$$F = w(\tfrac{1}{2}l - x), \qquad M = -\tfrac{1}{2}wx(l-x). \quad \text{..................(4)}$$

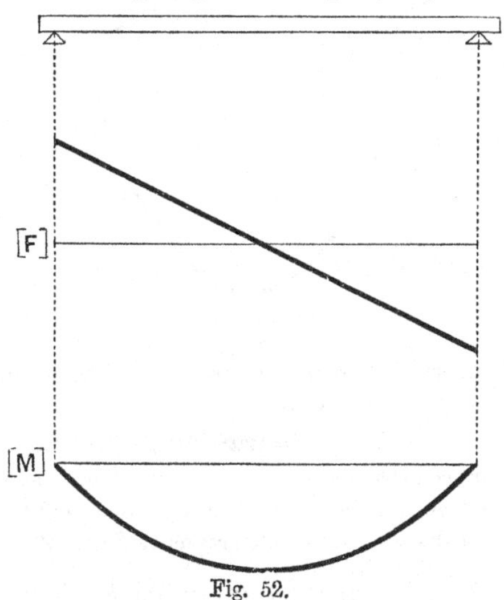

Fig. 52.

These results might have been obtained directly by resolving and taking moments, as in the next example.

Ex. 2. Let the supports C, D be at equal distances a $(<\frac{1}{2}l)$ from the centre O.

To take advantage of the symmetry we choose the origin at O, and consider only the portion OB of the beam. If x be the coordinate of a point P between D and B, we have, resolving vertically the forces on the segment PB, and taking moments about P,

$$F = w(\tfrac{1}{2}l - x), \qquad M = \tfrac{1}{2}w(\tfrac{1}{2}l - x)^2; \quad \text{.....................(5)}$$

for the weight of this segment is $w(\frac{1}{2}l - x)$, and its centre of gravity is of course at its middle point.

Again, at the point O we have obviously $F=0$, whilst

$$M_0=-\tfrac{1}{2}wla+\tfrac{1}{8}wl^2, \quad\text{.............................(6)}$$

as is found by taking moments about O of the forces acting on the half OB. These consist of the upward pressure $\frac{1}{2}wl$ of the support D, and the weight

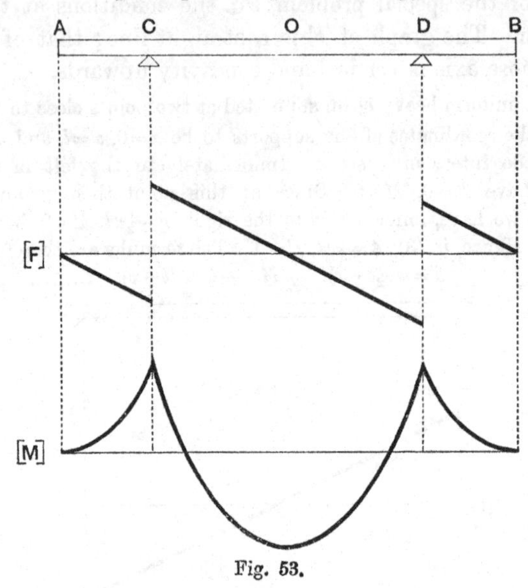

Fig. 53.

$\frac{1}{2}wl$ acting at a distance $\frac{1}{4}l$ from O. Hence if P lies between O and D, we find, considering the equilibrium of OP,

$$F=-wx, \qquad M=\tfrac{1}{2}wx^2-\tfrac{1}{2}wla+\tfrac{1}{8}wl^2. \quad\text{.................(7)}$$

It should be noticed that the relations (1) are verified, and that the two formulæ for M agree when $x=a$, whilst F has a discontinuity of the proper amount. The figure shews the complete graphs of F and M.

It is to be remarked that in any problem of the above kind the values of F and of M, due to different loads, will give by addition the values corresponding to the combined load. For instance, the case of Fig. 51 might have been obtained by superposition from three cases of the type of Fig. 50.

29. Sliding Friction.

The usually adopted laws of sliding friction have been stated for the case of a particle in Art. 9. The natural generalization is as follows:

If a body be in contact with a rough surface or surfaces at one or more points, equilibrium is possible if, and only if, the statical conditions can be satisfied by the assumption of reactions at the various points whose inclination to the common normal does not, in each case, exceed a certain value λ. This limiting value λ is connected with the coefficient μ of friction by the relation $\tan\lambda = \mu$. The values of μ, and therefore of λ, may of course be different for the different pairs of surfaces which are in contact.

Ex. A bar AB rests, like a ladder, with the upper end A against a vertical wall, which we will suppose to be smooth, and the lower end B on a rough horizontal plane.

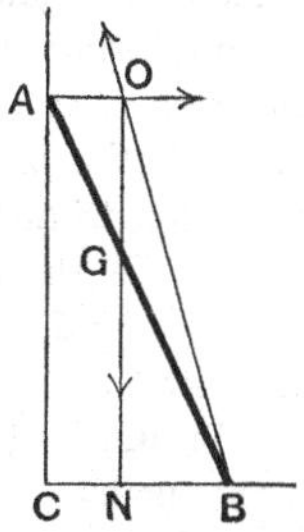

Since there are only three forces acting on the bar, viz. its weight, the horizontal pressure at A, and the reaction of the ground at B, these must be concurrent, say in the point O. Hence if G be the centre of gravity of the bar, the angle GOB in the figure cannot exceed λ, the angle of friction at B, consistently with equilibrium. If θ be the inclination of the bar to the vertical, we have, in the figure,

Fig. 54.

$$\tan\theta = \frac{CB}{AC}, \qquad \tan GOB = \frac{BN}{NO} = \frac{BN}{AC},$$

and therefore
$$\tan\theta = \frac{CB}{BN}\tan GOB = \frac{AB}{BG}\tan GOB. \quad \text{.................(1)}$$

Hence, for a given position of G, the greatest admissible inclination of the bar is given by

$$\tan\theta = \frac{AB}{BG}\tan\lambda = \mu . \frac{AB}{BG}. \quad \text{........................(2)}$$

For a given inclination θ, we must have

$$BG \not< \mu . AB . \cot\theta. \quad \text{..............................(3)}$$

If with the point of contact P of two surfaces as vertex we describe a right circular cone of semi-angle λ about the common normal as axis, the direction of the resultant reaction at P must lie within this cone, which is accordingly called the 'cone of friction' at P. In two-dimensional problems, such as we deal with here, this cone is represented by two opposite generating lines, viz. the two lines in the plane of the figure which make angles λ on either side with the normal.

The existence of friction often leads to cases of statical indeterminateness, which however become physically determinate (although not always easily calculable) when account is taken of the elastic properties of the bodies concerned. Cf. Art. 25.

Ex. A bar AB rests on two inclined planes which face each other.

The case where the planes are smooth has been solved in Art. 23, Ex. 3. When friction is to be taken into account, we draw through each of the points A, B a pair of straight lines making angles with the normal equal to the corresponding angle of friction. In the case illustrated in Fig. 55, these angles are less than the inclinations of the respective planes.

Since the directions of the reactions at A, B must lie within the angular spaces thus defined, it is necessary for equilibrium that the point of concurrence of the three forces acting on the bar should lie within the quadrilateral area which is common to these spaces. Hence if the vertical through G crosses this area, the position is one of equilibrium, but the precise position of the point of concurrence, and consequently the directions and magnitudes of the reactions, cannot be determined on purely statical considerations.

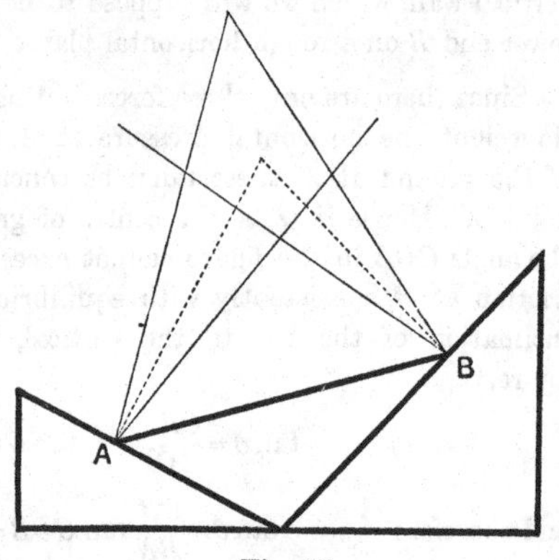

Fig. 55.

If the vertical in question passes through the extreme points of the quadrilateral, to the right or left, the reactions have their greatest possible obliquity to the normals, and the position is one of 'limiting equilibrium.' In this case the reactions are determinate. The formulæ for the extreme inclinations of the bar are obtained from Art. 23, (3) if we replace α and β in the one case by $\alpha+\lambda$ and $\beta-\lambda'$, and in the other by $\alpha-\lambda$ and $\beta+\lambda'$, respectively, where λ, λ' are the two angles of friction.

The indeterminateness, when it exists, is explained by the fact that the bar when placed in position is subject to a stress which varies with the manner in which the operation is performed. It is different, for instance, when the bar is gently laid on the planes, and when it is firmly pressed into its position. This implies a variable constituent in the reaction of each plane.

30. Friction at a Pivot. Rolling Friction.

The pivot is supposed to be cylindrical, and to fit almost exactly into a cylindrical hole, of slightly greater radius. We will assume that the contact takes place along a generating line of the cylinder*, which is represented in the figure by the point A. The resultant action of the pivot on the socket is then represented by a single force whose inclination to the normal AO cannot exceed a certain value λ. Hence if we make the angle OAB equal to λ, AB will be one of the two extreme directions of the force. Drawing OB at right angles from the centre O of the pin, we have $OB = OA \sin\lambda = b \sin\lambda$, if b denote the radius of the pin. Hence the line of the reaction cannot fall outside a circle of radius $b \sin\lambda$ described with O as centre. This is called the 'circle of friction.' In the case of well-lubricated joints it is usually small.

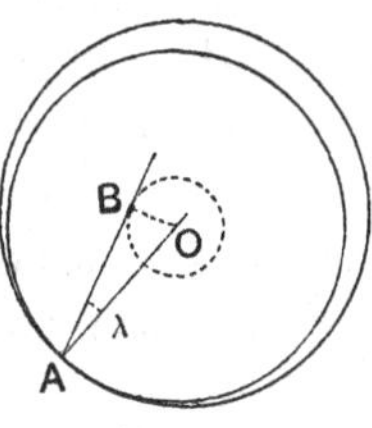

Fig. 56.

In the case of a bar jointed at each end, the reactions of the two pins must be equal and opposite, if we neglect the weight of the bar itself. The extreme positions of the line of action will therefore be given by the four common tangents to the two friction-circles. If the pins be small, and well-lubricated, it is sufficient for many purposes to assume that the line of action passes through the centres of the pins.

Ex. Suppose that a locomotive wheel of radius a carries a portion W of the load, and that the radius of the friction-circle is r. In limiting equilibrium the reaction between the axle and the wheel is tangential to the friction-circle, and since (if we neglect the weight of the wheel) it must be balanced by the reaction of the rail, it goes through the point of contact. If θ be its inclination to the vertical, we have therefore

$$\sin\theta = \frac{r}{a}, \quad \ldots\ldots\ldots\ldots\ldots(1)$$

and the horizontal component is

$$F = W\tan\theta = W\frac{r}{a}, \quad \ldots\ldots\ldots(2)$$

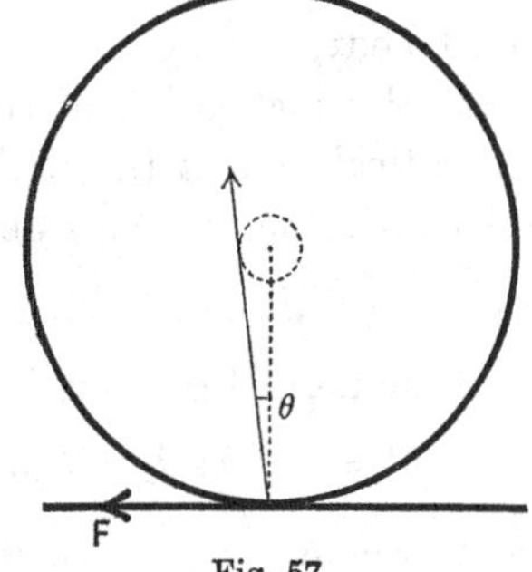

Fig. 57.

practically, since θ is usually small. The horizontal resistance is therefore less than in the case of sliding friction, in the ratio of r/a to μ. This shews the advantage of large wheels.

* This is not strictly correct as a representation of actual conditions, but may serve for purposes of general explanation.

So far, nothing has been said about 'rolling friction.' It is plain that as a matter of fact there must be some influence which opposes the rolling of one body over another, in addition to the force F of sliding friction already considered. For take the case of a body supported on a horizontal plane by cylindrical or spherical rollers, on which it rests at their highest points. If the action at each point of contact reduces to a single force, it is evident from a consideration of the equilibrium of a roller that this force must be vertical. Hence there would be no resistance to a horizontal pull on the body.

The usual empirical assumption is that at the place of contact there is a resisting *couple*, in addition to the tangential friction F, and that the maximum moment (M) of this couple is proportional to the normal pressure R, viz. we have

$$M = Rr, \qquad (3)$$

where r is a constant length depending on the nature of the surfaces and the curvature.

On this hypothesis the scheme of forces acting on the roller would be as shewn in Fig. 58, if we neglect the weight of the roller itself. It is assumed that there is a horizontal force on the supported body, urging it to the right. If θ denote the inclination of the reactions S to the vertical, and a the radius of the roller, we have, taking moments about the centre,

$$2Sa \sin\theta = M_1 + M_2. \qquad (4)$$

Since the normal pressure is $S\cos\theta$, we have

$$M_1 = r_1 S\cos\theta, \quad M_2 = r_2 S\cos\theta, \qquad (5)$$

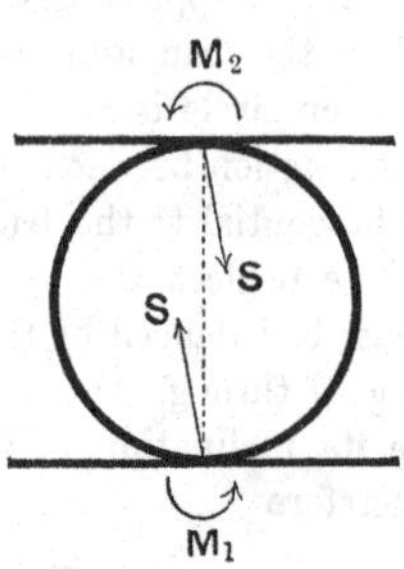

Fig. 58.

and therefore

$$\tan\theta = \tfrac{1}{2}(r_1 + r_2)/a. \qquad (6)$$

The origin of the couple is doubtless to be sought in the fact that the contact does not take place at a point, but over a certain area, the surfaces being deformed by the pressure, but into this we cannot enter. It appears at any rate that r is smaller the harder the materials. The values found by experiment are in many cases of the order of a fraction of a millimetre. The value of θ, and consequently the resistance, is therefore very small, unless the radius a be itself very minute*.

EXAMPLES. IV.

(General Theory.)

1. Shew how to construct geometrically the resultant of two forces whose lines of action meet outside the paper.

2. Prove the following construction for the preceding problem:

Let AA', BB' be straight lines representing the given forces completely. Through A, B draw AC, BC parallel to $A'B$, AB', respectively; and through A', B' draw $A'C'$, $B'C'$ parallel to AB', $A'B$, respectively. Then CC' will represent the resultant completely.

3. Prove that if a plane system of forces has equal moments about three points which are not collinear it reduces to a couple.

4. If three forces P, Q, R act along the sides BC, CA, AB, respectively, of a triangle ABC, the resultant will pass through the centres of the in- and circumscribed circles if

$$\frac{P}{\cos B - \cos C} = \frac{Q}{\cos C - \cos A} = \frac{R}{\cos A - \cos B}.$$

5. $ABCD$ is a convex quadrilateral; prove that if four forces represented completely by AB, BD, DC, CA are in equilibrium, AD must be parallel to BC.

6. Forces p.AB, q.BC, p.CD, q.DA act in the corresponding sides of a quadrilateral $ABCD$ in the senses indicated; prove that they are equivalent to a single force unless $p = q$, or unless $ABCD$ is a parallelogram.

7. AD, BE are the perpendiculars from the angles A, B of a triangle to the opposite sides, and a force P acts along DE. Replace it by three forces acting along BC, CA, AB, respectively. [$P\cos A$, $P\cos B$, $-P\cos C$.]

8. Two forces P, Q act through fixed points A, B. If they be turned through any the same angle about these points, in the same sense, so as to make a constant angle with one another, prove that their resultant R turns about a fixed point C (called the 'astatic centre'), and that

$$\frac{P}{BC} = \frac{Q}{AC} = \frac{R}{AB}.$$

Deduce the existence of an astatic centre for any plane system of forces acting through fixed points.

* The empirical laws of sliding and rolling friction were formulated by the French physicists C. A. Coulomb (1821) and A. Morin (1831–3).

For information as to more recent experiments the reader is referred to books on technical Mechanics.

EXAMPLES. V.

(Problems.)

1. If the addition of ·01 gramme in one scale-pan of a balance makes the pointer move over 5 mm. of its scale, find the depth of the centre of gravity of the beam below the plane of the three knife-edges, having given that the length of the beam between the outer knife-edges is 30 cm., its weight 180 grammes, and the length of the pointer 15 cm. [·25 mm.]

2. A lamina whose weight is W can turn in a vertical plane about a fixed point O. If weights P, Q be suspended from points A, B on opposite sides of the line OG joining O to the centre of gravity G, prove that the angle θ which OG makes with the vertical is given by

$$\tan\theta = \frac{Pa\sin\alpha - Qb\sin\beta}{Pa\cos\alpha + Qb\cos\beta + Wk},$$

where $a = OA$, $b = OB$, $k = OG$, and α, β denote the angles AOG, BOG.

3. A uniform ladder whose weight is 50 lbs. and length 10 ft. rests with its upper end against a smooth wall, and its foot is 4 ft. from the wall. A man whose weight is 150 lbs. stands on one of the rungs at a distance of 7 ft. from the lower end. Find, from a diagram drawn to scale, or by calculation, the horizontal thrust of the end on the ground. [56·7 lbs.]

4. AC and BC are two light rods hinged at A and B to two fixed points in the same vertical, and AC is horizontal. If $AC = 4$ ft., $AB = 2$ ft., find in lbs. the tension or thrust in each rod when a weight of 20 lbs. is suspended from C. [40 lbs., 44·7 lbs.]

5. A uniform rod AB of weight W hangs from a fixed point O by a string OA attached to the end A. If a couple of moment N be applied to the rod in a vertical plane, find its inclination to the vertical when in equilibrium.

6. A uniform heavy rod of given length is to be supported in a given position, with its upper end resting at a given point against a smooth vertical wall, by means of a string attached to the lower end of the rod, and to a point of the wall. Find by a geometrical construction the point in the wall to which the string must be attached.

7. A uniform beam of length l hangs from a fixed point by two ropes of lengths a, b; find their tensions.

$$\left[\frac{Wa}{\surd(2a^2 + 2b^2 - l^2)},\quad \frac{Wb}{\surd(2a^2 + 2b^2 - l^2)}.\right]$$

8. A sphere rests between two smooth inclined planes (α, α') whose line of intersection is horizontal. Find the pressures on the planes.

$$\left[\frac{W\sin\alpha'}{\sin(\alpha + \alpha')},\quad \frac{W\sin\alpha}{\sin(\alpha + \alpha')}.\right]$$

9. A uniform bar $APQB$ of given weight W rests with its lower end A on a smooth horizontal plane, and is in contact with smooth pegs at P and Q, of which the former is above and the latter below the bar. Find the pressures on the horizontal plane and on the pegs, having given the distance PQ, the distance of the centre of gravity from A, and the inclination of the line PQ to the horizontal.

10. A beam of weight W rests against a smooth horizontal rail, with its lower end on a smooth horizontal plane. Find what horizontal force must be applied to the lower end in order that the beam may be in equilibrium at a given inclination θ to the horizontal, having given the height h of the rail above the horizontal plane, and the distance a of the centre of gravity of the beam from the lower end. [$Wa/h \,.\, \sin^2\theta \cos\theta$.]

11. The ends A, B of a uniform bar can slide without friction along horizontal and vertical lines OA, OB, respectively; and the bar is maintained in equilibrium by a string connecting a given point P of it to O. Find the tension of the string, having given the inclinations of the string and the bar to the horizontal.

In what cases is equilibrium impossible?

12. In a 'wheel and axle' the diameter of the wheel is 3 ft., and that of the axle 6 in. A rope wrapped in opposite ways round the wheel and axle carries a pulley weighing 12 lbs. What force must be applied at right angles to a crank 18 in. long in order to maintain equilibrium?

13. Two uniform rods AB, BC are freely jointed at B, and are moveable about A, which is fixed. At what point must a prop be placed below BC, so that the rods may rest in a horizontal line?

[The prop must divide BC in the ratio $W' : W + W'$, where W, W' are the weights of the rods.]

14. A uniform beam AB can turn freely about a hinge at A, and to the end B is attached a string which passes over a small smooth pulley at C, vertically above A, and carries a weight P hanging freely. Prove that in equilibrium

$$BC = \frac{2P}{W} \,.\, AC,$$

where W is the weight of the beam.

15. A uniform rectangular plate $ABCD$, of weight W, can turn freely in a vertical plane about its highest point A, which is fixed; and the side BC rests on a smooth peg at its middle point, making an angle α with the horizontal. Give a diagram shewing the arrangement of the forces acting on the plate; and prove that the pressure on the hinge is $W \sin\alpha \sec\beta$, where β is the angle ACB.

16. A uniform rectangular plate $ABCD$ of weight W hangs from A. If a weight W' be suspended from B, the inclination θ of the diagonal AC to the vertical is given by

$$\tan\theta = \frac{2W'ab}{W(a^2+b^2)+2W'a^2}$$

where $a = AB$, $b = AD$.

17. A rectangular picture-frame hangs from a perfectly smooth peg by a string of length $2a$ whose ends are attached to two points on the upper edge at distances c from its middle point. Prove that if the depth of the picture exceeds $2c^2/\sqrt{(a^2-c^2)}$ the symmetrical position of equilibrium is the only one.

18. A uniform elliptic plate of weight W rests on two smooth pegs P, Q at the same level, its plane being vertical. Prove that a principal axis must be vertical, unless the pegs are at the extremities of conjugate radii.

Prove that in the latter case the pressures on the pegs are

$$\frac{CQ}{PQ}.W \quad \text{and} \quad \frac{CP}{PQ}.W,$$

respectively, where C denotes the centre of the plate.

19. A uniform rod can turn freely about one end, which is fixed, and is supported in an inclined position by a string attached to the other end. Determine graphically the reaction of the joint, and the tension of the string, the inclinations of the rod and string being given.

Find also for what direction of the string the reaction is least, the position of the rod being given; and shew that the reaction is then at right angles to the rod.

20. Two fixed smooth rods OA, OB are inclined downwards at equal angles α to the horizontal. Two equal uniform rods AC, CB, hinged together at C, have rings at the extremities A, B, which can slide on the fixed rods. Prove that in equilibrium the inclination θ of the moveable rods to the horizontal is given by

$$\tan\theta = \tfrac{1}{2}\cot\alpha.$$

21. Two uniform rods AB, BC, each of length $2a$, are smoothly jointed at B, and rest in a vertical plane on two smooth pegs at a distance $2c$ apart. Prove that they are in equilibrium if each rod makes with the vertical an angle θ given by

$$\sin\theta = \sqrt[3]{(c/a)}.$$

22. Two equal circular disks of radius a, with smooth edges, are placed on their flat sides in the corner between the smooth vertical sides of a box, and touch each other in the line bisecting the angle. Prove that the radius of the least disk which may be pressed between them without causing them to separate is $\cdot 414a$.

23. Two spheres are connected by a string passing over a smooth peg, and hang in contact. Prove that the distances of their centres from the peg are inversely proportional to their weights.

24. A chain is composed of n equal symmetrical links, each of weight W, and is attached at the ends to two fixed points. If α, β be the inclinations of the end links to the horizontal, prove that the horizontal pull on either support is

$$\frac{(n-1)\,W}{\tan\alpha+\tan\beta}.$$

25. An equilateral triangle ABC formed of three uniform bars, each of weight W, is suspended from the corner A; prove that the reaction at B is equal to $\cdot 764\,W$, and makes an angle of about 41° with the horizontal.

26. If AC, BC represent the jib and tie-rod of a crane, find the stresses in them when a weight W is suspended from C, neglecting the weights of AC, BC themselves.

If the weights of AC, BC be W_1, W_2, the centres of gravity being at the middle points, prove that the action at C on the jib AC consists of a force $\frac{1}{2}W_1$ upwards, and a force

$$\{W+\tfrac{1}{2}(W_1+W_2)\}\frac{AC}{AB}$$

in the direction from C to A.

What are the corresponding results for the rod BC?

27. A frame of three rods smoothly jointed together so as to form a triangle ABC is in equilibrium under three forces perpendicular to the sides at their middle points, and proportional to the sides. Prove that the reactions at the joints are all equal, and that their directions are tangential to the circle ABC.

28. A link polygon is acted on by forces perpendicular to the sides, at the middle points, which are respectively proportional to those sides, and are directed all inwards or all outwards. Prove that if the polygon is in equilibrium a circle can be described through its angular points, and that the reactions at the joints are all equal, and tangential to this circle.

EXAMPLES. VI.

(Bending Moments.)

1. Prove that the bending moment at any point P of a beam, due to a concentrated load at Q, is equal to the bending moment at Q due to an equal load at P.

2. Construct a diagram shewing the variation of the bending moment at any point P of a beam AB, supported at A and B, for different positions of a concentrated load.

3. A bar AB is supported at two points C, D equidistant from the ends. Draw the diagrams of shearing force and bending moment when equal weights W are suspended from A, B.

4. If AB be a horizontal beam supported at the ends, prove that the curves of bending moment for all distributions of the same load which have a resultant in the same vertical line will have a common tangent at A, and also at B.

Illustrate by the cases where the beam is (1) uniformly loaded from A to O (the middle point), and (2) subject to the same load concentrated at the middle point of AO.

5. A beam is confined laterally between stops at A and B, and has a stanchion passing through it at right angles at a point C. If a couple be applied at C about the axis of the stanchion, find the distribution of shearing force and bending moment; and illustrate by figures.

6. A light rod AB is held in place by two smooth rings at the ends. Two equal and opposite forces are applied at right angles to the rod at points C, D of it. Draw the diagrams of shearing force and bending moment.

What is the connection between this and the preceding question?

7. A uniform beam is supported, as in Fig. 53, p. 62, at two points equidistant from the ends; find where these must be situate in order that the maximum bending moment (without regard to sign) may be as small as possible.

Also, in this case, find the ratio of the maximum bending moment to what it would be if the beam were supported at the ends.

[The distance between the supports is ·586 of the length of the beam; the ratio of the maximum bending moments is ·171.]

8. A uniform beam ABC rests on two supports at A and B, and carries a weight, equal to one-fourth its own weight, at C. Draw carefully the curve of bending moments, having given $AB = 10$ ft., $BC = 6$ ft.

9. A beam $ABCD$, supported at A and D, carries a uniform load from B to C; sketch the diagrams of shearing force and bending moment.

Explain the relations between the discontinuities (if any) in the two diagrams.

How are the diagrams altered (1) if the load be concentrated at its middle point, (2) if it be equally divided between the points B and C?

10. A uniform load of given length PQ travels along a horizontal beam supported at the ends A, B. Prove that the bending moment at any given point O is a maximum when

$$PO : OQ = AO : OB.$$

11. Draw the diagrams of shearing force and bending moment for the case of two uniform heavy beams AB, BC, hinged at B, and resting in a horizontal position on three supports at A, D, C. (Take $AB = BD = \frac{1}{2}DC$.)

12. A uniform beam $AOCB$ consists of two portions smoothly jointed at C; it is supported at the ends A, B, and at the middle point O. If $AO = OB = a$, $OC = b$, prove that the pressures on the three supports are

$$\tfrac{1}{2}w(a-b), \quad w(a+b), \quad \tfrac{1}{2}w(a-b).$$

Prove that if O be taken as the origin of x, the values of F and M for the two portions AO, and OCB are

$$F = -wx - \tfrac{1}{2}w(a+b), \quad M = \tfrac{1}{2}w(x+a)(x+b),$$

and

$$F = -wx + \tfrac{1}{2}w(a+b), \quad M = \tfrac{1}{2}w(x-a)(x-b),$$

respectively.

Draw the corresponding diagrams.

13. If the sectional area of a cantilever of length l, projecting horizontally from a wall, varies as $l^2 - x^2$, where x denotes distance from the wall, prove that the shearing force and bending moment are given by

$$F = \frac{w}{3l^2}(l-x)^2(2l+x), \qquad M = \frac{w}{12l^2}(l-x)^3(3l+x),$$

where w is the weight per foot at the wall.

14. A curved rod in the form of an arc of a circle has its ends connected by a tense string. Prove that the bending moment at any point varies as $\sin\frac{1}{2}\alpha \sin\frac{1}{2}\beta$, where α, β are the angular distances of the point in question from the ends.

EXAMPLES. VII.

(Friction.)

1. A circular hoop hangs over a rough peg; shew how to find the position of equilibrium when a given weight is suspended by a string tangentially from the rim.

If the weight of the hoop be 10 lbs., what is the greatest weight that can be so suspended, the angle of friction being 30°? [10 lbs.]

2. A uniform rod of length $4a$, bent into the form of a square, hangs with one of its sides over a rough peg. If μ be the coefficient of friction, prove that the greatest possible distance of the peg from the centre of the side is $\frac{1}{2}\mu a$.

3. A uniform plank, whose thickness may be neglected, rests in a horizontal position across a fixed circular cylinder of radius a. If l be the length of the plank, and λ the angle of friction, prove that the greatest weight which can be attached at one end without causing the plank to slip is

$$\frac{a\lambda}{\frac{1}{2}l - a\lambda} \cdot W,$$

where W is the weight of the plank.

4. A light rod rests in a horizontal position with one end A against a rough wall, being supported by a string which joins a point O of the rod to a point of the wall vertically above A. Prove that a weight may be suspended from any point of the rod within distances

$$\frac{\mu}{\tan\alpha \pm \mu} \cdot AO$$

from O, without causing the end A to slip, α denoting the inclination of the string to the horizontal.

Examine the case of $\tan\alpha < \mu$.

5. A circular cylinder rests with its axis horizontal on a rough plane of inclination $\alpha\,(>\lambda)$, and is supported by a string tangential to the cylinder and perpendicular to its axis. Prove that the greatest admissible inclination of the string to the horizontal is given by

$$\cos\theta = \frac{\sin(\alpha - \lambda)}{\sin\lambda}.$$

6. A uniform bar AB is supported at an inclination α to the horizontal, with the end B on a rough horizontal plane, by a string attached to A. Prove that the extreme inclinations (θ) of the string to the vertical are given by

$$\cot\theta = \frac{1}{\mu} \pm 2\tan\alpha.$$

7. A uniform bar AB rests on a plane of inclination α, along a line of greatest slope. If the lower end A be gently raised by means of a string attached to it, and making an angle θ with AB, prove that the end B will, or will not, begin to slip according as

$$\cot\theta \gtrless 2\tan\alpha + \mu.$$

How do you reconcile this result with the fact that when $\theta = 0$ the bar will not slide unless the tension of the string exceeds a certain value?

8. A uniform ladder rests in limiting equilibrium with its lower end on a rough horizontal plane and its upper end against a smooth wall. If θ be the inclination of the ladder to the vertical, prove that $\tan\theta = 2\mu$.

9. A uniform ladder of length l and weight W rests with its foot on the ground (rough) and its upper end against a smooth wall, the inclination to the vertical being θ. A force P is applied to it horizontally at a distance c from the foot, so as to make the foot approach the wall. Prove that P must exceed

$$\frac{Wl}{l-c}(\mu + \tfrac{1}{2}\tan\theta).$$

10. A uniform rod passes over one rough peg and under another, the coefficient of friction in each case being μ. The pegs are at a distance a apart, and the line joining them makes an angle α with the horizontal. Shew that equilibrium is not possible unless the length of the rod exceeds

$$a\left(1 + \frac{\tan\alpha}{\mu}\right).$$

11. A horizontal lamina has two pins A, B projecting from it which can slide in two straight slots in a horizontal table, at right angles to one another. If a couple be applied to the lamina, prove that it will not move unless the sum of the angles of friction at A and B falls short of 90°.

12. A bracket can slide along a fixed vertical rod, and supports a weight W suspended from its arm at a distance c from the rod. Prove that the bracket will not slip down if $c > AB/2\mu$, where A, B denote the points of contact, and μ is the coefficient of friction at each of these points.

13. A rectangular window-sash of breadth a and depth b, fitting somewhat loosely in its frame, is supported by two weights. If one of the cords be broken, find the relation between a, b, and the coefficient of friction μ, in order that the sash may not run down.

14. A uniform rod $ABCD$ rests across two horizontal rails at the same level at B and C, and $AB = CD$. A gradually increasing horizontal force is applied at A at right angles to the rod; determine whether the rod will begin to slip at B or at C first, the coefficients of friction at these points being the same.

15. A bar rests on two pegs, and makes an angle α with the horizontal. The centre of gravity is between the pegs, at distances a, b from them. Prove that for equilibrium

$$\tan\alpha \not> \frac{\mu_1 b + \mu_2 a}{a + b},$$

where μ_1, μ_2 are the coefficients of friction at the pegs.

Prove that if $\tan\alpha$ is less than the above limit the friction at each peg is indeterminate; and give the physical explanation.

16. A bar rests with its extremities on two inclined planes facing each other, as in Art. 23, Ex. 3; prove that if the inclination of each plane to the horizontal is less than the corresponding angle of friction, every position is one of equilibrium.

17. A bar rests on two planes of inclinations α, β which face each other, and λ, λ' are the corresponding angles of friction. Examine geometrically the condition of equilibrium when $\alpha < \lambda$, $\beta > \lambda'$.

Find the inclination of the bar to the vertical, in limiting equilibrium.

18. A uniform bar AB rests with its ends on two planes of equal inclination α which face each other, in a plane perpendicular to the line of intersection. If the angle λ of friction at each end be less than α, prove that the limiting inclination θ of the bar to the horizontal is given by

$$\tan\theta = \frac{\sin 2\lambda}{2\sin(\alpha - \lambda)\sin(\alpha + \lambda)}.$$

19. A rectangular table stands on a plane of inclination a, one pair of edges being horizontal. The lower pair of feet are fitted with castors, so that the friction there may be neglected. If the coefficient of friction (μ) for the upper pair be greater than $\tan a$, prove that there will be equilibrium provided the vertical through the centre of gravity cuts the plane at a distance from the line of the lower feet which lies between $l/\mu \,.\, \tan a$ and l, where l is the distance between the upper and lower feet.

What is the corresponding result if the upper feet (alone) are fitted with castors?

20. If in the preceding example there be friction at all four feet, the coefficients being μ for the upper and μ' for the lower feet, prove that if $\mu > \tan a$, $\mu' < \tan a$, the distance from the line of the lower feet of the point where the vertical through the centre of gravity cuts the plane must exceed

$$\frac{\tan a - \mu'}{\mu - \mu'} \,.\, l.$$

21. If in Ex. 8 the wall and the ground be equally rough, prove that the extreme inclination of the ladder to the vertical is twice the angle of friction.

22. A carriage is on an inclined plane, the axles being horizontal. Prove that it can be prevented from running down most easily by braking the front or the hind wheels according as the point where the vertical through the centre of gravity meets the ground is nearer to the line of contact of the front wheels or to that of the hind wheels.

23. A circular cylinder of radius a and weight W rests with its axis horizontal in a V-shaped groove whose sides are inclined at equal angles a to the horizontal, and a gradually increasing couple is applied to it in a plane perpendicular to its axis. If $a < \lambda$ prove that the cylinder will begin to roll up one side of the groove when the couple exceeds $Wa \sin a$. If $a > \lambda$ prove that the least couple which will make the cylinder turn about its axis is

$$\tfrac{1}{2} Wa \sec a \sin 2\lambda.$$

CHAPTER IV

GRAPHICAL STATICS

31. Force-Diagram and Funicular Polygon.

The graphical method of reducing a plane system of forces, which is now to be explained, involves the construction of two figures, viz. a 'force-diagram' and a 'funicular polygon'*.

In the force-diagram no regard is had to the precise configuration of the lines of action of the various forces. It is constructed by drawing in succession a series of vectors AB, BC, CD, ... to represent the given forces in magnitude and direction, and by joining the vertices of the polygon $ABCD$... to an arbitrary 'pole' O.

In the funicular, or 'link' polygon as it is sometimes called, we start with the lines of action of the various forces as given, but their magnitudes are not in evidence. Its vertices lie on the lines of action, and its sides are respectively parallel to the lines drawn from O in the force-diagram. More particularly, the two sides

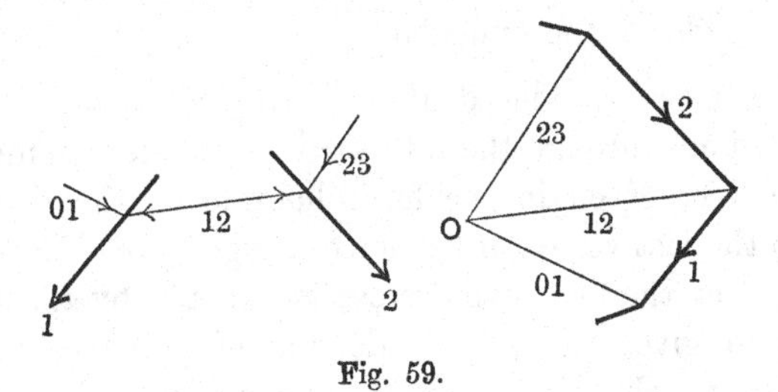

Fig. 59.

* The method was given by C. Culmann (1821–81), professor of engineering at Zürich, in his book *Die graphische Statik*, 1st ed., 1864.

which meet at any vertex are respectively parallel to the lines drawn from O to the ends of that side of the force-polygon which represents the corresponding force. In drawing the funicular we may start at any point on the line of action of one of the forces, and the rest of the figure is then determinate.

The relation between the two diagrams will be understood from the annexed Fig. 59, where corresponding lines in the force-diagram on the right, and the funicular on the left, are numbered similarly.

The sides of the force-polygon may be arranged in the first instance in any order, and the force-diagram can then be completed in a doubly-infinite number of ways, owing to the arbitrary position of the pole O. And in the case of each force-diagram we can draw a simply-infinite number of funiculars, in which corresponding sides are of course parallel.

32. Graphical Reduction of a Plane System of Forces.

The two diagrams being supposed constructed, it is seen that the force in any one of the lines of action in the funicular can be replaced by two components acting in the sides which meet on that line, and that the magnitudes and directions of these components are given by a corresponding triangle in the force-diagram. Thus the force numbered 1 in the preceding figure is equivalent to two forces represented by 01 and 12. When this process of replacement is complete, each terminated side of the funicular is the seat of two forces which neutralize each other, and there remain only two uncompensated forces, viz. those which act in the first and last sides of the funicular.

If these sides (produced if necessary) intersect, the final resultant will act through the intersection, and its magnitude and direction will be given in the force-diagram by the line joining the first to the last vertex of the force-polygon; see Fig. 60, where the resultant of the four given forces is denoted by x. Thus the force-diagram gives the magnitude and direction of the single resultant to which the system in general reduces, whilst the funicular determines a point on its line of action.

It will be noticed that the construction can be made to give not only the final resultant, but also the resultant of any group of consecutive forces. For this purpose we have only to make use of those portions of the diagrams which relate to this group.

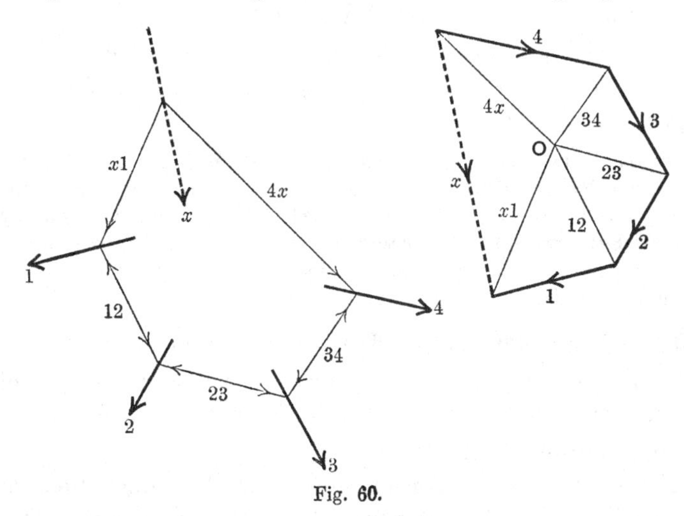

Fig. 60.

As a special case it may happen that the force-polygon is 'closed,' i.e. its first and last points coincide. The first and last sides of the funicular will then in general be parallel, and the two uncompensated forces will form a couple. It will be noticed that here also, as in Art. 19, a couple appears as the equivalent of an infinitely small force acting in an infinitely distant line.

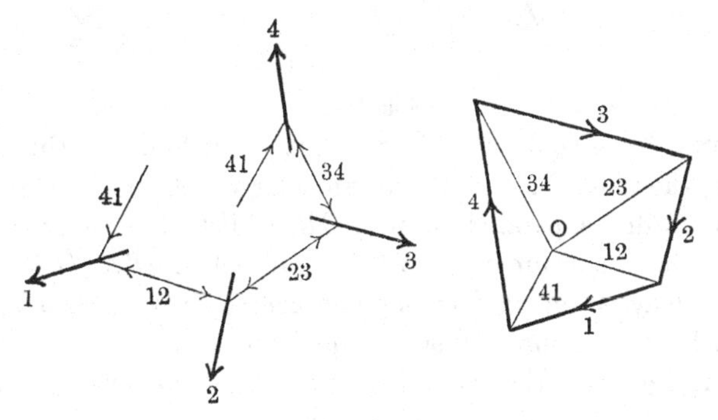

Fig. 61.

If, however, the force-polygon being closed, the first and last sides of the funicular are coincident, instead of parallel, the two outstanding forces will neutralize one another, and we have equilibrium. This is illustrated by Fig. 60, if we imagine the force x to be reversed and included in the system of given forces.

Hence the graphical conditions of equilibrium of a plane system of forces are that the force-polygon and the funicular polygon should both be closed.

A system of bars, smoothly jointed where they meet, having the shape of the funicular polygon, would evidently be in equilibrium under the action of the given forces, supposed applied at the joints. Moreover any bar in which the stress is of the nature of a tension, as opposed to a thrust, might be replaced by a string. This explains the origin of the names 'link-polygon' and 'funicular.' Cf. Art. 11.

33. Properties of the Funicular Polygon.

The figures obtained by taking two distinct positions O, O' of the pole in the force-diagram, and drawing the respective funiculars, have various interesting properties.

In the first place we may prove that the intersections of corresponding sides of the two funiculars will be collinear*. This is, essentially, a theorem of pure Geometry, but a simple statical

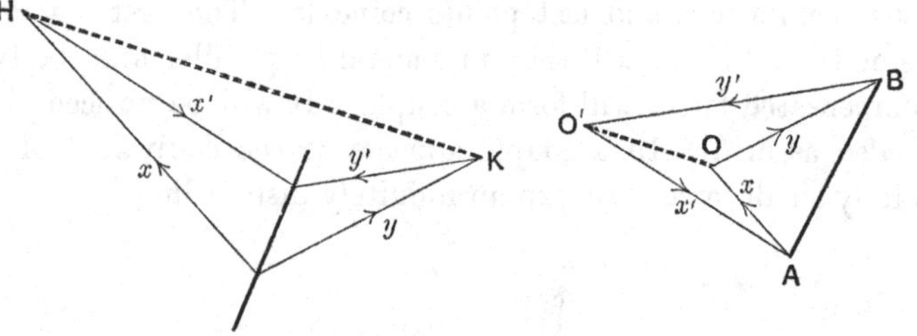

Fig. 62.

proof may be given. If AB represent any force in the force-polygon, the pair of sides of one funicular which meet on its line of action will be parallel to OA, OB, whilst the corresponding sides of the other funicular will be parallel to $O'A$, $O'B$. The force AB may be replaced by the two components marked x, y in the left-hand diagram, whilst an equal and contrary force BA in the same line of action is equivalent to the components marked

* C. Culmann, *l.c. ante* p. 77.

x', y'. The four forces x, y, x', y' are therefore in equilibrium. Now x, x' have a resultant through the point H, represented as to magnitude and direction by O′O, whilst y, y' have a resultant through K, represented by OO′. Since these two resultants must balance, HK must be parallel to OO'. If L be the intersection of a consecutive pair of corresponding sides of the two funiculars, it appears in the same way that KL is also parallel to OO', and is accordingly in the same line with HK; and so on. This line HKL... may be called, for a reason which will appear presently, the 'axis of perspective' of the two figures.

This theorem enables us, when one funicular is given, to draw any other without further reference to the force-diagram.

By considering the figure for three forces in equilibrium we are led to the theorem that if the lines AA', BB', CC' joining corresponding vertices of two

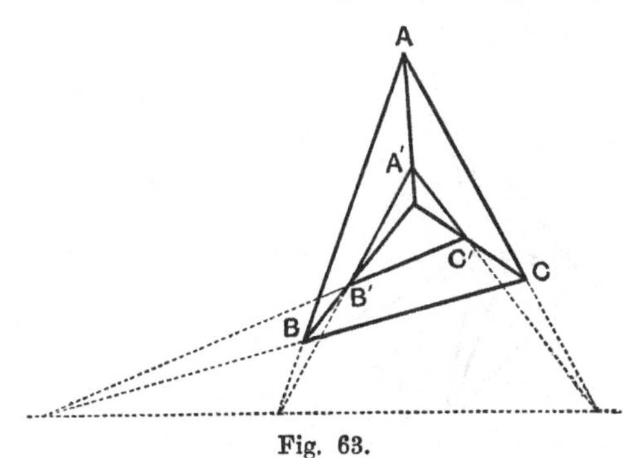

Fig. 63.

triangles ABC, $A'B'C'$ are concurrent, the intersections of corresponding sides will be collinear, and conversely. These are well-known theorems of Geometry. In a similar way we may infer from statical principles that if the sides of a closed polygon of $n+1$ sides pass each through a fixed point on a given straight line, and if n of its vertices describe given straight lines, the locus of the remaining vertex is also a straight line.

Again, it is obvious that in the case of a plane system of forces in equilibrium, the complete figure obtained by taking two poles O, O' in the force-diagram may be regarded as being the orthogonal projection of the edges of a closed plane-faced polyhedron. The polyhedron in question consists in fact of two pyramids, with vertices represented by O, O', standing on a common base represented by the force-polygon (see Fig. 64).

A similar statement holds with respect to the corresponding funicular diagram, with its two funicular polygons. For let p be the line on which pairs of corresponding sides of the two funicular polygons intersect, and imagine any two planes ϖ, ϖ' to be drawn through p. Through the vertices of the two funiculars

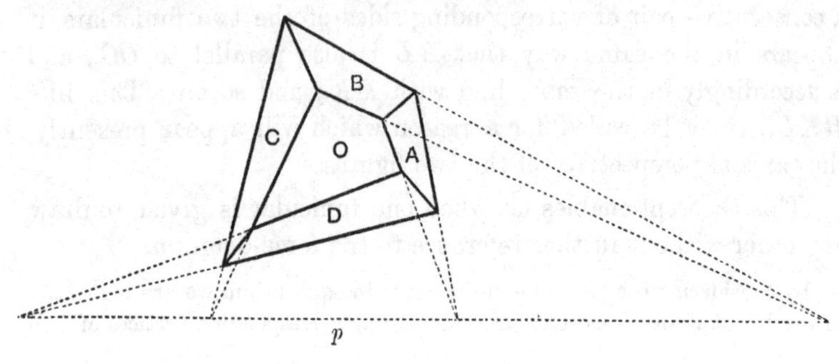

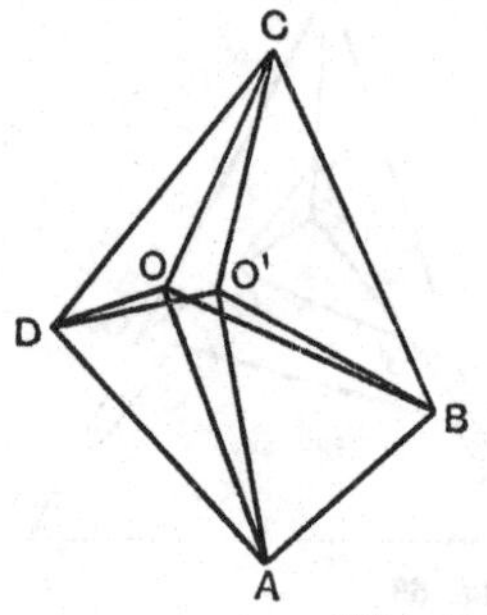

Fig. 64.

draw lines perpendicular to the plane of the paper, to meet in one case the plane ϖ and in the other the plane ϖ'. The points in ϖ and ϖ' thus determined will be the vertices of a polyhedron of which the funicular diagram is the orthogonal projection*.

34. Reciprocal Figures.

If we compare the two diagrams in Fig. 64 we notice that to each line in one there is a parallel line in the other, and that the lines which meet at a point in one diagram correspond to lines forming a closed polygon in the other. Two figures which are related in this way are said to be 'reciprocal,' since the properties

* L. Cremona, *Le figure reciproche nella statica grafica*, 1872.

of the first figure in relation to the second are the same as those of the second in relation to the first*. A simpler instance of two reciprocal figures is supplied by the funicular and force-diagram of three or more *concurrent* (or parallel) forces in equilibrium (Fig. 65).

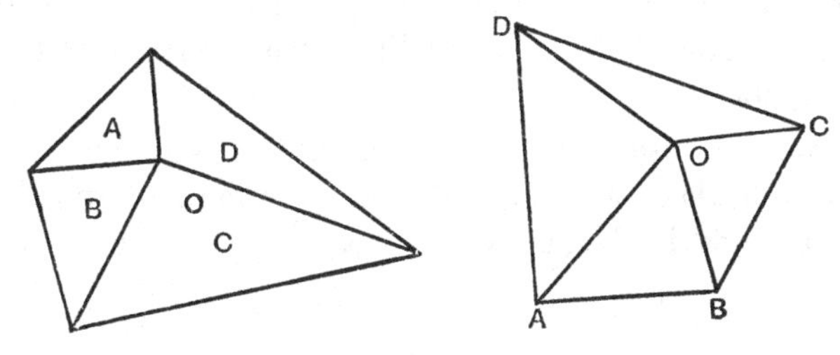

Fig. 65.

It is convenient to have a notation which shall put this reciprocal character in evidence. For this purpose we may designate the various *points* of one figure by letters $A, B, C, \ldots$, and the corresponding *polygons* in the other figure by the same letters. A line joining the points A, B in the former will then be parallel to the line which is common to the two polygons A, B in the latter. This notation was introduced by R. H. Bow† in connection with the theory of Frames (see Chap. v), where, also, reciprocal diagrams, or portions of such, present themselves. It is illustrated in Figs. 64, 65‡.

It is to be observed that a figure composed of points and straight lines joining them does not of necessity admit of a reciprocal. It may be shewn that a necessary and sufficient condition is that the figure should be the orthogonal projection of a closed polyhedron with plane faces. When this is fulfilled, a kind of reciprocal polyhedron can be constructed theoretically (in various ways), the orthogonal projections of whose edges form

* The theory of reciprocal figures, as thus defined, was first studied by J. Clerk Maxwell (1864), in a slightly different connection. The reciprocal relations of force-diagrams and funiculars were investigated by Cremona.

† *Economics of Construction in relation to Framed Structures*, London, 1873.

‡ The letter O, in the left-hand portion of Fig. 65, refers to the polygon formed by the four outer lines. In Fig. 64 the symbol O' in the first diagram is omitted. It corresponds to the outer quadrilateral.

the lines of the reciprocal plane diagram. The letters $A, B, C, \ldots$ in Bow's notation may be taken to refer to the *vertices* of one polyhedron and the *faces* of the other.

35. Parallel Forces.

When the given forces are all parallel, the force-polygon consists of segments of a straight line. This case has important practical applications.

Thus we may use the graphical method to find the resultant of any given loads on a girder. The construction is shewn, for the case of four loads, in the annexed figure.

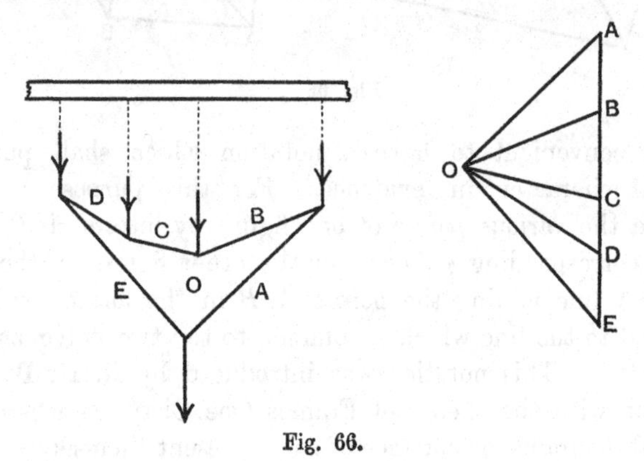

Fig. 66.

By a slight modification of the process we may resolve a system of parallel forces into two components acting in two given straight lines parallel to them. In this way we may find the pressures on the supports of a beam loaded in any given manner.

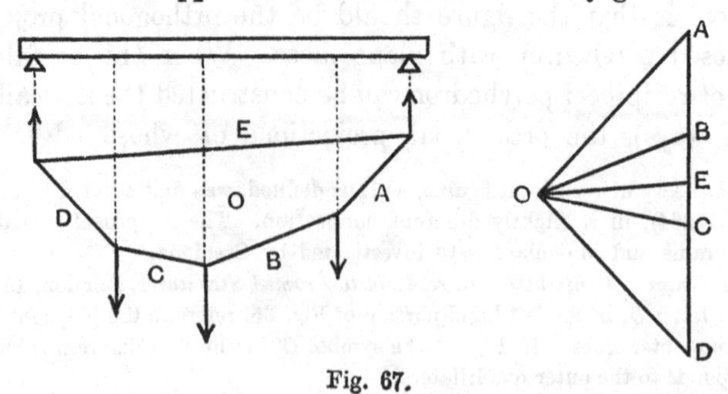

Fig. 67.

Thus if, in the force-diagram (Fig. 67), AB, BC, CD represent the given loads, we first construct the sides of the funicular which are parallel to OA, OB, OC, OD, and then join the points in which the extreme sides, corresponding to OA, OD, meet the vertical lines through the two supports. This joining line is called the 'closing line' of the funicular. If we draw OE parallel to it in the force-diagram, the segments DE, EA will represent the upward pressures exerted by the two supports on the beam. For the forces thus indicated, combined with the given loads, will constitute a system in equilibrium, since the force-polygon ($ABCDEA$) and the corresponding funicular are both closed.

The graphical method can be applied to find the 'centre' of a system of parallel forces acting through given points A_1, A_2, ... (Art. 22). For if we construct the line of action of the resultant for each of two given directions of the forces, the intersection of these lines will be the required point. The most convenient plan is to take the two directions at right angles. It is unnecessary to construct a second force-diagram, since the sides of the second funicular are sufficiently determined by the fact that they are respectively perpendicular to those of the first. The construction may be used to find the centre of gravity of a plane system of particles situate at A_1, A_2, ... whose masses are proportional to the several forces.

The method is in use for finding the mean centre of irregular plane areas. These are divided into strips, whose individual mean centres are found by estimation, the areas themselves being determined by the planimeter.

36. Bending Moments.

The funicular polygon can also be made to indicate the *moment* of a force, or of a system of forces, about any assigned point P.

For let F be a force, corresponding to AB in the force-diagram, and draw a parallel to it through P, meeting in H, K the sides of the funicular which correspond to OA, OB. Let R be the intersection of these sides, and draw OM perpendicular to AB. In taking the moment of F we may substitute for it its two components in

RH, KR, which are represented by AO, OB, respectively, in the force-diagram. Also, if we imagine these forces to act at H and K we need only take account of their components at right angles

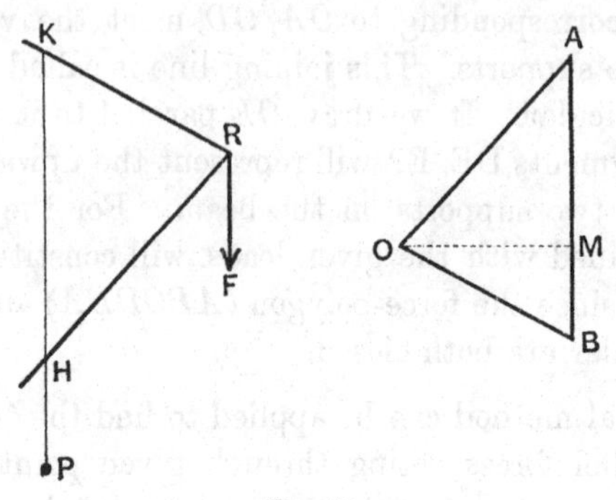

Fig. 68

to HK, which correspond to MO, OM, respectively. The moment is therefore given numerically by the product $OM.HK$. The sign to be attached to the result will depend on whether O lies on one side or other of AB, and on the sense of HK as compared with that of AB.

The result applies at once to any system of forces, provided F (and AB) refer to their resultant. The moment of the system about any point P is equal to OM multiplied by the length HK which the first and last sides of the funicular intercept on a line drawn through P parallel to the resultant.

The case of a system of parallel (say vertical) forces is specially simple, since OM is then a fixed length. The moment of such a system is then represented on a certain scale by the intercept made by the first and last sides of the funicular on the vertical through P. Moreover, by a proper convention, the sense of this intercept will determine the sign.

For instance, in the case of a beam in equilibrium under given loads and the reactions at the supports, if the funicular polygon, which is now closed, be constructed as in Fig. 67, the vertical through any point P will intersect two of its sides, which are the first and last sides of that portion of the funicular which belongs to the group of forces lying to the left (say) of P. The length intercepted will accordingly be proportional to the moment of this group about P, and therefore to the *bending moment* at P. The complete funicular will therefore serve as a diagram of bending moments. If the closing line were horizontal, it would be identical with the diagram employed in Art. 27 *.

Ex. 1. Let it be required to find the pressures on the supports of a beam, and to construct the diagram of bending moments, for the case shewn in Fig. 69. This is a variation on the case of Fig. 67.

We distinguish the given loads by the numerals 1, 2, 3, and the unknown pressures by x and y †. We first draw the part of the force-diagram corresponding to the given forces 1, 2, 3, assuming an arbitrary pole O. This enables us to construct the sides $x1$, 12, 23, $3y$ of the funicular, and thence the closing line xy. A parallel to this through O (not shewn in the figure) will determine the pressures x, y.

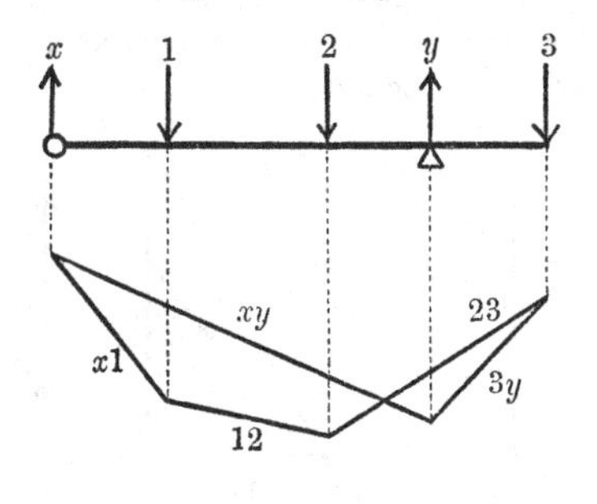

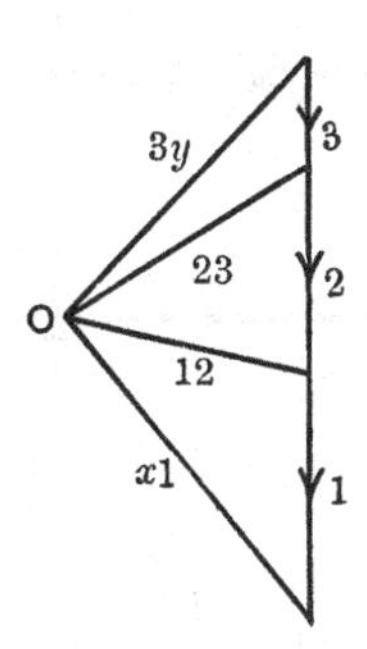

Fig. 69.

It will be noticed that in this example the funicular polygon intersects itself, indicating that the bending moment vanishes and changes sign at the corresponding point of the beam. Hence the results are the same as if the beam had consisted of two portions hinged together at this point.

A girder or bridge which is so long as to require more than two points of support is sometimes made in sections hinged together. In this way it is easy to arrange that the pressures of the supports shall be statically determinate, and accordingly independent of the elastic properties of the structure, and unaffected

* When one funicular has been drawn, it is easy to arrange another so that the closing line shall be horizontal. The above theory is due to Culmann, *l.c.*

† Bow's notation might have been used in this case also, but it is a little more difficult to indicate to which compartments of the funicular diagram the several letters belong.

by changes of temperature, or by a slight settlement of the foundations.

Ex. 2. We take the case of a beam composed of two portions hinged together, and resting on three supports, as shewn in Fig. 70. For simplicity we suppose the loads on each portion to be replaced by their resultant; this makes no essential difference so far as the method is concerned. The resultant loads are numbered 1 and 2, and the pressures of the supports are denoted by x, y, z. The force-diagram gives at once the directions of the sides $x1$, 12, $2z$ in the funicular, and the portion of the funicular relating to the left-hand portion of the beam is completed by drawing the side xy, which must intersect the side 12 on the vertical line through the hinge, since the bending moment vanishes there. Since the extremities of the remaining side yz of the funicular are now determined, the figure can be completed. The lines drawn through the pole of the force-diagram parallel to xy and yz then fix by their intersections with the vertical side the magnitudes of the reactions x, y, z. In the case figured, the bending moment changes sign at a certain point on the right-hand beam, as well as at the hinge.

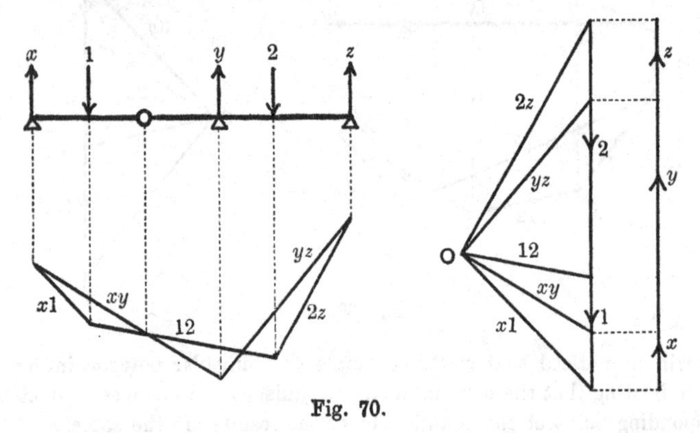

Fig. 70.

It is sometimes required to construct bending-moment diagrams for a series of different positions of a moveable load, or system of loads*, on a beam. For this purpose it is convenient to regard the funicular for the given system as fixed, and to shift the lines of action of the pressures of the supports relatively to it, keeping these lines, of course, at the proper distance apart. The only change is then in the position of the closing line; see Fig. 71, where corresponding lines of support are denoted by aa, bb, cc, ...

* For instance, the pressures exerted by the wheels of a locomotive on a girder bridge.

It follows from a known theorem of Geometry that the different positions of the closing line envelope a parabola whose axis is vertical, and which touches the extreme sides of the funicular*.

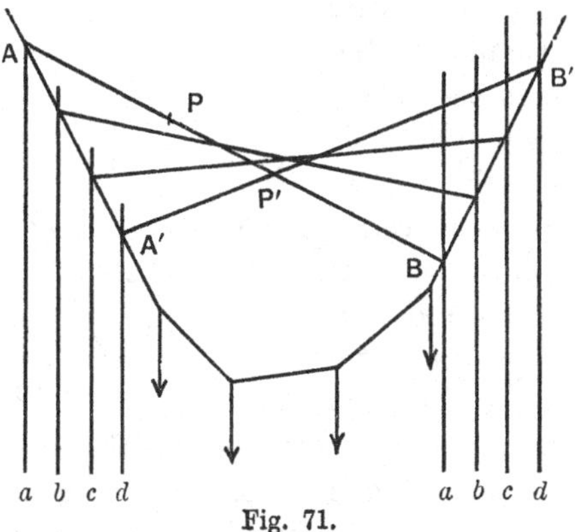

Fig. 71.

Hence if AB be any position of the closing line, touching the parabola at P, the figure included by AB and the funicular gives the distribution of bending moment for a given position of the travelling load. This figure admits, moreover, of another interpretation. If $A'B'$ be any other tangent to the parabola, meeting AB in P', the length $A'P'$ which is intercepted by two given tangents will have a constant horizontal projection, equal to that of AP. Hence P and P' may be supposed to refer to *the same point of the beam*, in the two positions relative to it of the travelling load. The figure bounded by AB and the funicular will therefore represent the variation of the bending moment at the point corresponding to P, for different positions of the travelling load†.

The preceding statements require modification when one or more of the loads passes off the beam. Instead of a single parabola we have then a series of parabolic arcs, consecutive arcs having a common tangent where they meet

37. Continuous Load.

The effect of a continuous load may be approximated to by

* The line joins homologous points of two 'similar' rows, and its limiting directions are vertical.

† A line constructed with the bending moment *at a given point* as ordinate, and the position of the travelling load as abscissa, is called the 'influence line' of that point.

dividing the beam into small segments, and supposing the load on each of these to be replaced by a single force through its centre of gravity. In the limit the funicular polygon becomes a curve; and the tangents at the extremities of any arc are parallel to the lines drawn through the pole O in the force-diagram to the ends of that segment of the vertical line which represents the corresponding portion of the load. If this portion were replaced by a single force through its centre of gravity, the arc in question would be replaced by the two tangents, and the force in question would act through the intersection of these tangents. Hence in the approximate construction referred to, the true curve touches the sides of the funicular polygon obtained. This method of drawing the curve is in practical use.

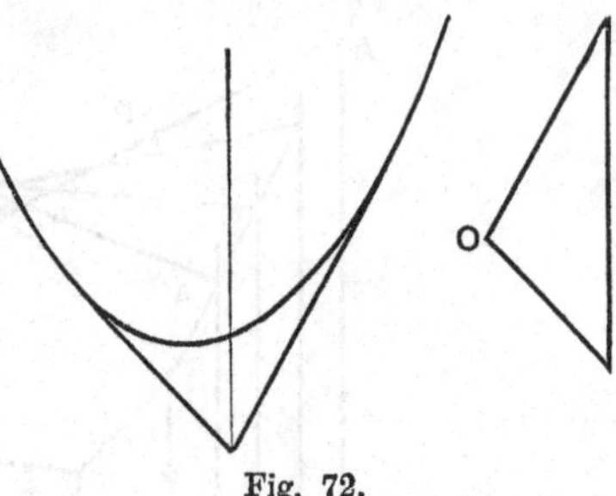

Fig. 72.

We have seen (Arts. 12, 28) that in the case of a *uniform* load the funicular curve is a parabola with vertical axis.

EXAMPLES. VIII.

1. Shew that the graphical method of constructing the resultant of two parallel forces agrees with the ordinary rule.

2. Explain how the construction of Art. 32 for the resultant of a plane system fails if the pole O be taken on the closing line of the force-polygon.

3. Prove that if the lines of action of four forces be given, it is in general possible to assign the magnitudes of the forces so that they shall be in equilibrium; and that the ratios of the forces are determinate.

What is the failing case?

4. Prove that any closed plane polygon is a possible funicular polygon of a system of forces acting through the vertices, the directions of all of which, except one, are given; and that the ratios of the forces are determinate.

5. Given the lines of action of three forces in equilibrium, construct a funicular polygon so that each side shall pass through an assigned point.

6. Given a figure consisting of four points in a plane and the lines joining them, prove that the circumcentres of the four triangles formed by these lines determine a figure which, when turned through a right angle, is reciprocal to the former one.

7. Construct the diagram of bending moments for the following case, the loads being proportional to the numbers given:

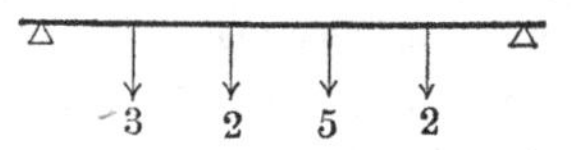

8. Also for the case where the loads (in order) are as the numbers 1, 2, 3, 4.

9. A rod is confined between stops as shewn, and is acted on by two equal and opposite forces P, at right angles; construct the diagram of bending moments.

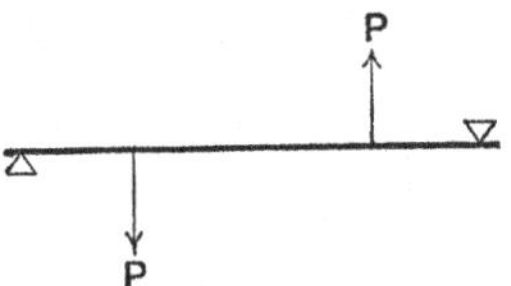

10. A beam supported symmetrically as shewn carries (1) equal and (2) unequal loads at the ends. Construct the diagram of bending moments for the two cases.

11. Construct the diagram of bending moments in the following case, the loads being proportional to the numbers given.

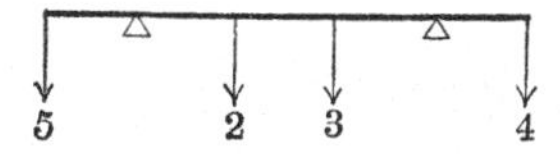

12. Two beams AB, BC, jointed at B, are supported horizontally at the points A, O, C, and are loaded as shewn. Construct the diagram of bending moments. (Take $AB = 2BO = \frac{2}{3}OC$.)

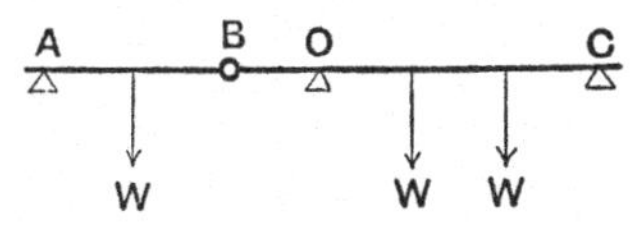

13. Three beams AB, BC, CD, jointed at B and C, are supported horizontally as shewn, and carry loads at the respective centres. Construct the diagram of bending moments, the loads being as indicated.

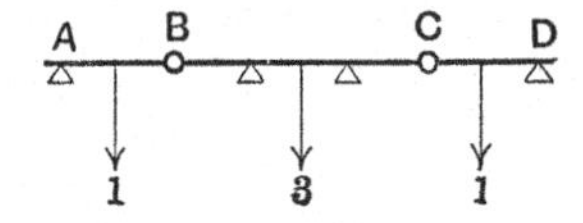

14. A beam AB, supported at A, B, carries another horizontal beam EF resting on it at E and F. Construct the diagrams of bending moment for AB and EF when a given system of loads (say three) is placed on EF, shewing that both diagrams can be included in one figure.

15. Find by a funicular polygon the reactions at the ends of a beam supported as in Fig. 39, p. 50, due to a given system of vertical loads.

Can the resulting figure be used as a diagram of bending moments?

16. Prove that if the length of a beam between its two supports be divided into n equal parts, and $n-1$ equal loads be placed at the points of division, the vertices of the diagram of bending moments lie on a parabola whose axis is vertical.

CHAPTER V

THEORY OF FRAMES

38. Condition for Rigidity.

The theory of framed structures claims attention on several grounds. It has, of course, important applications in practice, in the design of roofs and bridges; whilst it appeals to the theoretical student by the interesting exemplifications of statical principles which it affords, and by the elegance of the geometrical and other methods which have been devised to meet the various problems which it presents.

By a 'frame' is meant a structure made up of rigid pieces, or 'members,' each of which has two 'joints' connecting it with other members. In a two-dimensional frame, each joint may be conceived as consisting of a small cylindrical pin fitting smoothly into holes drilled through the members which it connects. This supposition of perfect smoothness is a somewhat ideal one, and in practice is often only roughly fulfilled.

We shall assume, in the first instance, that external forces act on the frame at the joints only, i.e. we shall suppose them applied to the *pins*. Hence for equilibrium it is necessary that the reactions on any member at its two joints should be equal and opposite. This combination of equal and opposite forces will be called the 'stress' in the member; it may be of the nature of a *tension* or a *thrust*.

The shapes of the members may be any whatever, but for diagrammatic purposes a member is sufficiently represented by a straight line terminating at two points which represent the joints; these lines will be referred to as the 'bars' of the frame.

In technical applications it is generally necessary that the frame should be 'stiff' or 'rigid,' i.e. it must be incapable of deformation without alteration of length of at least one of its bars. It is said to be 'just rigid' if it ceases to be rigid when any one of its bars is removed. If it has more bars than are essential for rigidity it is said to have 'superfluous' members, or to be 'over-stiff.' In this book we contemplate only frames whose bars are all in one plane, and the terms above defined have reference only to the possibility of deformations in this plane.

There is a definite relation between the number of joints and bars in a plane frame which is just rigid. Let the number of joints be n. Suppose one bar, with its two joints, to be fixed; this will by hypothesis fix the frame. The positions, relative to this bar, of the remaining $n-2$ joints will involve $2(n-2)$ coordinates (Cartesian or other); and these must be completely determined by the equations which express that the remaining bars have given lengths. These equations must therefore be $2n-4$ in number, i.e. the total number of bars must be $2n-3$. Otherwise: if the positions of the joints be referred to any axes whatever in the plane, their $2n$ coordinates will determine the position as well as the shape of the frame. Since the position depends on *three* independent quantities (Art. 13), the shape will involve $2n-3$ independent relations.

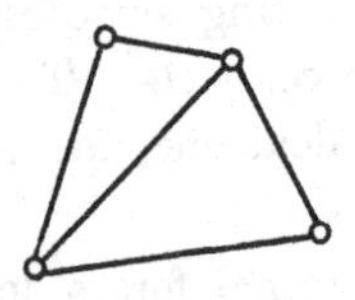
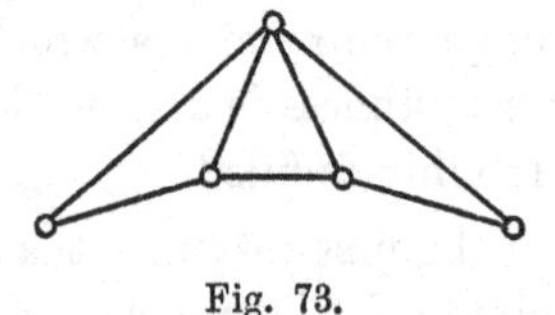
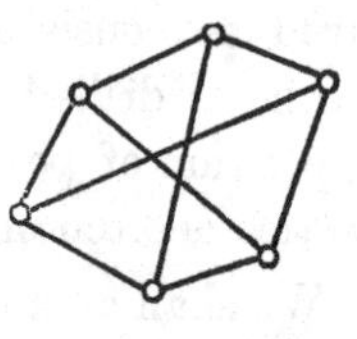

Fig. 73.

A joint where two bars only are connected is called a 'single' joint; one where three bars are connected is called a 'double' joint; and so on. If we count the number of bars which meet at each joint of the frame, and add the results, each bar will be reckoned twice, so that in the case of a just rigid frame the number thus obtained will be $4n-6$. The average number of bars at a joint is therefore $4-6/n$, and is accordingly less than 4. Hence a just rigid frame must have at least one single or one double joint; and if $n<6$, there must be at least one single joint.

39. Determinateness of Stresses.

When a frame which is just rigid is subject to a given system of forces in equilibrium, acting on the joints, the stresses in the various bars are in general uniquely determinate. For the conditions of equilibrium of the forces at the joints furnish $2n$ equations, viz. two for each pin, which are linear in respect of the stresses and the external forces. These equations must involve the three conditions of equilibrium of the frame as a whole, which are by hypothesis already satisfied. There remain, therefore, $2n-3$ independent relations which determine in general the $2n-3$ unknown stresses. The frame is then said to be 'statically determinate.' In particular, when there are no external forces the frame will in general be free from stress.

The same argument shews that the stresses in an over-rigid frame are indeterminate, since the number of unknowns exceeds the number of independent equations, and in particular that the frame may be in a state of stress independently of the action of external forces. Such a frame is accordingly said to be 'statically indeterminate.'

The physical explanation of the indeterminateness is of the same kind as in Art. 25. If we start with a frame which is just rigid, and add a superfluous bar between two of its joints A, B, stresses will be at once introduced unless the natural length of this bar (between the centres of the eyeholes) be exactly equal to the original distance between the centres of the pins at A, B. Thus if the bar be too short, it will exert a tension T depending on the elastic properties of the bar and the original frame, and on the amount of the slight discrepancy in length. Stresses proportional to T will accordingly be produced in the bars of the original, just rigid, frame, viz. those due to equal and opposite extraneous forces T acting at A and B. It is evident that each superfluous bar added in this way will produce its own degree of indeterminateness. Even if an over-rigid frame were accidentally free from internal stress, such stresses would be evoked by a change of temperature if the expansions were not uniform.

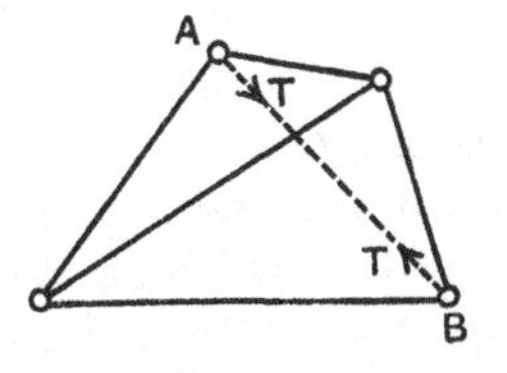

Fig. 74.

It may happen that, owing to some special relation between the lengths of the bars, a frame of n joints and $2n-3$ bars may admit of an *infinitesimal* deformation*. It may be shewn that in this case also the frame may be self-stressed, and consequently that the stresses due to given external forces are indeterminate. The theory of such 'critical forms' is, however, deferred for the present. (See Chap. VI, Arts. 55, 56.)

40. Graphical Determination of Stresses.

The analytical method sketched in the preceding Art., and already partially exemplified in Art. 26, is of course always possible, but when the number of joints is at all large the system of equations becomes troublesome to manipulate, and accidental errors are not easily detected.

The graphical method which was initiated by Maxwell† has great advantages. It proceeds in a regular manner, step by step; some unknown quantity is determined at each stage; and an error usually betrays itself at once by some inconsistency in the drawing Moreover the final diagram gives a complete view of the distribution of stresses in the frame which is much more readily intelligible and more instructive than any table of numerical results.

In practice the external forces acting on the frame consist partly of known forces, e.g. the loads on the various parts of the structure, and partly of the reactions of the supports, which have, as a first step, to be ascertained. It is assumed that this question is a determinate one. This imposes a certain limitation on the nature of the supports; they must involve *three* degrees of constraint and no more. If the number were to exceed three, the reactions would be indeterminate (Art. 25), and the frame, though itself statically determinate, might be in a state of stress independently of the loads. If the number were less than three the structure would not be fixed. A frequent arrangement is that one end of the structure is hinged to a pier, whilst the other rests on

* It is hardly necessary to do more than mention the remaining case of exception, where a frame having the proper number of bars fails to be rigid owing to some parts being over-stiff whilst others are deformable.

† James Clerk Maxwell (1831–79), professor of experimental physics at Cambridge (1871–79). The paper referred to was published in 1864. The systematic application to roofs and bridges was explained by F. Jenkin and M. Taylor, *Trans. R. S. Edin.* 1869.

rollers carried by a second pier (see Fig. 39, p. 50), so that the direction of the reaction there is prescribed. This is seen to be included in the general scheme of Fig. 22, p. 29, if we imagine two of the links to be attached to the same point of the body. A structure supported in this way can expand freely with change of temperature, and the stresses are due solely to the loads.

We assume then that the reactions have been determined, by means of a force-diagram and a funicular polygon, or otherwise. The problem now is to evaluate the stresses in a statically determinate frame subject to a given system of external forces in equilibrium acting at the joints. The methods to be adopted will depend to some extent on the 'structure' of the frame*.

A frame which can be built up from a single bar by successive steps, in each of which a new joint is annexed by two new bars meeting there, is called a 'simple' frame. Since each additional joint involves the addition of two new bars, the total number of bars is $2(n-2)+1=2n-3$, if n be the total number of joints. It is otherwise obvious that a frame built up in this way is just rigid, and that the critical case referred to in Art. 39 will not arise unless two adjacent bars are in the same line.

The stresses in a simple frame can be found graphically by considering the equilibrium of the various joints in a proper succession. We begin at a 'single' joint, e.g. the joint last added in the process of construction above explained. At this point we have three forces in equilibrium, viz. the known external force, and the tension or thrust of each of the two bars which meet there. These latter forces can accordingly be found from a triangle of forces. We may next imagine the bars in question to be removed, their place being supplied by the tensions or thrusts which they exert at their other extremities, and which must now be reckoned as known external forces acting on the 'simple' frame which remains. This frame, again, has at least one single joint; the directions of all the forces which act at this joint are given, and the magnitudes of all but two of them are now known. Hence

* Two frames are said to possess the same 'structure' when they differ only in the lengths of the various bars. Thus the structure of the frames in Figs. 75, 76 is the same.

the polygon (or triangle) of forces for this joint can be constructed, and the stresses in the two bars which meet there can be determined. The process can be continued, until all the stresses in the frame have been found.

41. Maxwell's Method.

The procedure above explained is straightforward, but it involves the construction of a number of disconnected polygons, and the stress in each bar is represented twice over. An important simplification was introduced by Maxwell, who shewed that in a great variety of cases the various polygons can be fitted together into a single diagram, in which the lines corresponding to the external forces form a closed polygon.

The method will be best understood from a few examples, and after a little practice will be found very easy. We will suppose that (as is usually the case) the external forces act at joints which are situate on the external contour of the frame; and we will further assume for the present that there is no crossing of bars.

If we distinguish the several compartments of the frame, and also the compartments into which the surrounding space is divided by the lines of action of the external forces, by letters $A, B, \ldots$, these compartments will be represented by points in the force-diagram, and any line AB in this diagram will be parallel to the line separating the compartments A, B in the frame-diagram. This is in fact Bow's notation (Art. 34), as applied to the kind of question where it was first used.

Ex. 1. A quadrilateral frame stiffened by a diagonal bar is subject to equal and opposite forces at the joints not on this diagonal.

We begin in each case by drawing the polygon of external forces, the sides being taken in the order of the forces as they occur round the contour. In the present instance this polygon consists of the two vectors XY, YX in the second diagram. The triangle of forces for the upper joint, viz. XYA, is constructed by drawing lines XA, YA through X and Y, respectively, parallel to the corresponding bars of the frame; and it is to be observed that this triangle indicates the *sense* as well as the magnitude of the forces at this joint, viz. these are represented by the vectors XY, YA, AX*. In the same

* The student should draw the diagram step by step, following the indications in the text. It is useful to mark the sense of the various forces at a joint by arrowheads in the frame-diagram, as we proceed.

way we construct the triangle YXB corresponding to the opposite joint, the forces there being represented by YX, XB, BY. Again, the left-hand joint is

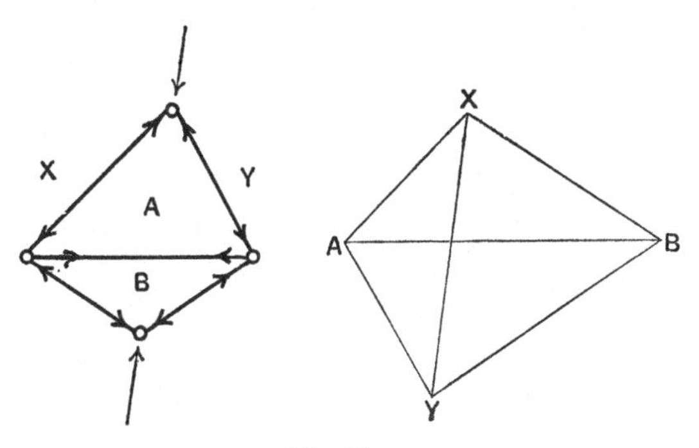

Fig. 75.

in equilibrium under three forces, of which two are already represented in the force-diagram, viz. by the vectors BX, XA. The remaining force, viz. that exerted by the diagonal bar, must accordingly be represented by AB. The forces at the remaining joint are then given by BA, AY, YB.

Incidentally the process indicates which bars are in tension and which in thrust. In the present instance the diagonal bar acts as a tie, and the remaining bars as struts.

It is important to notice that the construction of the figure would have been impossible if the conditions of equilibrium of the external forces had not been fulfilled. For instance, if the two external forces, though represented by the vectors XY, YX, had not been in the same line, the line AB in the force-diagram would not have come out parallel to the diagonal bar.

Ex. 2. The frame in Fig. 76 (p. 100), which resembles a common form of roof-truss, but is purposely drawn unsymmetrical, is assumed to be in equilibrium under three forces acting as shewn.

The triangle XYZ of the extraneous forces having been constructed, we can at once form the triangles ZXA and ZBY corresponding to the two lower joints. Proceeding next to the upper joint, we notice that three sides, viz. AX, XY, YB, of the corresponding polygon of forces have been found; the remaining force at this joint is therefore represented by BA. Hence if the data and the drawing are correct, this line in the force-diagram must necessarily come out parallel to the diagonal bar in the frame. The triangle of forces for the remaining joint is ABZ.

The success of the method depends as before on the accuracy of the data; in particular it is essential that the lines of action of the three extraneous forces should be concurrent, or parallel.

It will be found on examination that the two outer bars in the figure are in thrust and the others in tension, as indicated by the arrowheads.

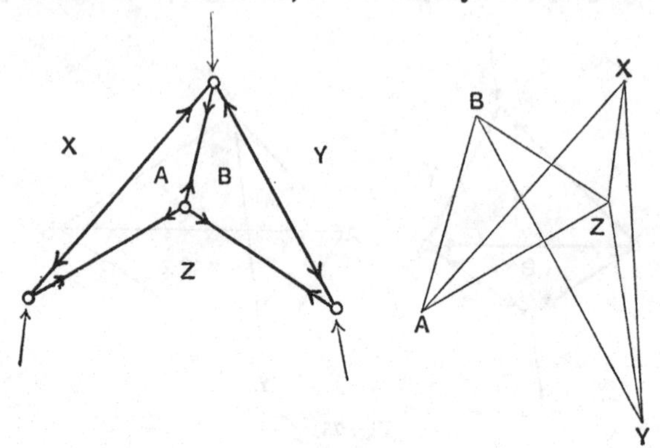

Fig. 76.

42. Further Examples. Parallel Forces.

When the external forces are parallel, as in the case of a roof or a girder loaded vertically at one or more joints, and resting freely on two supports, the polygon of forces consists of segments of a vertical line, and the force-diagram simplifies.

Ex. 1. The frame in Fig. 77 is supposed to be loaded, and supported, symmetrically.

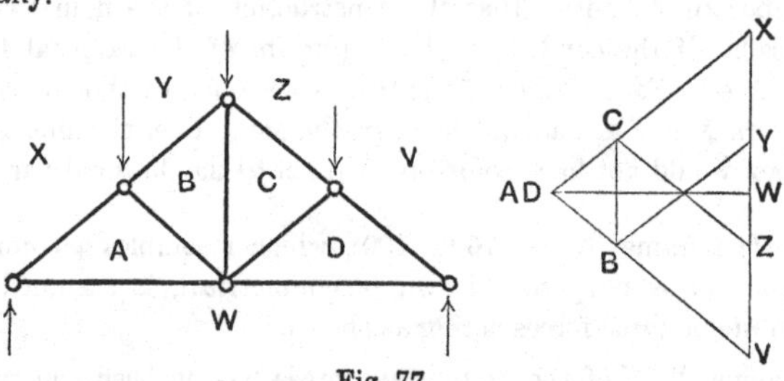

Fig. 77.

The external forces are represented by XY, YZ, ZV, VW, WX in the force-diagram. The triangles *WXA*, *VWD* are first constructed. It is found that the points *A* and *D* coincide, but this is an accident, due to the symmetry of the frame and the loads. The polygon *XYBA*, corresponding to the second joint from the left, can then be completed, by drawing lines through the points *Y* and *A* parallel to the directions of the two unknown stresses at this joint. Similarly for the polygon *ZVDC*. The figure is completed by joining *CB*.

Ex. 2. The first diagram of Fig. 78 represents a bridge loaded unequally at two upper joints.

If the loads are given by the vectors XY, YZ, the pressures of the supports will be represented by ZV, VX, where V is some point in XZ. The precise position of this point can be found by taking the moments of the given loads about the points of support, or graphically by taking a pole O in the force-diagram and constructing a corresponding funicular polygon (Art. 35). The success of the subsequent construction will depend on V being placed correctly. The triangles VXA, ZVD, DVC can then be drawn in this order. The polygon $YZDCB$ is next completed, and it only remains to join AB. If the loads are equal, and the three upper bars equal in length, the points B and C will coincide. This indicates, what is otherwise obvious, that there is then no stress in the diagonal bar.

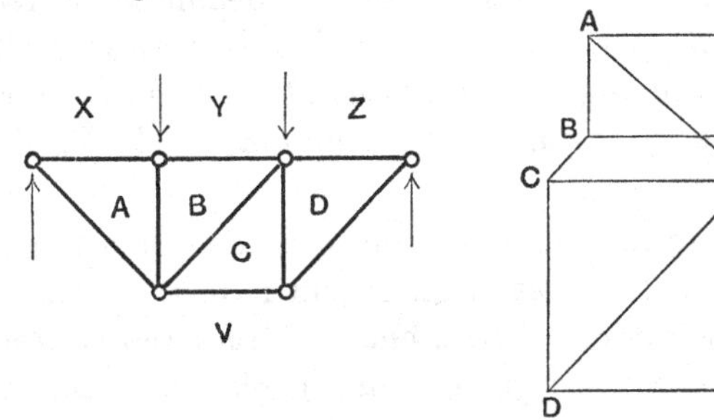

Fig. 78.

If two bars cross, it is convenient to imagine them pinned together at the intersection. This introduces one new joint and two new bars, so that the condition for determinateness (Art. 39) is not impaired. In the diagram of forces the stresses in the four bars which meet at this new joint will be represented by the sides of a parallelogram*.

Ex. 3. The annexed frame is assumed for simplicity to be symmetrical, and symmetrically loaded. The external forces are given by XY, YZ, ZV,

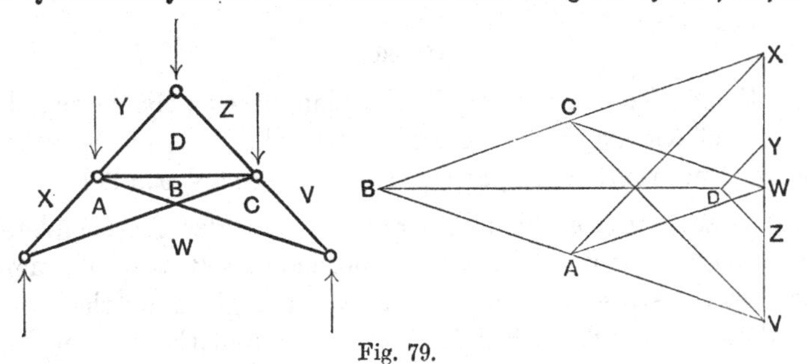

Fig. 79.

* This artifice does not avail when three or more bars cross at the same point, since the introduction of a new joint there would make the frame over-rigid.

VW, WX. We construct the triangles WXA, VWC, YZD, and complete the parallelogram $WABC$, which gives the stresses at the crossing point. Finally we draw BD.

More elaborate examples of 'simple' frames will be found in books on Graphical Statics and Technical Mechanics, but the method of treatment is the same in all cases.

43. Reciprocal Figures.

It will be noticed that in each of the above examples the two diagrams fulfil to a certain extent the definition of reciprocal figures given in Art. 34. The relation is however not quite complete; thus in Fig. 77, for instance, there are in the frame no closed polygons answering to the points X, Y, Z, V, W of the force-diagram.

Two complete reciprocal diagrams are however obtained if we take a pole O in the force-diagram, join it to the corners of the polygon of the external forces, and construct the corresponding funicular in the frame-diagram, as in Fig. 80. Each figure is now

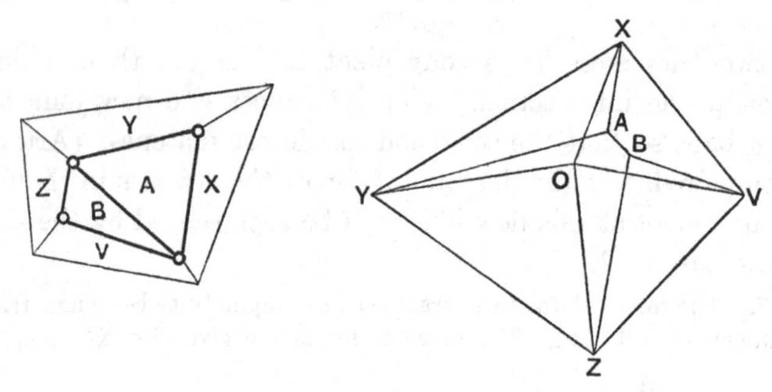

Fig. 80.

the orthogonal projection of a closed plane-faced polyhedron; the vertices of one polyhedron correspond to the faces of the other, and the lines representing corresponding edges are parallel*.

It is seen that either figure, and in particular the completed frame-diagram, may be taken to represent a self-strained frame, and that the stresses in the several bars are given by the corresponding lines in the other figure. This is remarkable, since the

* L. Cremona, *l.c. ante*, p. 82. The letter O corresponds to the quadrilateral formed by the four outer lines in the left-hand figure.

relation between the numbers of bars and joints, in the completed frame-diagram, is that which ordinarily ensures statical determinateness. We have in fact here hit upon a method of obtaining 'critical forms' in endless variety. (See Art. 39.)

44. Method of Sections.

Whenever a frame possesses a 'single' joint, the stresses in the bars which meet there can be found graphically, and if the frame which remains when these two bars are removed also possesses a single joint, the process can be continued as explained in Art. 40.

In the case, however, of a frame which is not 'simple,' in the sense of Art. 40, we arrive at length at a form which possesses no single joints. The method then comes to a stop; at a double joint, for instance, we should require to construct a force-polygon of which three sides are known only in direction. Various methods of meeting this difficulty have been devised, some of them special to particular types of frame.

In some cases recourse may be had to the 'method of sections.' This is applicable whenever the frame can be regarded as made up of two distinct portions which are connected by *three* bars. For a reason given in Art. 15 (p. 32) we will suppose that the lines of these bars are not concurrent or parallel.

If an ideal section be drawn across any frame so as to divide it into two parts, the extraneous forces on either portion must be in equilibrium with the forces exerted on it by the bars cut across*.

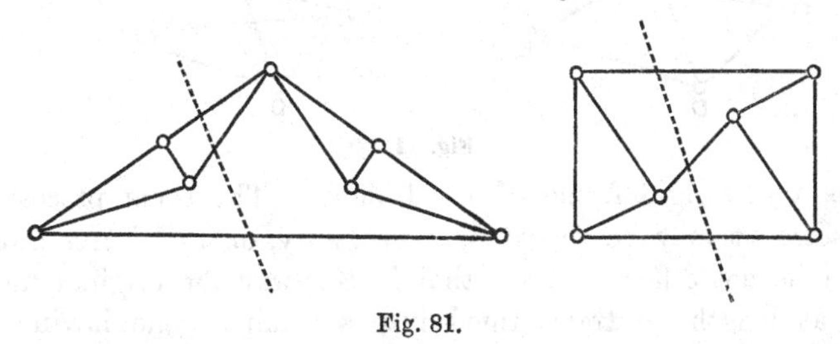

Fig. 81.

* Hence in the force-diagram of a 'simple' frame the lines representing the external forces on either side, together with those corresponding to the stresses in the bars cut across, must form a closed polygon. This gives a test of the correctness of the drawing.

And if a section can be drawn meeting three bars only, the forces which they exert can be uniquely determined, by Art. 21, so as to be in equilibrium with the resultant of the external forces on either side. In this way the stresses in the three bars in question are ascertained*; and if each portion of the frame (apart from these bars) is itself a simple frame, the solution can be completed by the previous method.

45. Method of Interchange.

It is not, however, always possible to draw a section fulfilling the above condition. A more general method† rests on the fact that a just rigid frame, of any structure, can be converted into a simple frame by a series of operations, each of which consists in removing a bar and replacing it by another occupying a different position in the frame.

We will suppose that we have a just rigid frame of n joints, none of which are single. There must, by Art. 38, be at least one joint A where only three bars meet. If one of these bars, say AB, be removed, the frame becomes deformable, but it may be made rigid again by inserting a bar of suitable length connecting two other joints, say C and E. The joint A is now a single one, and the two remaining bars which meet there contribute nothing to the rigidity of the rest of the structure, which therefore now

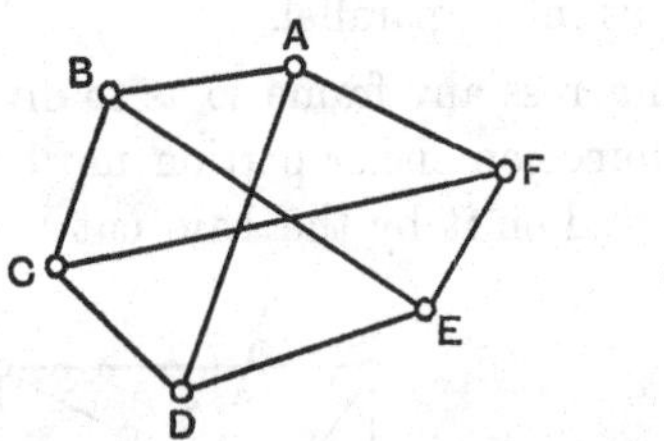

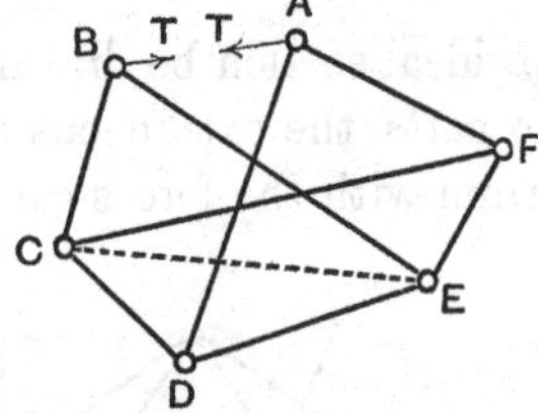

Fig. 82.

forms a just rigid frame of $n-1$ joints. The same process of replacement may be repeated, if necessary, on this latter frame, and it is not difficult to see that in this way the original frame will at length be transformed into a 'simple' one having its

* This method is due to A. Ritter (1863). He determines the stress in each of the three bars by taking moments about the intersection of the lines of the other two.

† Due to L. Henneberg, *Statik der starren Systeme*, 1886.

n joints in the same relative positions, but a different arrangement of bars.

We will confine ourselves, however, for simplicity, to cases where *one* operation of the above kind suffices. An interesting example is that of a frame whose bars form the sides of a hexagon $ABCDEFA$ and the three diagonals AD, BE, CF (Fig. 82).

In the process devised by Henneberg we begin by finding, on Maxwell's plan, the stresses produced in the *modified* frame by two distinct systems of external forces. The first system is that of the actual external forces (not indicated in the figure). Suppose that these produce in the inserted bar CE a tension P, and in the remaining bars tensions Q, R, ... The second system consists merely of two equal and opposite *unit* forces at A and B, tending to produce approach of these points. Let the tensions thus evoked be p in CE, and q, r, ... in the remaining bars. We infer that the tensions produced by the combined action of the given extraneous forces, and of any two equal and opposite forces T urging the points A and B towards one another, will be $P+pT$ in CE, and $Q+qT$, $R+rT$, ... in the remaining bars. The value of T is as yet arbitrary; we choose it so that

$$P+pT=0. \qquad \text{.........................(1)}$$

The bar CE is now free from stress and may be removed without affecting the equilibrium. We have in fact determined the stresses which are consistent with equilibrium in the *original* form of the frame; viz. the tension in AB is

$$T=-P/p, \qquad \text{.........................(2)}$$

and those in the remaining bars are given by the expressions

$$Q-qP/p,\ R-rP/p, \ldots \qquad \text{..................(3)}$$

It will be noticed that the process fails if $p=0$. The tensions are then infinite, unless also $P=0$, in which case they are indeterminate. The vanishing of p is in fact an indication that we are dealing with a 'critical form' (Art. 39).

A method of determining the stress in any one bar of a just rigid frame, independently of the stresses in the remaining bars, will be explained in the next Chapter, where also the subject of critical forms will be more fully considered.

46. Case of External Forces acting on the Bars.

Up to this point the external forces acting on the frame have been supposed to act exclusively at the joints, viz. on the pins. When external forces are applied to the bars themselves, the stress in each bar will no longer consist of a purely longitudinal tension or thrust. To find the reactions at the joints we may proceed as follows.

The resultant W of the external forces on any bar may be replaced by two components P, Q in lines through the centres of

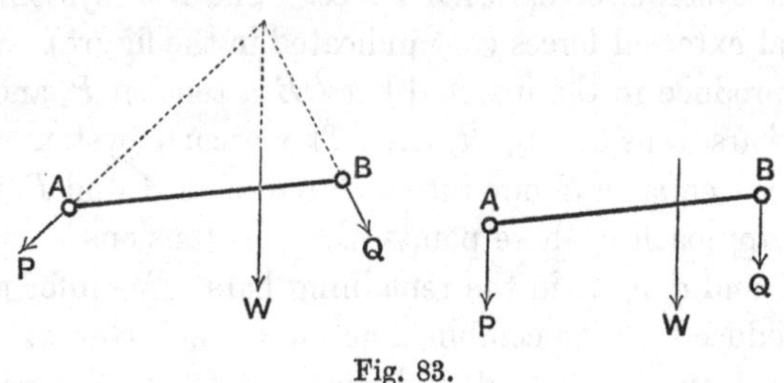

Fig. 83.

the pins at its extremities. This can be done in an infinite number of ways, but in practical cases the forces in question are generally due to gravity, and therefore vertical, and the two components P, Q are then most conveniently taken to be vertical also.

We first alter the problem by transferring these forces P, Q from the bars *to the pins.* The stresses in the bars, in the problem as thus modified, are supposed to be found as before, by a reciprocal diagram or otherwise. We have next to infer from the results thus obtained the directions and magnitudes of the reactions in the original form of the question. To find the pressure exerted by any bar AB on the pin at A we compound with the force in AB, given by the auxiliary diagram, a force equal to P; for in this way the conditions of equilibrium of each pin will obviously be satisfied. Conversely, to find the pressure exerted by the pin A on the bar AB we combine with the force given by the diagram a force equal and opposite to P.

This question arises in practice in connection with 'three-jointed' structures, of which a rudimentary form is shewn in Fig. 84. It is required to find the magnitudes and directions of the reactions at the joints when given loads are placed anywhere on the structure.

For the present purpose the structure is sufficiently represented by two bars AC, CB, and we may suppose the loads on the two members to be replaced by their resultants W_1, W_2. These resultants may again be replaced, as above explained, by P_1, Q_1 acting vertically at A, C, and P_2, Q_2 acting at C, B. The auxiliary diagram in the figure is a portion of the reciprocal diagram constructed on the hypothesis that the loads act on the pins. In particular XY represents the force Q_1+P_2 acting on the pin C, YZ the thrust exerted on this pin by the bar BC, and ZX the thrust exerted on it by the bar AC. If we compound with ZX the vector XV, which represents Q_1, we get the actual pressure ZV exerted by the bar AC on the pin C. The directions and magnitudes of the true reactions at A and B are then easily determined from the consideration that the three forces on each bar must be concurrent, and must be related by a triangle of forces.

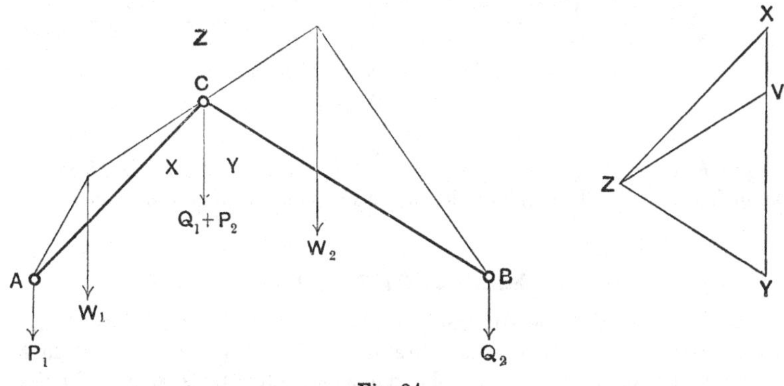

Fig. 84.

On account of the practical importance of this problem several other graphical methods of solution have been devised. The following construction (see Fig. 85) may be noticed as an application of the geometrical property of funiculars proved in Art. 33.

Let W be the resultant of W_1 and W_2, its line of action being determined graphically or otherwise. It is evident that any triangle having its vertices on the lines of action of W_1, W_2, W, respectively, will be a possible form of funicular triangle for the three equilibrating forces W_1, W_2, $-W$. As a particular case, the lines of the three reactions at A, B, C will constitute such a triangle. The problem therefore resolves itself into the construction of a funicular triangle whose sides shall pass through three given points A, B, C. If we take any point L' on the line of action of W, and draw AL', BL', meeting the lines of action of W_1, W_2 in H' and K', respectively, we obtain an auxiliary funicular triangle $H'K'L'$. By the theorem referred to, the intersections of the sides of this triangle with the corresponding sides of the desired funicular must be collinear. We therefore produce $H'K'$, AB to meet in X, and through X draw XC meeting the lines of action of W_1, W_2 in

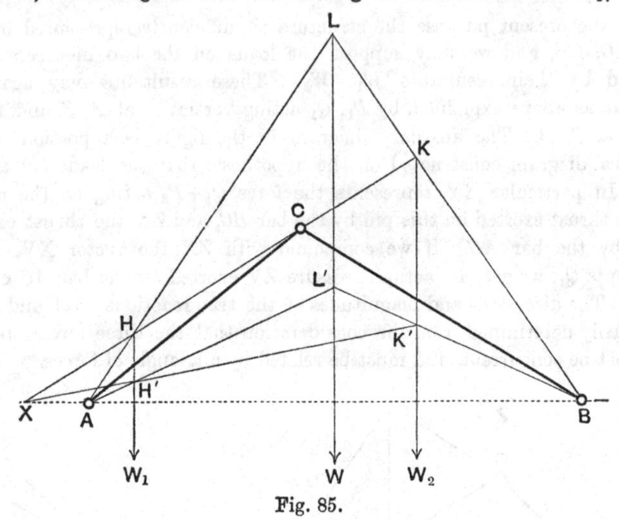

Fig. 85.

H and K, respectively. If we now draw AH, BK to meet in L, HKL is the triangle required. The magnitudes of the reactions are then easily found.

EXAMPLES. IX.

1—9. Determine the stresses in the following frames, when loaded and supported as shewn. Where there are several loads these may in the first instance be taken as equal; but a diagram should afterwards be made for the case of unequal loads.

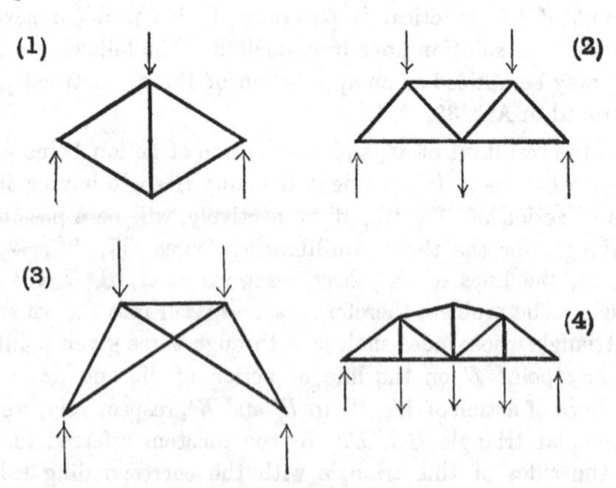

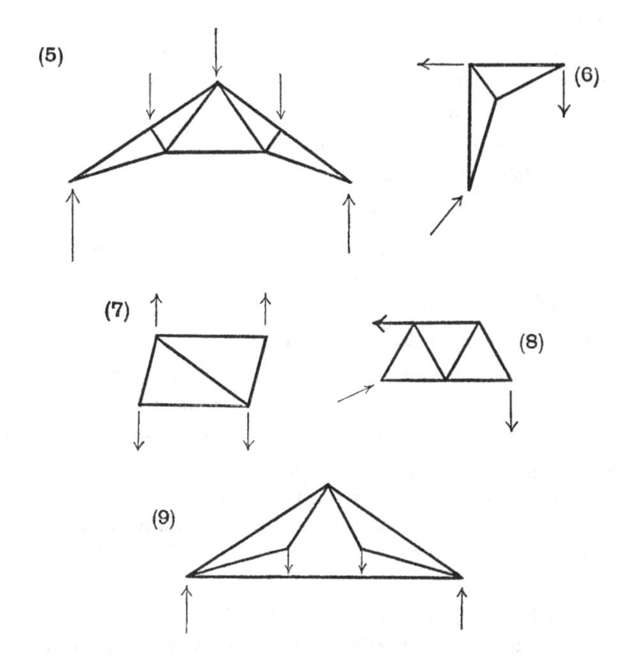

10. A frame is composed of five light rods, hinged together, forming a quadrilateral $ABCD$ with the diagonal BD. Find, by a diagram drawn to scale, the stress in BD when the frame hangs from A, and a weight of 50 lbs. is suspended from C, having given

$$AB=18, \quad BC=24, \quad CD=15, \quad DA=24, \quad BD=16.$$

11. A plane frame consists of four sides AB, BC, CD, DE of a regular hexagon, and the five bars connecting the vertices A, B, C, D, E to the centre O. The corners A, E are pulled outwards with equal and opposite forces P; find the stress in each member, and indicate which members are in tension and which in thrust.

12. Determine graphically the ratio of two forces in equilibrium acting at the two inverse points of a Peaucellier linkage (*Infinitesimal Calculus*, Fig. 107).

13. Four bars are jointed so as to form a quadrilateral $ABCD$, and are in equilibrium under the action of four forces applied to the joints. The lines of action of these forces are produced so as to form another quadrilateral $PQRS$; prove that the diagonals of $PQRS$ pass respectively through the intersections of opposite sides of $ABCD$.

14. Four rods jointed at their extremities form a cyclic quadrilateral. If the opposite joints be connected by strings prove that the tensions of these strings are inversely proportional to their lengths.

15. *ABC* is a triangle of jointed bars, and three links connect the corners *A*, *B*, *C* with the orthocentre. Prove that if the frame thus formed be self-strained, the stress in each bar is proportional to the length of the opposite bar, i.e. the one which it does not meet.

16. If n_1 be the number of single joints in a just rigid frame, and n_2 the number of double joints, prove that

$$2n_1 + n_2 \nless 6.$$

17. If p joints of a plane frame of n joints are attached to fixed points in the same plane by links, how many bars must the frame consist of in order that the whole structure may be just rigid?

Sketch the cases $n=4$, $p=4$; $n=5$, $p=4$.

18. Prove that if a chain formed of a number of uniform links of equal weight hangs freely between two fixed points, the tangents of the angles which successive links make with the horizontal are in arithmetic progression.

19. A rhombus *ABCD* formed by four uniform rods each of weight *W*, jointed at their extremities, is suspended from *A*, and is prevented from collapsing by a light horizontal strut *BD*. Find the thrust in *BD*, and the reaction at *C*, having given the inclination a of *AB* to the vertical.

[$2W \tan a$, $\frac{1}{2} W \tan a$.]

20. Six equal uniform bars, each of weight *W*, are jointed together so as to form a regular hexagon *ABCDEF* which hangs from the point *A* and is kept in shape by strings *AC*, *AD*, *AE*. Find the tensions of these strings.

[$\sqrt{3}\,W$, $2W$, $\sqrt{3}\,W$.]

CHAPTER VI

WORK AND ENERGY

47. Work of Forces on a Particle.

The 'work' done by a force acting on a particle, in any infinitely small displacement, is defined as the product of the force into the orthogonal projection of the displacement on the direction of the force. Hence if F be the force, δs the displacement, and θ the angle between the directions of F and δs, the work is $F.\delta s \cos\theta$. Since this may be written as $\delta s.F\cos\theta$, we may also say that the work is the product of the displacement into the component of the force in the direction of the displacement. If the displacement is at right angles to the force the work vanishes*.

The total work done by two forces acting on a particle in any infinitely small displacement is equal to the work of their resultant. For, let the vectors OA, OB represent the two forces, OC their resultant, and let OH be the direction of the displacement δs. The proposition follows at once from the fact that the sum of the orthogonal projections of OA, OB on OH is equal to the projection of OC. It is not essential to the proof that δs should be in the same plane with OA and OB. The result can obviously be extended to the case of any number of forces acting in any given directions.

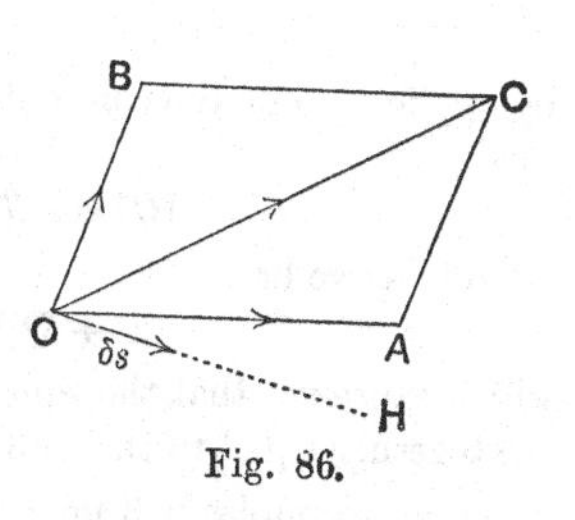

Fig. 86.

* The notion of 'work,' under one designation or another, was developed gradually. The *name* ('travail') appears to have been definitely established by G. Coriolis (1829).

In the same way we see that the total work done by a force in two or more successive small displacements of a particle is equal to the work done by the same force in the resultant displacement.

These relations have a concise expression in the notation of the theory of Vectors. Given any two vectors **P**, **Q**, the product of the absolute value of either into the projection of the other on its direction is called the 'scalar product*,' and is denoted by (**PQ**), or simply by **PQ**. Thus if P, Q be the absolute values, and θ the angle between the directions of the vectors, we have

$$\mathbf{PQ} = PQ \cos\theta. \qquad (1)$$

In particular the scalar square of a vector is the square of its absolute value; whilst the scalar product of two perpendicular vectors is zero.

Since, by the definition,

$$\mathbf{PQ} = \mathbf{QP}, \qquad (2)$$

we see that scalar multiplication of vectors follows the commutative law. Moreover, the theorem above proved shews that

$$\mathbf{P}(\mathbf{Q} + \mathbf{R} + \ldots) = \mathbf{PQ} + \mathbf{PR} + \ldots, \qquad (3)$$

so that the distributive law also holds. The scope of Vector Algebra is thus greatly extended.

For example, referring to Fig. 2, p. 3, and writing AB = **P**, AD = **Q**, and therefore AC = **P** + **Q**, we have

$$\begin{aligned}(\mathbf{P} + \mathbf{Q})^2 &= (\mathbf{P} + \mathbf{Q})\,\mathbf{P} + (\mathbf{P} + \mathbf{Q})\,\mathbf{Q} \\ &= \mathbf{P}^2 + 2\mathbf{PQ} + \mathbf{Q}^2, \qquad (4)\end{aligned}$$

by application of the commutative and distributive laws. This is equivalent to

$$AC^2 = AB^2 + 2AB.AD \cos DAB + AD^2. \qquad (5)$$

Again, we have

$$(\mathbf{P} + \mathbf{Q})^2 + (\mathbf{P} - \mathbf{Q})^2 = 2(\mathbf{P}^2 + \mathbf{Q}^2), \qquad (6)$$

which expresses that the squares on the diagonals of the parallelogram $ABCD$ are together equal to the sum of the squares on the four sides.

Other examples will present themselves in Chap. VIII.

48. Principle of Virtual Velocities.

It follows from the preceding theorem that when the forces acting on a particle are in equilibrium their total work in any

* It was called by H. Grassmann (1844) the 'inner' product.

infinitesimal displacement is zero. In symbols, if P denote any one of the forces, and δp the component of the displacement in the direction of P, we have

$$\Sigma(P.\delta p)=0. \qquad (1)$$

This is an equation between infinitesimals, and is to be understood on the ordinary conventions of the Differential Calculus. The sum $\Sigma(P.\delta p)$ vanishes, not because the quantities δp themselves tend to the limit zero, but in virtue of the *ratios* which these quantities ultimately bear to one another. The equation therefore holds if the resolved displacements δp are replaced by any finite quantities having to one another the ratios in question.

For instance they may be replaced by the resolved *velocities* of the particle in the directions of the several forces, in any imagined motion of the particle through the position of equilibrium. Thus dividing by the element of time δt, we have

$$\Sigma\left(P.\frac{dp}{dt}\right)=0, \qquad (2)$$

where the quantities dp/dt are the resolved velocities in question. They were called by the older writers*, the 'virtual' or effective velocities of the particle; and the theorem embodied in the equation (1) or (2) is accordingly known as the principle of 'Virtual Velocities.'

The physical significance of the principle can hardly be apprehended without some reference to Dynamics. The equation (2) expresses that whenever a particle in motion passes through a possible position of equilibrium, the *rate* at which the forces are doing work upon it is for the moment zero. By a theorem of Dynamics it follows that the kinetic energy is at that instant stationary in value. An example is furnished by a simple pendulum consisting of a particle connected with a fixed point by a light rod. There are two positions of equilibrium, and in the passage through these the kinetic energy is in one case a maximum, in the other a minimum.

* The first complete statement appears to be due to John Bernoulli (1717). By the time of Lagrange (1788) the word 'virtual' had come to be used in a different sense, as indicating the *hypothetical* character of the displacements δp. Many recent writers, following G. Coriolis (1834), have designated the principle, in the form (1), as that of 'virtual work,' with the same altered meaning of the adjective.

It is to be noticed that the equation (1) is not only a consequence of the ordinary conditions of equilibrium of a particle, but includes these, if regard be had to the arbitrary character of the displacements. In other words, if the total work done by a system of forces acting on a particle is zero for all infinitesimal displacements from a given position, that position is one of equilibrium. For if there were a resultant, the work would not vanish for a displacement in its direction. Moreover, it is sufficient that the equation should hold for two, or three, independent displacements, according as the question is one of two or three dimensions; for the most general displacement can be compounded of these.

In applying the principle it is often convenient to take the displacement at right angles to the direction of one of the forces, in which case the corresponding δp vanishes. In this way an unknown reaction may be eliminated. The process is of course equivalent to resolving the forces in the direction of the displacement.

Ex. 1. A weight W is maintained in equilibrium on a smooth plane of inclination a by a force P which acts (1) horizontally, or (2) up the plane.

(1) If we imagine the weight to receive a displacement δs up the plane, the projections of δs on the directions of the forces P and W are $\delta s \cos a$ and $-\delta s \sin a$. Hence

$$P\delta s \cos a - W\delta s \sin a = 0, \qquad (3)$$

the work of the normal pressure being zero, since it is at right angles to δs. Hence

$$P = W \tan a, \qquad (4)$$

(2) In the second case we have

$$P\delta s - W\delta s \sin a = 0,$$

or

$$P = W \sin a. \qquad (5)$$

Again, it is sufficient if the resolved displacements in (1) be calculated accurately to the *first order* of small quantities. Thus in the case of a force P acting from a fixed point O, the corresponding term in the equation is $P\delta r$, to the first order, if r denote distance from O. For if

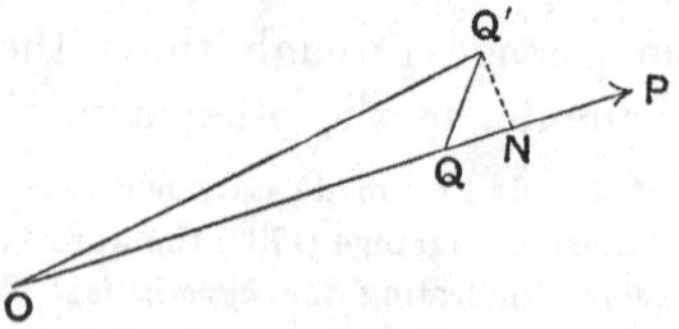

Fig. 87.

the displacement be from Q to Q', and $Q'N$ be drawn perpendicular to OQ, we have

$$\delta p = QN = ON - OQ = OQ' - OQ = \delta r,$$

to the first order.

Ex. 2. A particle of weight W is constrained to lie on a smooth vertical circle, and is urged towards the highest point A by a constant force P, for example by means of a string passing over a smooth pulley at A and carrying a weight P which hangs vertically; to find the positions of equilibrium.

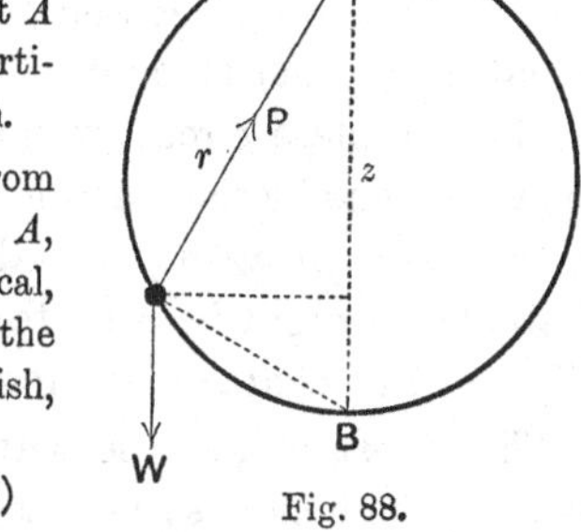

Fig. 88.

If r denote the distance of the particle from A, z its vertical depth below the level of A, θ the angle which r makes with the vertical, then in an infinitesimal displacement along the arc the work of the normal pressure will vanish, and we shall have

$$-P\delta r + W\delta z = 0. \quad \text{.........(6)}$$

Now, if c be the diameter,

$$r = c\cos\theta, \quad z = r\cos\theta = c\cos^2\theta,$$

$$\delta r = -c\sin\theta\,\delta\theta, \quad \delta z = -2c\cos\theta\sin\theta\,\delta\theta,$$

whence

$$(P - 2W\cos\theta)\sin\theta = 0. \quad \text{.........................(7)}$$

This is satisfied by $\theta = 0$, as we should expect, and again by

$$\cos\theta = P/2W, \quad \text{..............................(8)}$$

but the latter position is imaginary if $P > 2W$.

49. Potential Energy.

When a particle subject to given forces undergoes a *finite* displacement, the total work is to be found by integration of the amounts done on it in traversing the various elements δs of its path.

Let us suppose that we have a particle moveable in a certain 'field of force.' By this is meant that some of the forces, at all events, which act upon it are definite functions of its position. Besides these, we contemplate the action of other forces which may be applied arbitrarily, and which we distinguish as 'extraneous.' Let us suppose that the particle is guided by suitably adjusted extraneous forces so as to pass from some standard position A to another position P, along some prescribed path. We may imagine the operation to be performed with infinite slowness, so that the extraneous forces are always sensibly in equilibrium with those due to the field. The work done by the extraneous forces can then be calculated, and is definite for the particular path.

If the constitution of the field be quite arbitrary, the amount of work required may, and in general will, be different for different paths from A to P. If it be less for one path than for another, then if the particle were made to pass from A to P by the former and return to A by the latter, the net amount of work required for the complete circuit would be negative. The work done by the reaction of the particle on the external agency would therefore on the whole be positive. The process might be repeated indefinitely, and we should have in a field of this character an inexhaustible source of work.

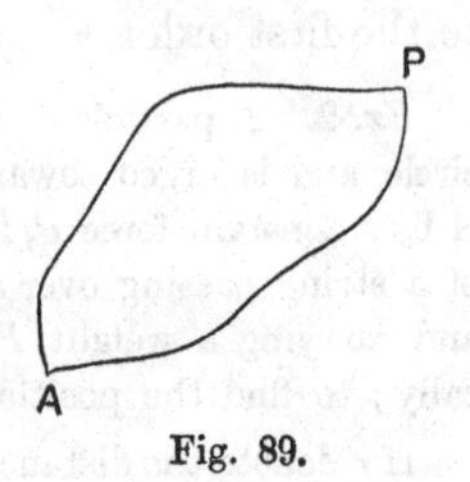

Fig. 89.

The quest of the 'perpetual motion' has, however, long been given up as hopeless; and it is accordingly recognized that in the case of any *natural* field of force the work required to bring a particle from rest in one position to rest in another, in the manner above explained, must be independent of the particular path taken. A field fulfilling this condition is said to be 'conservative*.

In a conservative field, then, the work required to bring a particle from rest in a standard position A to rest in any other position P is a function of the position of P only. This definite capacity for doing work against external resistance, which the particle possesses when at P, over and above what it possesses when at A, in virtue of its position only, is called its 'statical or 'potential†' energy. The particular position A which is our zero of reckoning is of course arbitrary; a change in it has merely the effect of adding a constant to the energy.

The simplest instance of a conservative field is that of ordinary gravity. If W be the weight of a particle, and z its altitude above a fixed horizontal plane, the work required to move it through a space δs is $W\delta z$, where δz is the projection of δs on the upward vertical. Hence by integration the potential energy is

$$V = Wz + C, \qquad \text{.........................(1)}$$

* Since the law of Conservation of Energy can be shewn to hold in it.

† This name was introduced in 1853 by W. J. M. Rankine (1820–72), professor of engineering at Glasgow 1855–72.

where the constant depends on what is regarded as the standard position.

Again, if a particle is urged always towards a fixed point O by a force $\phi(r)$ which is a function of the distance r only, the work required to give it an infinitesimal displacement (in any direction) is, by Art. 48, $\phi(r)\,\delta r$. Hence for the potential energy V we have the formula

$$V = \int_a^r \phi(r)\,dr, \qquad (2)$$

where the lower limit refers to the arbitrary standard position.

Ex. 1. In the case of an attractive force varying as the distance, putting $\phi(r)=\mu r$, we have

$$V=\tfrac{1}{2}\mu r^2 + C. \qquad (3)$$

Again, for an attraction varying as the inverse square of the distance, putting $\phi(r)=\mu/r^2$, we find

$$V=-\frac{\mu}{r}+C. \qquad (4)$$

In each case the potential energy increases with increasing distance. The reverse would be the case if the forces were repulsive.

It is an immediate consequence of the above definition that the work done by the forces of a conservative field in any displacement whatever of a particle is equal to the decrement of the potential energy. Hence, by the principle of virtual velocities, the necessary and sufficient conditions of equilibrium, if there are no extraneous forces, are summed up in the formula

$$\delta V = 0. \qquad (5)$$

In words, the potential energy must be stationary for all infinitesimal displacements.

Ex. 2. A particle is attracted to several centres of force O_1, O_2, ... by forces $\mu_1 r_1$, $\mu_2 r_2$, ... proportional to the distances r_1, r_2, ... from these points, respectively.

We have
$$V=\tfrac{1}{2}(\mu_1 r_1^2+\mu_2 r_2^2+\dots)+C. \qquad (6)$$

It will appear later (Chap. VIII) that there is only one position of the particle satisfying the condition $\delta V=0$, viz. the mass-centre of a system of particles with masses proportional to μ_1, μ_2, ..., situate at O_1, O_2, ..., respectively.

The formula (5) will also hold even if there are frictionless constraints, provided the displacements be such as are consistent with the constraints. For instance, if a particle subject to gravity be constrained to lie on a smooth surface, the possible positions of equilibrium are where the tangent plane is horizontal.

If there are extraneous forces, i.e. forces in addition to those of the field, acting on the particle, and if we denote these generically by P, the principle of virtual velocities takes the form

$$\Sigma (P . \delta p) - \delta V = 0; \quad \text{.....................(7)}$$

i.e. the work done by the extraneous forces in any infinitesimal displacement from a position of equilibrium is equal to the increment of the potential energy.

50. Application to a System of Particles.

The principle of virtual velocities can be extended at once to any system of particles provided we take account of all the forces acting on each. These forces must include those which are due to the mutual actions of the particles, as well as the forces extraneous to the system.

The infinitesimal displacements contemplated may be quite arbitrary and independent for the several particles, but it is usually convenient so to arrange them that some unknown reaction or reactions shall disappear from the equation.

The work done by the mutual action between two particles depends on the change of length of the line (r) joining them. If two points A, B be slightly displaced so as to occupy the positions A', B', and if α, β be the orthogonal projections of A', B' on AB, we have

Fig. 90.

$$B\beta - A\alpha = \alpha\beta - AB = A'B' \cos\theta - AB, \quad \text{........(1)}$$

where θ is the small angle between AB and $A'B'$. This is equal to $A'B' - AB$, or δr, to the first order of small quantities. Hence if we have two particles at A and B exerting equal and opposite forces R on one another, the total work of these in the small displacement will be

$$R.A\alpha - R.B\beta = -R\delta r, \quad \text{..................(2)}$$

to the first order, provided the forces R be reckoned positive when attractive.

The mutual action therefore disappears from the equation of virtual velocities if the displacements be adjusted so that the distance AB is unaltered. In particular, we can eliminate in this way the tension of a straight string connecting the particles, whether the string be physically 'inextensible' or not. A similar inference can easily be made in the case of two particles connected by a string passing over a smooth peg, or a smooth surface of any form, provided we assume that the tension is the same throughout the length*, and provided the displacements be such as the particles could undergo if the string were inextensible.

In calculating the work done by gravity, we may imagine the whole mass of the system to be collected at the centre of gravity, and to receive the displacement of this point. For if W_1, W_2, ... be the weights of the several particles, and z_1, z_2, ... their depths below some fixed horizontal plane of reference, the work done by gravity in a small displacement will be

$$\begin{aligned} W_1\delta z_1 + W_2\delta z_2 + \ldots &= \delta(W_1 z_1 + W_2 z_2 + \ldots) \\ &= \delta\{(W_1 + W_2 + \ldots)\bar{z}\} \\ &= (W_1 + W_2 + \ldots)\,\delta\bar{z}, \quad \text{.........(3)} \end{aligned}$$

if $\bar{z}$ be the depth of the centre of gravity.

Hence if gravity be the only force which does work in the displacement considered, we shall have $\delta\bar{z} = 0$. Thus in the case of the funicular polygon (Fig. 20, p. 23) where a number of weights are attached at various points of a string whose ends are fixed, the depth of the centre of gravity of the whole system is stationary for all small displacements which do not involve change of length of the strings.

* A formal proof of this is given in Chap. x.

Ex. 1. Two weights P, Q rest on two smooth planes whose inclinations are α, β, being connected by a string which passes over a smooth pulley, the free portions being parallel to the planes*.

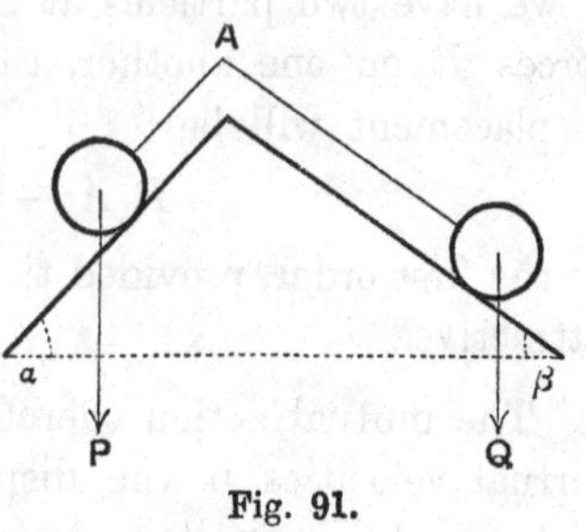

Fig. 91.

If the particles receive equal displacements δs along the planes, P downwards (say) and Q upwards, the projections of these on the downward vertical will be $\delta s \sin\alpha$ and $-\delta s \sin\beta$, respectively. Hence for equilibrium we must have

$$P\delta s \sin\alpha - Q\delta s \sin\beta = 0, \quad \ldots(4)$$

or

$$P \sin\alpha = Q \sin\beta. \quad \ldots\ldots(5)$$

The normal reactions of the planes, and the tension of the string, are omitted from the equation for the reasons explained.

Ex. 2. Two particles P, Q are connected by an inextensible string passing over a smooth peg S. Q hangs freely, whilst P rests on a smooth curve in a vertical plane through the peg. To determine the form of the curve so that P may be in equilibrium in all positions on it.

Let r denote the distance SP, z the depth of P below some fixed horizontal line. In a small displacement in which the string is unextended, Q will ascend through a space δr, whilst P will descend through a space δz. Hence for equilibrium

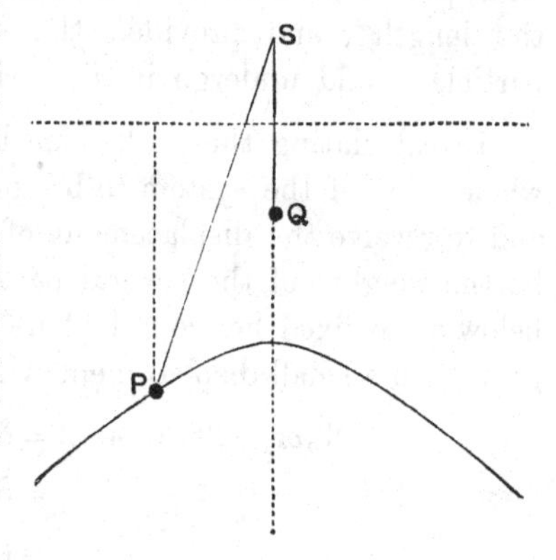

Fig. 92.

$$P\delta z - Q\delta r = 0, \ldots\ldots\ldots(6)$$

if the displacement of P be along the curve. Since this relation is to hold at every point of the curve, we find by integration

$$Pz = Qr + \text{const.},$$

or

$$r = \frac{P}{Q}(z - c), \quad \ldots\ldots\ldots\ldots\ldots\ldots\ldots\ldots(7)$$

where c is some constant. This is the equation of a conic whose focus is S and whose directrix is a horizontal line $z = c$, the eccentricity being P/Q.

* E. Torricelli (1664) quotes this problem as an example of his principle that the centre of gravity cannot descend.

51. Work of a Force on a Rigid Body.

The work of a force, acting on matter of any kind, in any infinitesimal displacement of its point of application, is defined as the product of the force into the orthogonal projection of this displacement on the direction of the force. Or, again, it is equal to the product of the displacement into the component of the force in the direction of the displacement.

It is important to notice that in the case of a body which is displaced without change of dimensions this definition will give the same result, to the first order, whatever point of the body in the line of action of the force be regarded as the point of application. For let A, B (Fig. 90) be any two points of the body in the line of action of a force R, acting (say) in the sense from A to B. If these points are displaced to A' and B', the work as calculated from the displacement of A is $R.A\alpha$, and as calculated from the displacement of B it is $R.B\beta$. We have seen that if $A'B'=AB$ these expressions are equal, to the first order.

It is evident, also, that the total work of two or more concurrent forces, in a small displacement, is equal to the work of the single force which is their geometric sum, acting at the point of concurrence, the proof being the same as in Art 47.

It follows from these results that the total work of any system of forces, in any infinitesimal displacement of a rigid body, is the same as that of any other system to which it is statically equivalent on the principles of Chapter III. For the only assumptions there made were those of the transmissibility of force, and of the geometrical addition of concurrent forces.

Thus in the case of a plane system of forces the work done in a pure translation $\delta\alpha$ is $X\delta\alpha$, where X is the sum of the components of the forces in the direction of $\delta\alpha$.

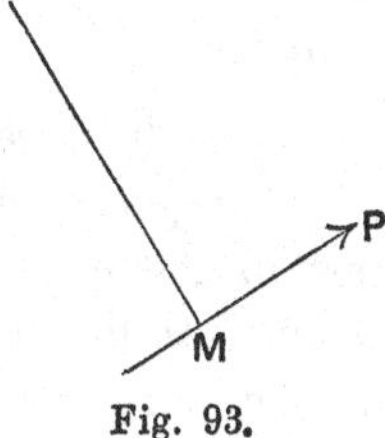

Fig. 93.

Again, the work done in a rotation $\delta\theta$ about any point O is $N\delta\theta$, where $\delta\theta$ is the angle of rotation, and N is the sum of the moments of the forces about O. For if P be one of the forces, and OM the perpendicular from O on its line of action, the displacement of the point M of the body will be $OM.\delta\theta$ at right angles to OM, and the work of P is therefore $P.OM.\delta\theta$; i.e. it is equal to the product of $\delta\theta$ into the moment of P about O. It is easy to see that this result will hold in all cases provided the usual conventions be observed as to the signs to be attributed to moments and angles of rotation.

The particular case of a couple is important. The work of a couple in a displacement of pure translation obviously vanishes. In the case of a rotation $\delta\theta$ about any point O it is equal to $N\delta\theta$, where N is the sum of the moments about O of the two forces constituting the couple, i.e. it is the constant moment of the couple (Art. 24).

52. Principle of Virtual Velocities for a Rigid Body.

It follows from Art. 51 that if a plane system of forces acting on a rigid body be in equilibrium, the total work in any infinitesimal displacement is zero. For the work is equal to that of the resultant force of the system, which by hypothesis vanishes.

Conversely, we can shew that if the work of a plane system of forces be zero for three independent small displacements of the body in the plane of the forces (and therefore for all such displacements) the forces must be in equilibrium. The three independent displacements may consist, for example, of pure translations in two assigned directions, together with a rotation about an assigned point. We learn that the sum of the resolved parts of the forces in each of the assigned directions is zero, and that the moment of the forces about the assigned point also vanishes. This is one form of the conditions of equilibrium investigated in Art. 23, and the other equivalent forms may be arrived at in a similar way.

As in the case of a system of discrete particles, it is possible by a proper choice of the arbitrary displacement to eliminate certain classes of reactions. Thus if a point of the body be constrained to lie on a smooth curve, the work of the normal reaction of the curve in any small displacement consistent with this constraint will be of the second order. Again, if the body can turn freely about a fixed point O, the reaction there does no work in a rotation about O.

Again, if the displacement be such that a surface fixed in the body rolls (without sliding) on a fixed surface, the reaction, whether normal or tangential, at the contact may be ignored in forming the equation of virtual velocities. For the displacement of that point (A') of the body which was originally in contact with the fixed surface (at A) is of the second order*.

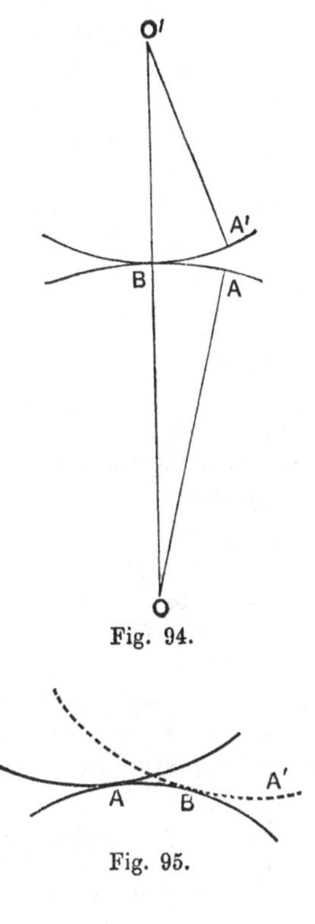

Fig. 94.

Fig. 95.

The *normal* component of the reaction will disappear even if the small displacement be such that one surface slides over the other, as well as rotates, provided it remain in contact. The point of contact now changes from its original position A to an adjacent position B, but if A' be the new position of that point of the body which was at A, the projection of AA' on the normal at A will be of the second order. This principle is specially useful in the case of contact between *smooth* surfaces.

Ex. 1. A bar AB rests with its ends on two smooth inclined planes (Fig. 40, p. 50).

If the bar be slightly displaced in a vertical plane so that its ends slide on the two inclines, the instantaneous centre I is at the intersections of the normals at A, B. The displacement of the centre of gravity G is therefore at right angles to IG. But the work of gravity must be zero, since no work is done by the remaining forces, viz. the reactions at A, B. Hence IG must be vertical. The analytical solution can be completed as in Art. 23, Ex. 3.

Ex. 2. A ladder rests with its upper end against a smooth wall and its foot on the ground (rough); to find the horizontal pressure at the foot.

* It is assumed here that the reaction is equivalent to a single force, the couple of rolling friction (Art. 30) being neglected.

Let l be the length of the ladder, θ its inclination to the vertical, a the distance of its centre of gravity G from the foot, W its weight, F the horizontal component of the reaction of the ground. The height of G above the ground is therefore $a\cos\theta$, and the distance of the foot from the wall is $l\sin\theta$.

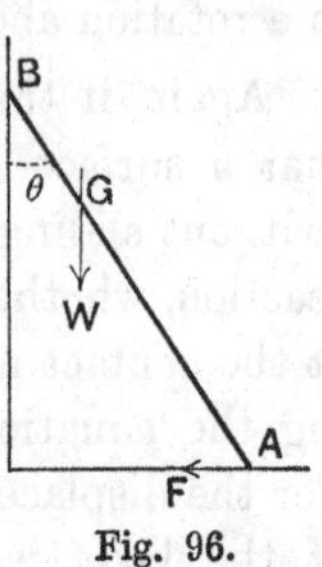

Fig. 96.

If the ladder be slightly displaced so as to remain in contact with the wall and the ground, the normal components of the reactions do no work, and the equation of virtual velocities reduces to

$$-W\delta(a\cos\theta)-F\delta(l\sin\theta)=0, \quad \text{.........(1)}$$

whence

$$F=\frac{Wa}{l}\cdot\tan\theta. \quad \text{.....................(2)}$$

Ex. 3. A bar AB rests on a smooth curve and the end A is restricted, by a frictionless constraint, to motion in a vertical line; to find the form of the curve in order that the bar may be in equilibrium in any position.

Since the height of the centre of gravity G must be stationary in all positions, the locus of G must be a horizontal line. The required curve is therefore the envelope of a straight line AG of constant length whose extremities move on two fixed straight lines at right angles to one another; i.e. it is one branch of a four-cusped hypocycloid (astroid).

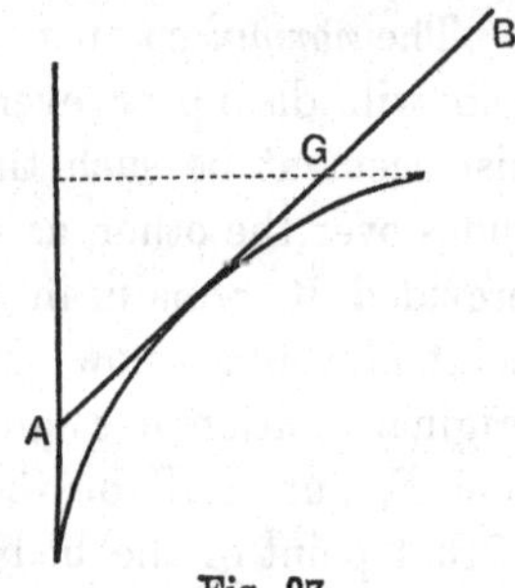

Fig. 97.

53. Extension to a System of Rigid Bodies.

The principle can be extended to any system of rigid bodies, connected together in any way, provided we take account of the mutual actions between the bodies. Any such action consists of two equal and opposite forces, each of which may contribute to the equation of virtual velocities.

As in Art. 50 the displacements of the various bodies may be arbitrary and independent, but it is often convenient to restrict them so that they shall be compatible with the various constraints and connections to which the system is subject.

Thus if two of the bodies be connected by a string or a light rod, and if the hypothetical displacement be such that the distance between the points of attachment is unaltered, the corresponding stress may be left out of account.

Again, the reaction (whether normal or oblique) between two bodies in contact will disappear from the equation, provided the displacement contemplated be such that one surface rolls without sliding relatively to the other. For the displacement may be supposed effected in two steps, in the first of which one body moves into its new position and the other moves with it, without *relative* displacement, whilst in the second step this latter body assumes its new position. It is evident that in the first step the work of the two equal and opposite pressures on the two bodies will cancel, since their points of application have the same displacement; and we have seen (Art. 52) that in the second step the work of one pressure is of the second order, whilst that of the other vanishes absolutely.

In the same way it is seen that the *normal* pressures between two surfaces will do on the whole no work even if there is a relative sliding of one surface over the other in the displacement considered. For this reason the reactions at *smooth* joints may be ignored in forming the equation of virtual velocities.

We are thus able to imagine a great variety of mechanical systems to which the principle of virtual velocities can be applied without any regard to the reactions introduced by the various connections or constraints, provided the hypothetical small displacements be such that none of these connections or constraints are violated. In many such cases the only forces that remain in the equation will be those of gravity, and we may then infer that for equilibrium the depth of the centre of gravity of the system must be 'stationary' (cf. Art. 50).

Ex. 1. A beam is supported on two equal cylindrical rollers, which rest with their axes horizontal on an inclined plane, the surfaces in contact being sufficiently rough to prevent slipping. To find the force F which must be applied to the beam, parallel to the plane, for equilibrium.

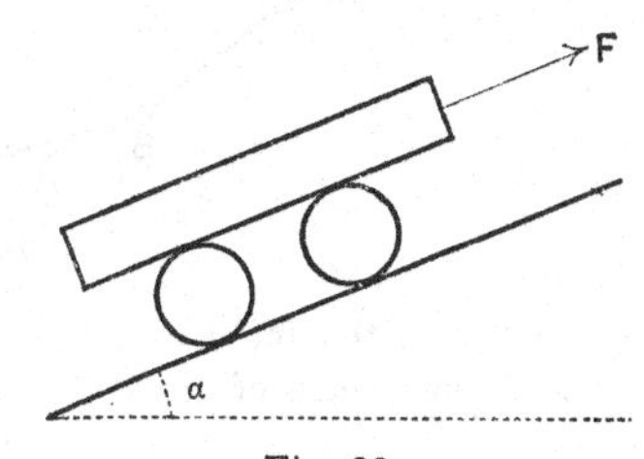

Fig. 98.

The instantaneous centre of each roller is at the point of contact with the plane. Hence if each roller turns through an angle $\delta\theta$, the displacement of the beam will be $2a\delta\theta$, where a is the radius of the roller, whilst the displacement of the centre of the roller will be $a\delta\theta$. Hence

$$F.2a\delta\theta - W.2a\delta\theta.\sin\alpha - W'.a\delta\theta.\sin\alpha = 0, \quad \text{..............(1)}$$

where W is the weight of the beam, and W' that of the pair of rollers; or

$$F = (W + \tfrac{1}{2}W')\sin\alpha. \quad \text{..............(2)}$$

Ex. 2. Four equal light bars are jointed together so as to form a rhombus $ABCD$; the whole hangs from a fixed point A, and is kept in shape by a light horizontal strut BD. A weight W hangs from C. To find the thrust in BD.

We may imagine the strut BD to be removed, provided we introduce two equal and opposite forces S acting outwards at B and D. The frame is now deformable, and if the inclination θ of the bars to the vertical were increased by $\delta\theta$, the distance BD would be increased by $\delta(2a\sin\theta)$, where a is the length of a side, whilst the increment of AC would be $\delta(2a\cos\theta)$. Hence

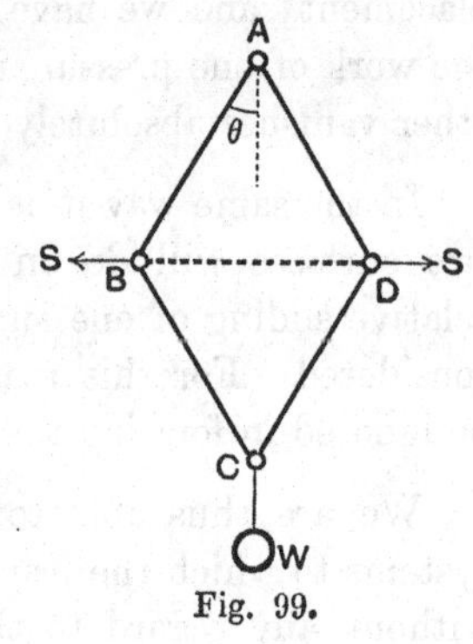

Fig. 99.

$$S.\delta(2a\sin\theta) + W.\delta(2a\cos\theta) = 0, \quad \text{......(3)}$$

or

$$S = W\tan\theta. \quad \text{..................(4)}$$

If the weights W' of the bars AB, BC, CD, DA are taken into account we must include in (3) a term $4W'\delta(a\cos\theta)$, since the depth of the centre of gravity of the whole below A is $a\cos\theta$. The result is

$$S = (W + 2W')\tan\theta. \quad \text{..............................(5)}$$

Ex. 3. A chain of four uniform links AB, BC, CD, DE, jointed at B, C, D, hangs from two points A, E at the same level. Let the weights of AB, DE be each $= W_1$, and those of BC, CD each $= W_2$. Also let $AB = DE = a$, $BC = CD = b$, so that the arrangement is symmetrical about the vertical through C. It is required to determine the form in which the chain hangs.

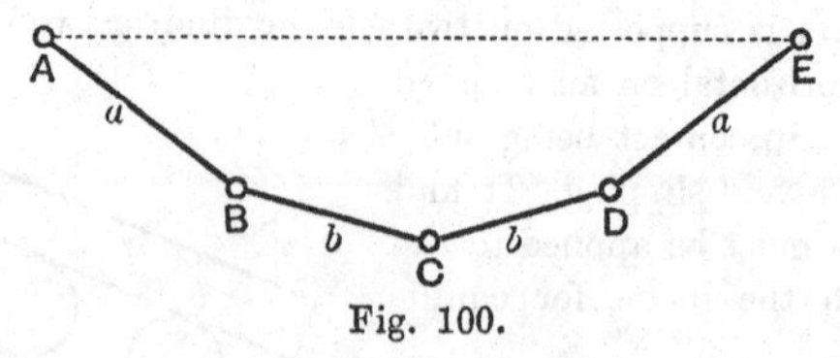

Fig. 100.

If θ, ϕ be the inclinations of AB, BC respectively to the vertical, the depth of the centre of gravity of the chain below AE is

$$\frac{W_1 \cdot \tfrac{1}{2}a\cos\theta + W_2(a\cos\theta + \tfrac{1}{2}b\cos\phi)}{W_1 + W_2},$$

If a small vertical displacement be given to the point C, the deformation will be symmetrical, and the condition that the depth of the centre of gravity should be stationary reduces to

$$(W_1+2W_2)\,a\sin\theta\,\delta\theta+W_2 b\sin\phi\,\delta\phi=0. \qquad\text{(6)}$$

The angles θ, ϕ are connected by the relation

$$2a\sin\theta+2b\sin\phi=c, \qquad\text{(7)}$$

where c is the distance AE, which is supposed fixed. Hence, differentiating, we find

$$a\cos\theta\,\delta\theta+b\cos\phi\,\delta\phi=0. \qquad\text{(8)}$$

Eliminating the ratio of $\delta\theta$ to $\delta\phi$ between (6) and (8), we find

$$(W_1+2W_2)\tan\theta=W_2\tan\phi. \qquad\text{(9)}$$

This relation, combined with (7), determines the angles θ, ϕ. If $W_1=W_2$, we have

$$\tan\theta=\tfrac{1}{3}\tan\phi. \qquad\text{(10)}$$

54. Application to Frames.

The principle of Virtual Velocities is specially convenient in the theory of frames, since the reactions at smooth joints and the stresses in inextensible bars may be left out of account.

In particular, in the case of a frame which is just rigid (Art. 38), subject to external forces at the joints only, the principle enables us to find the stress in any one bar independently of the rest. If we imagine the bar in question to be removed, the frame becomes deformable, but it will remain in equilibrium in its actual shape provided we introduce two equal and opposite forces S, to replace those which were exerted by the bar at the joints which it connected. In any infinitesimal deformation of the frame as thus modified, the work of the forces S, together with that of the original extraneous forces, must vanish. This determines S. The method has already been exemplified in Ex. 2 of Art. 53.

This procedure is particularly useful when the frame, although just rigid, is not 'simple' in the sense of Art. 40, and when, accordingly, the method of reciprocal diagrams is not available for a complete determination of the stresses in it. The analytical calculation would, however, usually involve a good deal of intricate trigonometrical work; and, to evade this, graphical devices have been introduced. For this purpose the infinitesimal displacements of the various joints are replaced by finite lines proportional to them, and therefore proportional to the velocities of the joints in some imagined motion of the frame (as rendered deformable by the suppression of a bar) through its actual configuration. This is really a reversion to the original notion of 'virtual velocities' (Art. 48).

The kinematical basis of the method* has been explained in Art. 15. Let I be the instantaneous centre of a bar CD which is connected to fixed points A, B by links AD, BC; and let s_1, s_2 be the virtual velocities of the points D, C. If these velocities were turned through a right angle, in the same sense, they would be represented on a certain scale by lines DD', CC', where C', D' are

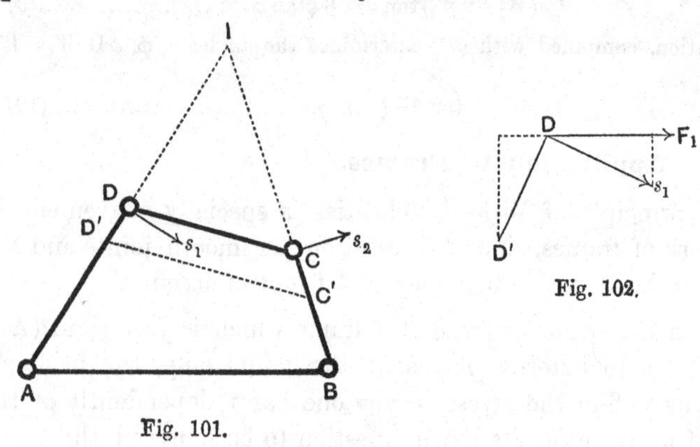

Fig. 101.

Fig. 102.

on IB, IA respectively, and $C'D'$ is parallel to CD. Further, if F_1 be any force acting on the joint D, its (rate of) work will be represented by the product of F_1 into the projection of s_1 on its line of action. The annexed Fig. 102 shews that this is the same as the moment of F_1 about D'. Hence the term which represents the work of F_1, in the equation of virtual velocities, is obtained simply by taking the moment of F_1 about D'.

Take, for example, the case of a frame of nine bars forming a hexagon with its three diagonals (Art. 45); and suppose that it is required to find the stress in CF due to a given system of external forces in equilibrium, acting on the joints. We imagine the bar CF to be removed, so that the frame becomes deformable. If AB be fixed, the instantaneous centre of CD will be at the intersection of AD and BC, so that if $C'D'$ be drawn parallel to CD, the lines CC', DD' may be taken to represent the virtual velocities of the points C, D, turned each through a right angle. Next, drawing $D'E'$ parallel

* Due to H. Müller-Breslau (1887).

to DE, we see that the virtual velocities of the points D, E will be represented on the same scale by the lines DD', EE', similarly turned through a right angle. Finally, if we draw $E'F'$ parallel to EF, the lines EE', FF'

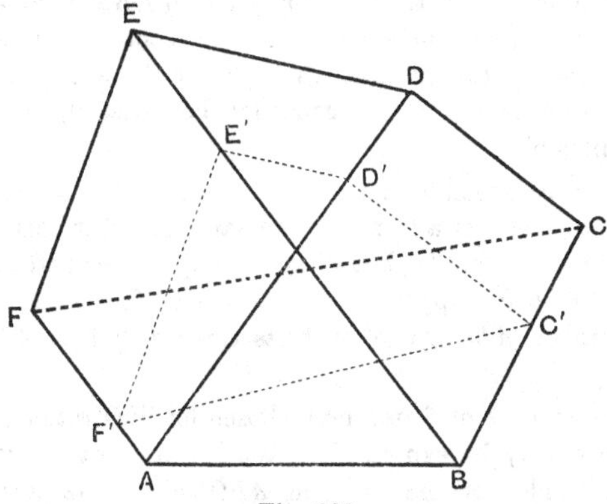

Fig. 103.

will indicate in the same way the virtual velocities of E and F. The equation of virtual velocities is then formed by taking moments about C', D', E', F' in the case of the external forces which act at C, D, E, F, respectively, and about C', F' in the case of the two equal and opposite forces S which we have supposed introduced at C and F to replace the effect of the removed bar CF.

55. Critical Forms.

The principle lends itself naturally to the discussion of the critical forms which have already been alluded to (Art. 39).

A 'critical form' is characterized by the property that, although the frame has a structure which is ordinarily sufficient, and just sufficient, for rigidity, yet in consequence of some special relation between the lengths of the bars it admits of an infinitely small deformation*. The simplest instance of all is that of a frame of three bars AB, BC, CA when the three joints A, B, C fall into a straight line. A small displacement of the joint B at right angles to AC would involve changes in the lengths of AB, BC which are of the second order, only, of small quantities. Another example is that of two rigid frames connected by three links which are concurrent or parallel, as in the annexed figure. (See Art. 15.)

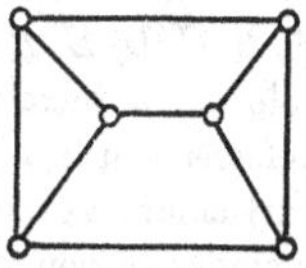

Fig. 104.

* Some questions of this type were treated by A. F. Möbius, *Lehrbuch der Statik*, 1837.

The method of Müller-Breslau explained in Art. 54 leads to a graphical criterion for the occurrence of a critical form. Thus in the hexagonal frame of Fig. 103, if an infinitesimal deformation is possible without removing the bar CF, the instantaneous centre of CF relative to AB will be at the intersection of AF and BC, and since CC', FF' represent the virtual velocities of the points C and F, turned each through a right angle, $C'F'$ must be parallel to CF. Conversely, if this condition be satisfied, an infinitesimal deformation is possible.

The result may be generalized into the statement that a just rigid frame has a critical form whenever a frame of the same structure can be designed with corresponding bars parallel, but without complete geometrical similarity. Thus if, in Fig. 103, $C'F'$ happens to be parallel to CF, a frame having the shape of the hexagon $ABC'D'E'F'$ with its three diagonals AD', BE', $C'F'$ fulfils the description.

In the particular type of frame here chosen for illustration the condition for a critical form may be expressed in another interesting way. Let the pairs of opposite sides of the hexagon $ABCDEF$ intersect in X, Y, Z, respectively, as shewn in Fig. 105. If the frame be slightly deformable,

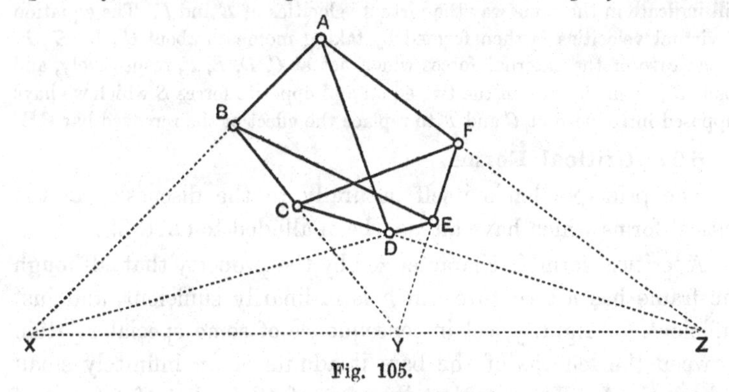

Fig. 105.

X will be the instantaneous centre of AD relatively to BE, Y that of BE relatively to CF, and Z that of CF relatively to AD. It follows from the kinematical theorem of the 'three centres' (Art. 17) that X, Y, Z must be collinear. By Pascal's theorem, this is the condition that the six points A, B, C, D, E, F should lie on a conic. Hence a frame whose bars form the sides of a hexagon and its three diagonals will admit of an infinitesimal deformation if, and only if, a conic can be described through its six joints. A particular case is where the hexagon has its opposite sides parallel; this includes of course the regular hexagon.

56. Self-stressed Frames.

When a frame has a critical form it may be in a state of stress independently of the action of external forces. Suppose as before that one of the bars is removed. If there are no external forces, the equation of virtual velocities reduces to

$$S.\delta s = 0, \qquad \text{.........................(1)}$$

where S is the tension in the bar removed, and δs is the diminution of distance between the joints which it connected. In a critical form we have $\delta s = 0$, and the equation is accordingly satisfied by an arbitrary value of S; and a consistent system of stresses in the remaining bars can then be found by preceding rules. A simple instance is that of the triangular frame with its three joints A, B, C in a straight line. The links AB, BC may be in tension, and AC in thrust, to any equal amounts.

The same conclusion may be reached by the graphical method of Art. 55. Thus in the case of Fig. 103 it is evident that if $C'F'$ be parallel to CF any two equal and opposite forces acting at C and F will have equal and opposite moments about C' and F', respectively, and will accordingly cancel in the equation of virtual velocities.

Again, we may verify that the condition that the hexagonal frame of Fig. 105 should be self-stressed is that the intersections X, Y, Z of pairs of opposite sides should be collinear If the bar AD is in stress, it is in equilibrium under four forces in the lines AB, AF and DC, DE, due to the stresses in the corresponding bars. Rearranging, we have certain forces in AB and DE in equilibrium with forces in AF and DC. The resultant of the first pair, which acts through X, must balance that of the second pair, which acts through Z. Again, the bar BE is in equilibrium under forces in BA, BC and ED, EF. Hence, rearranging, we have forces in BA and ED in equilibrium with forces in BC and EF. The resultant of the first pair, acting through X, must balance that of the second pair, acting through Y. The forces of the first pair are, however, exactly equal and opposite to those which act on the bar AD in the lines AB, DE, and whose resultant has been shewn to act in the line XZ. The points X, Y, Z must therefore be collinear. Conversely, it is easily seen that if this condition be satisfied, a consistent set of stresses in the bars of the frame can be assigned*.

We have seen in Art. 43 how self-stressed frames of n joints and $2n-3$ bars may be devised in abundance. It may be noticed that these all fulfil the graphical criterion (Art. 55) of a critical form; thus in Fig. 80 any number of funicular polygons can be

* The theory of self-stressed frames of n joints and $2n-3$ bars was originated by M. W. Crofton (1878) to whom the above example is due.

constructed with corresponding sides parallel, in virtue of Culmann's theorem (Art. 33).

The possibility of a state of self-stress implies that the stresses due to given external forces are indeterminate, since a distribution of internal stress of arbitrary amount may be superposed without altering the conditions of the problem.

It is more important to remark that the stresses in question are also in general theoretically infinite. To see this, imagine one of the bars of the frame to be removed as before, and replaced by two equal and opposite forces S at its ends. If P denote an external force, and δp the displacement of the joint on which it acts, resolved in the direction of P, the equation of virtual work takes the form

$$\Sigma (P.\delta p) + S.\delta s = 0, \quad \text{.....................(2)}$$

where the summation includes all the joints. If the frame were merely displaced as a rigid body we should have $\Sigma (P.\delta p) = 0$, since the external forces are assumed to be in equilibrium, but in the supposed deformation the sum $\Sigma (P.\delta p)$ will not in general vanish. It follows that the equation (2) cannot be satisfied by a finite value of S, since $\delta s = 0$ by hypothesis. This merely means that if the material of the frame were absolutely unyielding, no finite stresses in the bars could withstand the external forces. With actual materials, the frame would yield elastically, until its configuration is no longer exactly that of a critical form. The stresses in the bars would then be comparatively great, but finite. Frames which approximate to a critical form are of course to be avoided in practice, on account of the excessive stresses, and consequent abnormal extensions or contractions, which would be called into play.

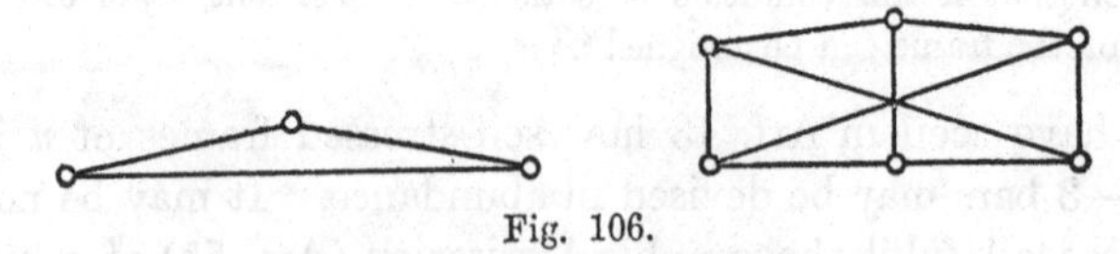

Fig. 106.

57. Potential Energy of a Mechanical System.

The idea of potential energy, which has been already developed in Art. 49 for the case of a single particle, may now be extended.

Suppose that we have a mechanical system of any kind, placed in a given field of force, and subject to given connections and constraints, as well as to other mutual actions between the various parts of it. If we consider the various ways in which the system might pass from rest in one configuration* A to rest in another configuration P, under the guidance of suitable extraneous forces, the amount of work required from the external agency might in general depend on the manner in which the transition is made. It is found, however, that if we exclude the action of frictional forces, the work required is in actual systems always the same. The system is then said to be 'conservative'; and the definite amount of work required to bring it from rest in some standard configuration A to rest in any other configuration P, is called its 'potential energy' in the latter configuration. This measures, as before, the capacity for doing work against external resistances which the system possesses in virtue of its configuration alone. The standard position is of course arbitrary, and the particular choice of it only affects the potential energy to the extent of an additive constant.

It is in general easy to decide whether a given imagined system is conservative or not. So far as ordinary gravity is concerned, we have already seen that the work required depends only on the initial and final configurations. The same thing holds as regards forces on the various particles which act in lines through fixed points, and are definite functions of the distances from these points. The expression for the work done then contains, as we have seen (Art. 49), terms of the type

$$\int_a^r \phi(r)\,dr, \qquad \text{.........................(1)}$$

whose value depends only on the initial and final distances a and r. We get a result of the same form in the case of a mutual action between any two particles which is a function of their distance. This appears from Art. 50 (2) if we put $R=\phi(r)$, $AB=r$. As regards unknown reactions, we have noticed in Art. 53 various types which may for the present purpose be ignored. It may be

* The word 'configuration' is here used to connote absolute as well as relative position.

well to mention, again, that friction, in the case of a relative sliding of two surfaces, is an exception.

If the extraneous forces which act on the system be denoted generically by P, the work which they do in a small displacement will be $\Sigma(P.\delta p)$, where δp is the resolved part of the displacement of the point of application of the force P in the direction of P. Since the work done by the forces intrinsic to a conservative system is $-\delta V$, the principle of virtual velocities takes the form

$$\Sigma(P.\delta p) = \delta V. \quad \text{.....................(2)}$$

If there are no extraneous forces, we have simply

$$\delta V = 0, \quad \text{.............................(3)}$$

i.e. for equilibrium the potential energy must be stationary for all infinitesimal displacements consistent with the constitution of the system.

58. Stability of Equilibrium.

The question as to *stability* of equilibrium belongs essentially to Kinetics, since it involves the consideration of what happens when a slight impulse, or a slight displacement, is given to the system. It appears that in the case of a conservative system the criterion of stability is that the potential energy must be a minimum, i.e. it must be less in the configuration of equilibrium than in any adjacent configuration. For example, whenever gravitational energy is the only form of potential energy involved, the height of the centre of gravity of the system must be a minimum. Thus the form assumed by a chain hanging between two fixed points is such that the centre of gravity is as low as possible.

We consider a few examples of the application of the above criterion.

Ex. 1. Take the case of a bar resting on two smooth inclines (Fig. 40). If the bar be displaced in a vertical plane so that its ends slide on the two inclines, the locus of G is an arc of an ellipse whose centre is in the intersection of the two planes (Art. 16, Ex. 2). Since this arc is convex upwards, the equilibrium is unstable.

Ex. 2. A uniform bar AB of weight W can turn freely about a fixed point A, whilst the end B is attached by a string passing over a small pulley C, vertically above A, to a hanging weight P. We will suppose that $CA > AB$.

Let $AB=2a$, $CA=c$, $CB=r$, and let θ denote the angle CAB. If l be the length of the string, the depth of P below C is $l-r$. The potential energy of the system is therefore

$$V = Wa\cos\theta + Pr + \text{const.} \qquad \text{............(1)}$$

Hence
$$\frac{dV}{d\theta} = -Wa\sin\theta + P\frac{dr}{d\theta}. \qquad \text{............(2)}$$

Since $r^2=c^2+4a^2-4ca\cos\theta$, $r\dfrac{dr}{d\theta}=2ca\sin\theta$, ...(3)

we have
$$\frac{dV}{d\theta} = \left(\frac{2Pc}{r} - W\right)a\sin\theta. \qquad \text{............(4)}$$

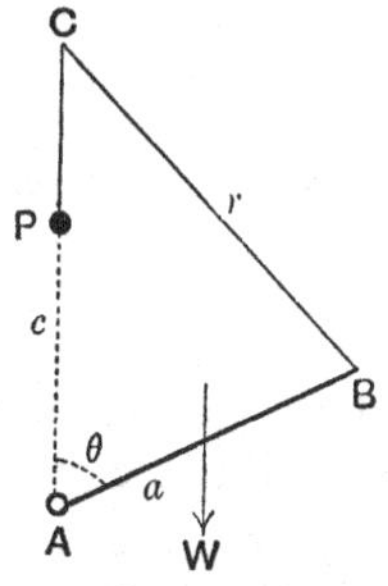

This vanishes for $\theta=0$, $\theta=\pi$, and again if

$$r=2Pc/W. \qquad \text{....................(5)}$$

Fig. 107.

There are therefore two vertical positions of equilibrium, and there may be an inclined one, but the latter is impossible if (5) makes

$$r > c+2a, \text{ or } < c-2a. \qquad \text{..........................(6)}$$

Hence, in order that the inclined position may be possible we must have

$$\frac{P}{W} < \frac{1}{2} + \frac{a}{c}, \text{ and } > \frac{1}{2} - \frac{a}{c}. \qquad \text{......................(7)}$$

To examine the stability, we differentiate again. Taking account of (3), we find

$$\frac{d^2V}{d\theta^2} = \left(\frac{2Pc}{r} - W\right)a\cos\theta - 4\frac{Pc^2a^2}{r^3}\sin^2\theta. \qquad \text{..............(8)}$$

When (5) is satisfied this is negative, and V is accordingly a maximum. Hence the inclined position of equilibrium, when it exists, is unstable. It is easily proved, further, that in this case the vertical positions are both stable, in virtue of the conditions (6). If however the inclined position is impossible owing to the ratio P/W being too great, the upper position ($\theta=0$) is stable and the lower ($\theta=\pi$) unstable. The contrary holds if the ratio P/W be too small.

59. Rocking Stones.

A general criterion for the case of a rigid body moveable parallel to a vertical plane, with one degree of freedom, when gravity is the only force to be taken into account, may be established as follows.

We have seen that the whole series of positions of the body which are kinematically possible is obtained if we imagine a certain curve fixed in the body to roll on another curve fixed in space, these curves being the 'pole-curves' of Art. 16. Since for equilibrium the altitude of the centre of gravity G must be stationary, G must lie in the same vertical with the point of contact I of the two curves. Again, it is known from the theory of 'roulettes' (*Inf. Calc.* Art. 165) that the locus of G will be concave or convex upwards according as

$$\frac{\cos\psi}{h} \gtrless \frac{1}{R} + \frac{1}{R'}, \qquad \text{......................(1)}$$

where R, R' are the radii of curvature of the two curves at I, ψ is the inclination of the common tangent at I to the horizontal, and h is the height of G above I. The signs of R, R' are to be taken positive when the curvatures are as in the standard case shewn in the figure. Hence for stability the relation (1) must hold, with the *upper* sign of inequality. This gives a limit to the height h of G above I.

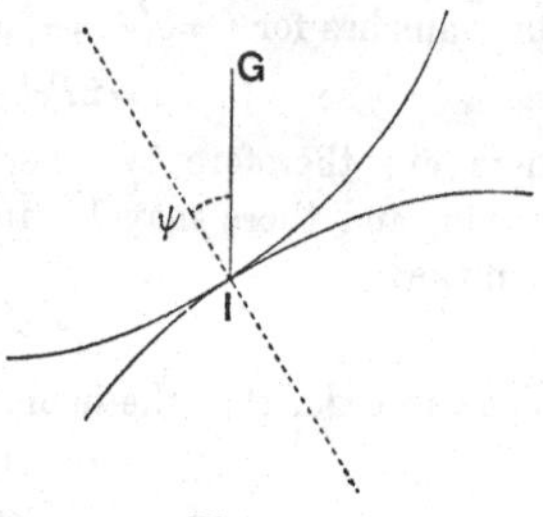

Fig. 108.

Since, by hypothesis, the reactions on the body, in any position, do no work in an infinitesimal rotation about I, they must be equivalent to a single force through I.

We may arrive at the same criterion in a more intuitive manner as follows. If the body be turned from the position of equilibrium through a small angle $\delta\theta$, say to the right, so that the point of contact changes from I to I', we have, by Art. 15,

$$\delta\theta = \left(\frac{1}{R} + \frac{1}{R'}\right)\delta s, \qquad \text{......................(2)}$$

where $\delta s = II'$. The consequent horizontal displacement of G will be $h\delta\theta$, and if this be less than the horizontal projection of δs, viz. $\delta s . \cos\psi$, the vertical through the new position of G will fall to the left of I', and the moment of gravity about I' will therefore tend to restore equilibrium. The condition for stability is therefore

$$\delta s . \cos\psi > h\delta\theta,$$

or

$$\frac{\cos\psi}{h} > \frac{1}{R} + \frac{1}{R'}. \qquad \text{......................(3)}$$

This gives the theory of 'rocking stones,' so far as it can be treated as a two-dimensional problem; the pole-curves are then identical with the profiles of the surfaces in contact.

Ex. 1. In the case of a body resting on an inclined plane we have $R' = \infty$, and the condition is

$$h < R\cos\psi, \quad \text{.....................(4)}$$

i.e. the centre of gravity of the body must be at a lower level than the centre of curvature of the surface in contact with the plane.

Ex. 2. A uniform bar AB hangs by two equal crossed strings from two fixed points C, D at the same level, such that $CD = AB$, and the bar is supposed restricted to the vertical plane through CD.

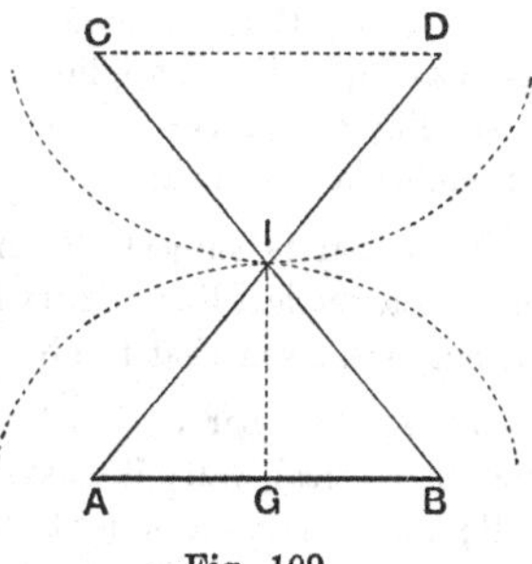

Fig. 109.

The circumstances are a little different from those assumed above, since it is now the *lower* pole-curve which is moveable; it is easily seen however that the condition of stability is

$$\frac{1}{h} < \frac{1}{R} + \frac{1}{R'}, \quad \text{...............(5)}$$

where h is the depth of the centre G of the bar below the point of contact I. If

$$AD = BC = 2a, \quad AB = CD = 2c,$$

the pole-curves are by Art. 16, Ex. 1, equal ellipses of semi-axes a, b, where $b = h = \sqrt{(a^2 - c^2)}$. Also $R = R' = a^2/b$. Hence (5) gives $c < b$, i.e. the depth of G below CD must exceed CD.

EXAMPLES. X.

(Virtual Velocities*.)

1. Two uniform rods AB, BC are freely jointed at B, and hang vertically from A. If a horizontal force P be applied at C, the inclinations θ, ϕ of AB and BC to the vertical are given by

$$\tan\theta = \frac{2P}{W_1 + 2W_2}, \qquad \tan\phi = \frac{2P}{W_2},$$

where W_1, W_2 are the weights of the two rods.

* Several of the examples at the ends of Chapters I and III are also suitable for treatment by this method; for instance, Examples II. (p. 25) 10, 13, 14, 15, and Examples V. (p. 68) 2, 10, 14, 16, 20, 21, 24.

2. A parallelogram $ABCD$ is formed of four rods jointed together. Equal and opposite forces P are applied inwards at A and C, and a string connects B and D. Prove that the tension of the string is equal to

$$P \times \frac{BD}{AC}.$$

3. Four equal uniform rods of length a are freely jointed together so as to form a rhombus. The system rests with a diagonal vertical, the two upper rods being in contact with two smooth pegs at a distance $2c$ apart, at the same level. Find the inclination of the rods to the vertical; and prove that if this inclination be 45° the pegs must be at the middle points of the upper rods.

4. Prove that a pressure F applied to the piston of a steam-engine produces a couple $F.OR$ applied to the fly-wheel, where OR is the intercept made by the line of the connecting rod on that diameter of the fly-wheel which is perpendicular to the axis of the cylinder. (See Fig. 25, Art. 15.)

5. A bar is subject to smooth constraints such that its upper end can move in a vertical line and its lower end on a given curve. Find the form of the curve in order that the bar may be in equilibrium in any position.

6. A light bar of length l rests with its lower end in contact with a smooth vertical wall; it passes over a smooth peg at a distance c from the wall; and carries a weight W suspended from the upper end. Find the inclination θ of the bar to the vertical in equilibrium, and the pressures on the wall and peg. [$\sin^3\theta = c/l$.]

7. Four equal uniform rods, each of weight W, are jointed so as to form a square $ABCD$; the side AB is fixed in a vertical position with A uppermost; and the figure is kept in shape by a string joining the middle points of AD, DC. Find the tension of the string. [$T = 5{\cdot}66\,W$.]

8. Six uniform bars, jointed together, hang from a fixed point A, and form a regular hexagon $ABCDEF$, which is kept in shape by light horizontal struts BF, CE. Prove that the thrusts in these are as 5 : 1.

9. Six equal bars are freely jointed at their extremities forming a regular hexagon $ABCDEF$ which is kept in shape by vertical strings joining the middle points of BC, CD and AF, FE, respectively, the side AB being held horizontal and uppermost. Prove that the tension of each string is three times the weight of a bar.

10. A frame consists of five bars forming the sides of a rhombus $ABCD$ with the diagonal AC. If four equal forces P act inwards at the middle points of the sides, and at right angles to the respective sides, prove that the tension in AC is

$$\frac{P\cos 2\theta}{\sin\theta},$$

where θ denotes the angle BAC.

11. The sides of a quadrilateral $ABCD$ of jointed rods are subject to a uniform normal pressure p per unit length, all directed inwards, the quadrilateral being kept in shape by a diagonal bar AC. Prove that the stress in AC, reckoned positive when a thrust, is

$$\tfrac{1}{2}p.AC(\cot B + \cot D).$$

12. Four bars are freely jointed at their ends so as to form a plane quadrilateral $ABCD$, and the opposite corners are connected by tense strings AC, BD. If T, T' be the tensions in these, prove that

$$\frac{T.AC}{OA.OC} = \frac{T'.BD}{OB.OD},$$

O being the point where the strings cross.

13. The middle points of opposite sides of a jointed quadrilateral are connected by light rods of lengths l, l'. If T, T' be the tensions in these rods, prove that

$$\frac{T}{l} + \frac{T'}{l'} = 0.$$

14. A plane frame of jointed rods is in a state of stress, but is subject to no external forces. If r be the length of any rod, and R the stress in it, reckoned positive when a tension, prove that $\Sigma(Rr) = 0$.

15. If in the preceding question, the frame be subject to given loads applied at the joints, and to the vertical pressures of supports, prove that

$$\Sigma(Rr) + \Sigma(Wy) = 0,$$

where y denotes the altitude of any joint above a fixed horizontal plane, and W the external force at that joint, reckoned as positive when it is a load, and negative when it is a supporting pressure.

EXAMPLES. XI.

(Potential Energy; Stability.)

1. Examine the stability of the various positions of equilibrium in the problem of Art. 48, Ex. 2.

2. A pendulum hangs by a string from the circumference of a horizontal circular cylinder of radius a, the string being wrapped round the cylinder; and l is the length of the free portion when vertical. Prove that the potential energy is, in dynamical measure,

$$\frac{1}{2}\frac{mg}{l}s^2 - \frac{1}{6}\frac{mga}{l^3}s^3 + \ldots,$$

where s is the arc described by the bob from the lowest position.

Hence shew that the tangential force necessary to maintain the bob in any position is

$$\frac{mg}{l}s - \frac{1}{2}\frac{mga}{l^3}s^2 + \ldots,$$

and verify this result independently.

3. A cylinder of circular section rolls along a horizontal plane, the centre of gravity not being on the axis. Apply the criterion of Art. 58 to distinguish between the stable and unstable positions of equilibrium.

4. A wheel with a cylindrical axle can turn about its axis of symmetry, but the wheel is loaded so that its centre of gravity is at an eccentric point. Given weights are suspended by strings wrapped round the circumferences of the wheel and the axle respectively. Find the locus of the centre of gravity of the system, and point out the stable and unstable positions, when equilibrium is possible.

5. An open cylindrical can whose height is 1 ft. and diameter 1 ft. is poised on the top of a sphere. Find the least diameter of the sphere consistent with stability. [9·6 in.]

6. A solid of uniform density is made up of a right cone and a hemisphere having a common base. Find the greatest admissible ratio of the height of the cone to the radius of the hemisphere in order that the solid may stand upright on a table. [1·732.]

7. One end of a uniform bar of weight W rests on a smooth horizontal plane, and the other on a smooth plane of inclination α. A string attached to this end passes over a small smooth pulley at the upper end of the inclined plane, and carries a weight P hanging freely. Prove that there is equilibrium in all positions of the bar provided

$$\sin\alpha = 2P/W.$$

8. A uniform plank of thickness $2h$ rests across the top of a fixed circular cylinder of radius a, whose axis is horizontal. Prove that the gain of potential energy when the plank is turned, without slipping, through an angle θ in a vertical plane parallel to its length is

$$W\{a\theta\sin\theta - (a+h)(1-\cos\theta)\},$$

and deduce the condition of stability.

9. A plank of thickness $2a$ rests horizontally across the top of a horizontal cylinder of radius a, so that the equilibrium is (to a first approximation) neutral. Prove that it is really unstable.

10. A cylinder whose section is an ellipse of semiaxes a, b rests with its axis horizontal on a rough plane of inclination α; prove that if

$$\sin\alpha \not> (a^2-b^2)/(a^2+b^2)$$

there are two positions of equilibrium, that one being stable in which the diameter of the cross-section which is parallel to the plane is greater than the vertical diameter.

11. One elliptic cylinder rests on another equal one, the lengths being horizontal, and the minor axes of the cross-sections in the same vertical plane. Prove that for stability the eccentricity of the cross-sections must exceed $1/\sqrt{2}$.

12. A uniform solid hemisphere rests with its curved surface in contact with a rough inclined plane; find the greatest admissible value of the inclination α. If α be less than this value is the equilibrium stable?

13. A uniform square plate rests in a vertical plane, with its two lower sides on two smooth pegs at a distance c apart in the same horizontal line. Prove that if c is less than one-fourth the diagonal of the square there is only one such position of equilibrium and that it is unstable. Also that if c exceeds the above value there are three positions of equilibrium, of which the symmetrical position is stable (for displacements in the vertical plane) and the others unstable.

14. A rectangular picture-frame hangs from a small perfectly smooth pulley by a string of length $2a$ attached symmetrically to two points on the upper edge at a distance $2c$ apart. Prove that if the depth of the picture is less than $2c^2/\sqrt{(a^2-c^2)}$ there are three positions of equilibrium, of which the symmetrical one is unstable.

If the depth exceed the above value the symmetrical position of equilibrium is the only one, and is stable.

15. A uniform elliptic plate of semiaxes a, b rests in a vertical plane on two smooth pegs P, Q at the same level. Prove that unsymmetrical positions of equilibrium are possible if PQ lies between $\sqrt{2}a$ and $\sqrt{2}b$.

Examine the stability of the various positions of equilibrium.

16. A cylinder whose section is a segment of a parabola rests on another whose section is an equal parabola, the generating lines being parallel and horizontal; and the axes of the sections are vertical and in the same straight line. Prove that if the upper cylinder be made to roll through any angle, every position will be one of neutral equilibrium, provided the centre of gravity be at the focus of a cross-section.

CHAPTER VII

ANALYTICAL STATICS

60. Analytical Reduction of a Plane System of Forces.

The general theorems relating to a plane system of forces have been proved in Chap. III in the manner which appears most direct and instructive. It remains to give the analytical treatment by the methods of Coordinate Geometry. This will not, of course, lead to anything new in the way of results, but the formulae obtained are of some interest in themselves, and are moreover important with a view to later applications in Dynamics.

Let (x_1, y_1), (x_2, y_2), ... be the rectangular coordinates of any points A_1, A_2, ... on the lines of action of the respective forces. The force at A_1 is supposed to be specified by its components (X_1, Y_1) parallel to the coordinate axes, that at A_2 by its components (X_2, Y_2), and so on. If we introduce at O two equal and opposite forces $\pm X_1$ along Ox, we see that the force X_1 at A_1 is equivalent to an equal and parallel force at O, together with a couple whose moment is $-y_1X_1$. In the same way the force Y_1 at A_1 is equivalent to an equal and parallel force at O together with a couple x_1Y_1.

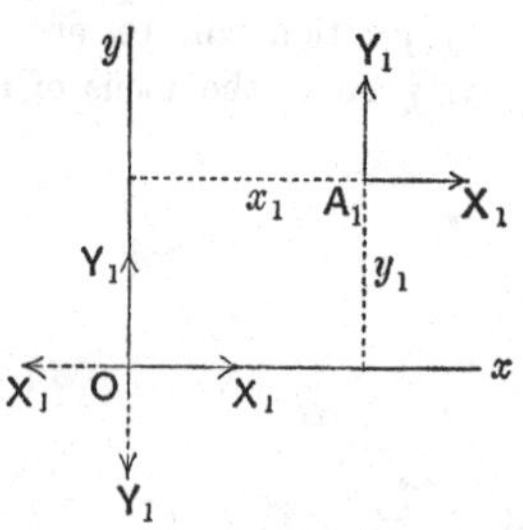

Fig. 110.

Since couples in the same plane are compounded by addition of their moments, we see that the force (X_1, Y_1) can be transferred from A_1 to O provided we introduce a couple of moment

$x_1 Y_1 - y_1 X_1$. This moment, it may be noticed, is the moment of the original force about O. If we deal with the remaining forces in the same way, the given system is finally replaced by two forces P, Q along the coordinate axes, and a couple of moment N, viz. we have

$$P = \Sigma(X), \quad Q = \Sigma(Y), \quad \text{.................(1)}$$

$$N = \Sigma(xY - yX). \quad \text{.....................(2)}$$

The two forces P, Q can be compounded into a single force R, in a direction making an angle θ with Ox, provided

$$P = R\cos\theta, \quad Q = R\sin\theta, \quad \text{...............(3)}$$

whence $$R^2 = P^2 + Q^2, \quad \tan\theta = Q/P. \quad \text{...............(4)}$$

The given system is thus reduced to a single force R acting through an arbitrary origin O, and a couple whose moment is the sum of the moments of the given forces about O. It is evident that the magnitude and direction of the force R are the same whatever point O in the plane be taken as origin (cf. Art. 21).

61. Conditions of Equilibrium.

For equilibrium we must have $R = 0$, $N = 0$, the former of which requires that $P = 0$, $Q = 0$, by Art. 60 (4). We are thus led to the *three* equations

$$\Sigma(X) = 0, \quad \Sigma(Y) = 0, \quad \Sigma(xY - yX) = 0, \quad \text{.........(1)}$$

which constitute the algebraic form of conditions stated in Art. 23.

When the force R does not vanish, we can by a change of origin secure that the couple shall vanish. For if O' be a point whose coordinates are (ξ, η), the moment of the couple, when the forces are referred to O' as a new origin, will be

$$N' = \Sigma\{(x - \xi)Y - (y - \eta)X\}$$

$$= N - Q\xi + P\eta. \quad \text{.........................(2)}$$

This will vanish provided O' lies on the straight line whose equation relative to the original axes is

$$Q\xi - P\eta = N, \quad \text{.......................(3)}$$

where ξ, η are now regarded as current coordinates. This is the equation of the line of action of the single resultant to which the system is always reducible unless $P = 0$, $Q = 0$ simultaneously (Art. 21).

If the given forces are all *parallel*, making (say) an angle θ with Ox, we may put

$$X_1 = S_1 \cos\theta,\quad Y_1 = S_1 \sin\theta,\quad X_2 = S_2\cos\theta,\quad Y_2 = S_2 \sin\theta,\ \ldots,$$

where $S_1, S_2, \ldots$ are the scalar quantities representing the given forces. Hence

$$P = \Sigma(S).\cos\theta,\quad Q = \Sigma(S).\sin\theta, \quad \ldots\ldots\ldots\ldots\ldots(4)$$

$$N = \Sigma(xS).\sin\theta - \Sigma(yS).\cos\theta, \quad \ldots\ldots\ldots\ldots\ldots\ldots(5)$$

and the equation (3) of the line of action of the resultant, when $\Sigma(S)$ does not vanish, becomes

$$\{\Sigma(xS) - \xi.\Sigma(S)\}\sin\theta - \{\Sigma(yS) - \eta.\Sigma(S)\}\cos\theta = 0. \quad \ldots(6)$$

This passes through the point whose coordinates are

$$\xi = \frac{\Sigma(xS)}{\Sigma(S)},\quad \eta = \frac{\Sigma(yS)}{\Sigma(S)}, \quad \ldots\ldots\ldots\ldots\ldots\ldots(7)$$

whatever the value of θ. This is the theorem as to the 'centre' of a system of parallel forces, of given intensities and acting at given points, which we have already met with in Art. 22.

62. Virtual Velocities.

Suppose that in consequence of an infinitesimal displacement the lines in a plane figure, of invariable form, which were originally coincident with the coordinate axes Ox, Oy, assume the positions $O'x'$, $O'y'$. Let $\delta\alpha$, $\delta\beta$ be the coordinates of O' relative to Ox, Oy; and let $\delta\theta$ be the angle through which the figure has been turned, i.e. the angle which $O'x'$ makes with Ox. Then, by projections, we find that

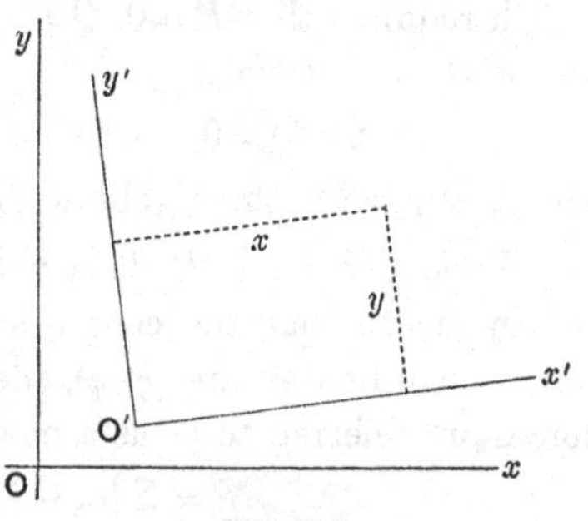

Fig. 111.

the coordinates of that point in the figure which was at (x, y) have become

$$\delta\alpha + x\cos\delta\theta - y\sin\delta\theta,\quad \delta\beta + x\sin\delta\theta + y\cos\delta\theta, \quad \ldots(1)$$

or

$$x + \delta\alpha - y\delta\theta,\quad y + \delta\beta + x\delta\theta, \quad \ldots\ldots\ldots\ldots\ldots(2)$$

to the first order. Hence the component displacements of the point in question are

$$\delta x = \delta\alpha - y\delta\theta,\quad \delta y = \delta\beta + x\delta\theta. \quad \ldots\ldots\ldots\ldots\ldots(3)$$

It may be noticed that by equating these to zero we should obtain the coordinates of the instantaneous centre (Art. 15).

If the point (x, y) is the point of application of a force (X, Y), the work of this force in the above displacement, being equal to the sum of the works of its two components (Art. 47), is

$$X\delta x + Y\delta y = X\delta\alpha + Y\delta\beta + (xY - yX)\,\delta\theta. \quad \text{......(4)}$$

Hence the total work of a plane system of forces acting on a rigid body, in an infinitesimal displacement parallel to the plane of the forces, is given by the expression

$$\Sigma(X\delta x + Y\delta y) = \Sigma(X).\delta\alpha + \Sigma(Y).\delta\beta + \Sigma(xY - yX)\,\delta\theta, \quad (5)$$

$$= P\delta\alpha + Q\delta\beta + N\delta\theta, \quad \text{........................(6)}$$

in the notation of Art. 60. We thus verify that the work is the same as that of the single force (P, Q) acting at O, and the couple N, to which the system was proved to be statically equivalent (cf. Art. 52).

If the forces are in equilibrium, the work vanishes. Conversely, if the total work is zero for three independent small displacements of the body, and therefore for all such displacements, we must have

$$P = 0, \quad Q = 0, \quad N = 0, \quad \text{..................(7)}$$

i.e. the forces must be in equilibrium. The principle of virtual velocities is therefore implied in, and implies, the ordinary conditions of equilibrium.

63. Scalar and Vector Products.

It may be worth while to point out the relations between some of the analytical expressions which have here presented themselves, and the theory of scalar and vector products.

If $\mathbf{r}$ denote the position-vector OA (Art. 5) of a point A relative to the origin, we may write

$$\mathbf{r} = x\mathbf{i} + y\mathbf{j}, \quad \text{..........................(1)}$$

where $\mathbf{i}, \mathbf{j}$ are two unit vectors parallel to the positive directions of x, y, respectively. As regards scalar products we have by Art. 47

$$\mathbf{i}^2 = 1, \quad \mathbf{j}^2 = 1, \quad \mathbf{ij} = 0. \quad \text{.....................(2)}$$

If $\mathbf{r} + \delta\mathbf{r}$ be the position-vector of an adjacent point A', we have

$$\text{AA}' = \text{OA}' - \text{OA} = \delta\mathbf{r}. \qquad (3)$$

Hence if A be a point on the line of action of a force $\mathbf{P}$, the work done in the displacement AA′ is given by the scalar product $\mathbf{P}\delta\mathbf{r}$, and the equation of virtual velocities may be written

$$\Sigma(\mathbf{P}.\delta\mathbf{r}) = 0. \qquad (4)$$

If X, Y be the rectangular components of $\mathbf{P}$, we have

$$\mathbf{P} = X\mathbf{i} + Y\mathbf{j}, \qquad (5)$$

and therefore

$$\mathbf{P}\delta\mathbf{r} = (X\mathbf{i} + Y\mathbf{j})(\delta x\mathbf{i} + \delta y\mathbf{j}). \qquad (6)$$

If we develope this product in accordance with the distributive law (Art. **47**), we find, in virtue of (2),

$$\mathbf{P}\delta\mathbf{r} = X\delta x + Y\delta y, \qquad (7)$$

as is otherwise known.

There is another kind of product which is recognized in vector analysis, and distinguished by a special notation. If we draw AB, BC to represent two vectors $\mathbf{P}$, $\mathbf{Q}$ in the plane of the paper, the 'vector product' of $\mathbf{P}$ into $\mathbf{Q}$ is defined as a certain vector $\mathbf{R}$ perpendicular to this plane, whose magnitude and sense are determined by the parallelogram $ABCD$ constructed on AB, BC as adjacent sides. The absolute magnitude is given by the area of the parallelogram, and the sense is towards or from the reader, according as a positive or negative rotation of less than two right angles would bring the direction of $\mathbf{P}$ into coincidence with that of $\mathbf{Q}$. This vector product is denoted by $[\mathbf{PQ}]$. If P, Q be the absolute magnitudes of $\mathbf{P}$, $\mathbf{Q}$, and θ the angle of rotation as above defined, the absolute value of $\mathbf{R}$, or $[\mathbf{PQ}]$, is equal to that of $PQ \sin\theta$, and the sense is towards or from the reader according as this expression is positive or negative.

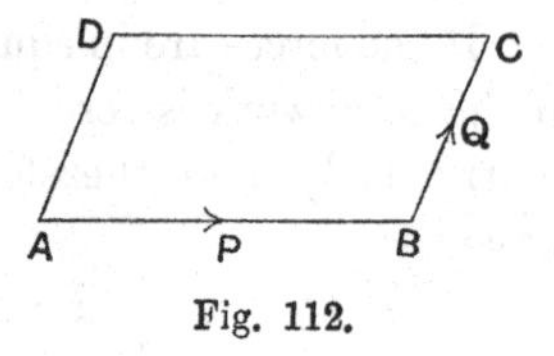

Fig. 112.

It follows that

$$[\mathbf{PQ}] = -[\mathbf{QP}], \qquad (8)$$

so that vector products do *not* follow the commutative law. Moreover, the vector square of any vector vanishes, since θ is then $=0$; or in symbols

$$[\mathbf{P}^2] = [\mathbf{PP}] = 0. \qquad (9)$$

It will be noticed that if $\mathbf{P}$ denote the position-vector of a point A relative to any given point O, and $\mathbf{Q}$ represent a force through A, the absolute magnitude of the vector $[\mathbf{PQ}]$ is equal to that of the *moment* of the force about O, and its sense is towards or from the reader according as this moment is positive or negative. In particular, the moment of any couple in the plane of the paper can be indicated in this way by means of a vector perpendicular to this plane; viz. if $\mathbf{Q}$ be either of the forces of the couple, and $\mathbf{P}$ the position-vector of any point on its line of action relative to any point on the line of action of the other force, the vector in question is $[\mathbf{PQ}]$. The theorem of Art. 24 shews that the law of composition of couples in the same plane is identical with that of addition of parallel free vectors.

It can be shewn that the distributive law

$$[\mathbf{P}(\mathbf{Q}+\mathbf{R})]=[\mathbf{PQ}]+[\mathbf{PR}] \qquad (10)$$

holds in all cases; but we are only concerned here with the case where the three vectors $\mathbf{P}$, $\mathbf{Q}$, $\mathbf{R}$ are all in the same plane, and the products are accordingly all vectors perpendicular to that plane. The proof is then mathematically equivalent to that of Art. 20, where it was shewn that the sum of the moments of two forces about any point is equal to the moment of their resultant.

Let us now take rectangular axes Ox, Oy, in the plane of the paper, in their usual relative positions, and let Oz be a third axis drawn normal to the paper and towards the reader. If $\mathbf{i}, \mathbf{j}, \mathbf{k}$ be three unit vectors respectively parallel to these axes, we have, by the definition,

$$[\mathbf{ij}]=-[\mathbf{ji}]=[\mathbf{k}], \quad [\mathbf{i}^2]=0, \quad [\mathbf{j}^2]=0. \qquad (11)$$

If $\mathbf{r}$ be the position-vector of a point (x, y) with respect to O, and $\mathbf{Q}$ be a force through (x, y), we may write

$$\mathbf{r}=x\mathbf{i}+y\mathbf{j}, \quad \mathbf{Q}=X\mathbf{i}+Y\mathbf{j}. \qquad (12)$$

Hence

$$[\mathbf{rQ}]=[(x\mathbf{i}+y\mathbf{j})(X\mathbf{i}+Y\mathbf{j})], \qquad (13)$$

or, developing the product by the distributive law, and making use of the relations (11),

$$[\mathbf{rQ}]=(xY-yX)\,\mathbf{k}, \qquad (14)$$

in agreement with the analytical expression for the moment obtained in Art. 60.

EXAMPLES. XII.

1. A lamina is moved in its own plane so that the point which was at the origin of rectangular coordinates becomes the point (α, β), and is turned through an angle θ; prove that the coordinates of the centre of rotation are

$$\tfrac{1}{2}(\alpha-\beta\cot\tfrac{1}{2}\theta), \quad \tfrac{1}{2}(\beta+\alpha\cot\tfrac{1}{2}\theta).$$

2. Let $O'x'$, $O'y'$ be the rectangular axes fixed in a moving lamina, and Ox, Oy fixed rectangular axes in the same plane with it. Let θ be the angle which $O'x'$ makes with Ox, and let the coordinates α, β of O' relative to Ox, Oy be given functions of θ. Prove that the coordinates of the instantaneous centre relative to Ox, Oy are

$$x=\alpha-\frac{d\beta}{d\theta}, \qquad y=\beta+\frac{d\alpha}{d\theta}.$$

3. A lamina receives infinitesimal rotations ω_1, ω_2, ... about points whose coordinates, relative to rectangular axes in its plane, are (x_1, y_1), (x_2, y_2), ..., respectively. Prove that the displacement is equivalent to a rotation $\Sigma(\omega)$ about the point whose coordinates are

$$\frac{\Sigma(\omega x)}{\Sigma(\omega)}, \quad \frac{\Sigma(\omega y)}{\Sigma(\omega)}.$$

If $\Sigma(\omega) = 0$, find the components of the equivalent translation.

4. Two forces P, Q acting through two given points A, B have a resultant R. Prove that if they be turned through any angle, in the same sense, about A, B, respectively, the resultant will also turn about a fixed point C (called the 'astatic centre'). Also that

$$P : Q : R = BC : CA : AB.$$

Deduce the existence of an astatic centre for any plane system of given forces acting through given points.

5. Prove that if the forces of a plane system be turned through 90°, in the same sense, about their respective points of application, the equation of the line of action of their resultant becomes

$$P\xi + Q\eta = M,$$

where $P = \Sigma(X)$, $Q = \Sigma(Y)$, $M = \Sigma(xX + yY)$.

6. Prove that if a plane system of forces in equilibrium be turned about their respective points of application through an angle a they will be equivalent to a couple of moment $M \sin a$.

7. Prove *analytically* the existence of an astatic centre (see Ex. 4) for any plane system of forces acting at given points, and shew that its coordinates are

$$\frac{MP + NQ}{P^2 + Q^2}, \quad \frac{MQ - NP}{P^2 + Q^2},$$

in the notation of Art. 60 and Ex. 5.

8. If the coordinates of the n joints of a plane frame be

$$(x_1, y_1), \quad (x_2, y_2), \quad \ldots, \quad (x_n, y_n),$$

and if the equations which express that the lengths of the bars are constant be

$$\phi = 0, \quad \phi' = 0, \quad \phi'' = 0, \quad \ldots,$$

prove that the principle of virtual velocities leads to n pairs of equations of the type

$$\left.\begin{aligned} X_r + \lambda\frac{\partial\phi}{\partial x_r} + \lambda'\frac{\partial\phi'}{\partial x_r} + \lambda''\frac{\partial\phi''}{\partial x_r} + \ldots = 0, \\ Y_r + \lambda\frac{\partial\phi}{\partial y_r} + \lambda'\frac{\partial\phi'}{\partial y_r} + \lambda''\frac{\partial\phi''}{\partial y_r} + \ldots = 0, \end{aligned}\right\}$$

where X_r, Y_r are the components of external force on the rth joint, and λ, λ', λ'', ... are undetermined multipliers.

9. Shew that the preceding equations include the conditions of equilibrium of the external forces, and so are equivalent to $2n-3$ independent relations.

Shew also that the equations are identical with those obtained by considering the equilibrium of each joint separately.

10. Deduce from the formulæ of Ex. 8 the analytical condition that a frame of n joints and $2n-3$ bars should be capable of self-stress; and shew that it is identical with the condition that the frame should admit of an infinitesimal deformation.

CHAPTER VIII

THEORY OF MASS-SYSTEMS

64. Centre of Mass.

We consider in the first instance a system of isolated points $P_1, P_2, \ldots, P_n$, not necessarily in one plane, with which are associated certain coefficients $m_1, m_2, \ldots, m_n$, respectively. In the application to Mechanics these coefficients are the masses of particles situate at the respective points, and it will accordingly be convenient to speak of them in this way; but it is to be noticed that so far as the geometrical theory is concerned there is no necessity for the coefficients to be all of one sign. We shall however exclude the case where the sum $m_1 + m_2 + \ldots + m_n$, or $\Sigma(m)$, vanishes*.

In such a system there is one and only one point G, called the 'mass-centre,' which is such that

$$\Sigma(m.\mathrm{GP}) = 0. \quad \text{......................(1)}$$

This is to be understood as a vector equation, and expresses that the geometric sum of the vectors drawn from G to the several points $P_1, P_2, \ldots, P_n$, multiplied respectively by the corresponding scalar quantities $m_1, m_2, \ldots, m_n$, is to be zero.

To prove the statement, we take any point O and construct by geometrical addition a vector OG such that

$$\mathrm{OG} = \frac{\Sigma(m.\mathrm{OP})}{\Sigma(m)}. \quad \text{......................(2)}$$

This determines a certain point G, and we then find

$$\begin{aligned}\Sigma(m.\mathrm{GP}) &= \Sigma\{m(\mathrm{GO}+\mathrm{OP})\} = \Sigma(m).\mathrm{GO} + \Sigma(m.\mathrm{OP}) \\ &= \Sigma(m).\mathrm{GO} + \Sigma(m).\mathrm{OG} = \Sigma(m).\mathrm{GG} = 0, \quad \text{......(3)}\end{aligned}$$

so that the condition (1) is fulfilled.

* This excluded case occurs in the theory of Magnetism.

Again, if G' also satisfies (1) we have

$$\Sigma (m).\mathrm{GG'} = \Sigma \{m(\mathrm{GP}+\mathrm{PG'})\} = \Sigma (m.\mathrm{GP}) - \Sigma (m.\mathrm{G'P}) = 0,$$

i.e. G' must coincide with G*.

It is easily seen that the position of the mass-centre is not altered, if for any group of particles in the system we substitute a single particle whose mass is equal to the total mass of the group, situate at the mass-centre of the group. Thus if H be the mass-centre of the group $m_1, m_2, \ldots, m_s$, and G that of the whole system, we have, by (2),

$$m_1.\mathrm{GP}_1 + m_2.\mathrm{GP}_2 + \ldots + m_s.\mathrm{GP}_s = (m_1 + m_2 + \ldots + m_s)\,\mathrm{GH}. \quad \text{.........(4)}$$

Hence in forming the sum on the left-hand side of (1), the particles $m_1, m_2, \ldots, m_s$ may be replaced by a single particle of mass $m_1 + m_2 + \ldots + m_s$ situate at H. This principle is used constantly in the determination of the mass-centres of particular systems.

If the particles be displaced in any way, so that m_1 comes to P_1', m_2 to P_2', and so on, and if G' be the new position of the mass-centre, then

$$\Sigma (m.\mathrm{PP'}) = \Sigma (m).\mathrm{GG'}. \quad \text{.................(5)}$$

This is obvious at once if we write

$$\mathrm{PP'} = \mathrm{PG} + \mathrm{GG'} + \mathrm{G'P'}, \quad \text{.................(6)}$$

and remember that in virtue of the fundamental definition

$$\Sigma (m.\mathrm{PG}) = 0; \quad \Sigma (m.\mathrm{G'P'}) = 0. \quad \text{...........(7)}$$

The theorem (5) has an important application in Dynamics.

Ex. 1. In the case of three equal particles situate at A, B, C, we have

$$\mathrm{GB} + \mathrm{GC} = 2\mathrm{GA'}, \quad \text{..............................(8)}$$

where A' is the middle point of BC. Hence, since by hypothesis

$$\mathrm{GA} + \mathrm{GB} + \mathrm{GC} = 0,$$

we have

$$\mathrm{AG} = 2\mathrm{GA'}, \quad \text{..................................(9)}$$

i.e. G lies in the line AA' at the point of trisection furthest from A.

* The definition of the mass-centre by means of the vector equation (2) was given by H. Grassmann, *Ausdehnungslehre* (1844). The more usual plan has been to define the centre of gravity of a system on the lines of Art. 22 *ante*, and to deduce the formula (1) as a consequence. This formula, in its statical interpretation (Art. 65), is attributed to Leibnitz.

More generally, if m_1, m_2, m_3 be the masses of any three particles situate at A, B, C, the mass-centre of m_2, m_3 is at a point A' in BC such that

$$m_2 . \mathrm{BA'} = m_3 . \mathrm{A'C}\,; \qquad (10)$$

and the mass-centre of m_1, m_2, m_3 is at a point G in AA' such that

$$m_1 . \mathrm{AG} = (m_2 + m_3)\,\mathrm{GA'}. \qquad (11)$$

Again, we have

$$\Delta GBC : \Delta ABC = GA' : AA' = m_1 : m_1 + m_2 + m_3, \qquad (12)$$

and therefore $\qquad \Delta GBC : \Delta GCA : \Delta GAB = m_1 : m_2 : m_3. \qquad (13)$

It is easily seen that by giving suitable values (positive or negative) to the ratios $m_1 : m_2 : m_3$ we can make G assume any assigned position in the plane of ABC. We have here the notion of the 'barycentric*' or 'areal' coordinates of Analytical Geometry. For instance, if $m_1 : m_2 : m_3 = a : b : c$, it appears from (13) that G is equidistant from the sides of the triangle ABC, i.e. it is at the centre of the inscribed circle. If $m_1 + m_2 + m_3 = 0$, G is at infinity.

Ex. 2. In the case of four equal particles at A, B, C, D, we have

$$\mathrm{GB} + \mathrm{GC} + \mathrm{GD} = 3 . \mathrm{GA'}, \qquad (14)$$

where A' is the mean centre (Art. 4) of B, C, D. Hence G lies in AA', and is such that $AG = \frac{3}{4} AA'$.

Again, if E, F be the middle points of any two opposite edges of the tetrahedron $ABCD$, say AB and CD, E is the mass-centre of the two particles at A, B, and F that of the two particles at C, D. Hence G coincides with the middle point of EF.

If the particles at A, B, C, D have any given masses m_1, m_2, m_3, m_4, respectively, and if A' be the mass-centre of m_2, m_3, m_4, the mass-centre of the whole system is at a point G in AA' such that

$$m_1 . \mathrm{AG} = (m_2 + m_3 + m_4)\,\mathrm{GA'}. \qquad (15)$$

It is readily found that the volumes of the four tetrahedra which have G as a common vertex, and the triangles BCD, ACD, ABD, ABC, respectively, as bases, are to one another as the masses m_1, m_2, m_3, m_4. Also by a suitable choice of the ratios $m_1 : m_2 : m_3 : m_4$ we can make the mass-centre assume any assigned position in space. In particular, if the masses are proportional to the areas of the opposite faces, G is at the centre of the sphere inscribed in the tetrahedron $ABCD$.

65. Centre of Gravity.

The formula (2) of Art. 64 may be interpreted as shewing that the resultant of a system of concurrent forces represented by $m_1 . \mathrm{P_1O}$, $m_2 . \mathrm{P_2O}$, ..., $m_n . \mathrm{P_nO}$ will be represented by $\Sigma(m) . \mathrm{GO}$; and the statement holds, of course, if the forces and the resultant

* The name given by the inventor, A. F. Möbius (1827).

be altered in any the same ratio. Now if we imagine the point O to recede to infinity in any direction, the vectors GO, P_1O, P_2O, ..., P_nO will ultimately be to one another in a ratio of equality; and we learn that a system of parallel forces proportional to $m_1, m_2, \ldots, m_n$, acting at $P_1, P_2, \ldots, P_n$, have a resultant proportional to $\Sigma(m)$ and acting always in a line through the mass-centre G, whatever be the direction of the forces.

We infer, as in Art. 22, that so far as the attractions of the Earth on the various particles of a body of invariable form can be regarded as a system of parallel forces proportional to the respective masses, they have a resultant equal to their sum acting always through the mass-centre. In other words, the mass-centre is under these conditions also a true 'centre of gravity.'

This proof of the existence of a centre of gravity, since it depends on an approximation, has of course its limitations. These are however practically quite unimportant, except in some questions of Physical Astronomy. The most interesting exceptional case is that of the Moon's attraction on the Earth. This may be reduced with sufficient accuracy to a single force, but there is no one point in the Earth through which the line of action always passes. It is true that the distance of this line from the mass-centre is never very great, amounting at most to a few hundred feet, but the effects of the deviation happen to be cumulative, and are responsible mainly for the slow change in the direction of the Earth's axis of rotation which makes itself apparent through the 'precession of the equinoxes.'

66. Formulæ for Mass-Centre.

If through $P_1, P_2, \ldots, P_n$ we draw a system of parallel planes meeting a straight line OX, drawn in a given direction through a fixed point O, in the points $M_1, M_2, \ldots, M_n$, respectively, the collinear vectors $OM_1, OM_2, \ldots, OM_n$ may be called (as in Art. 5) the projections of $OP_1, OP_2, \ldots, OP_n$ on OX. Let these projections be denoted algebraically by $x_1, x_2, \ldots, x_n$, the sign being positive or negative according as the direction is that of OX or the reverse. Since the projection of a vector-sum is the sum of the projections of the several vectors, we infer from Art. 64 (2) that

$$\bar{x} = \frac{\Sigma(mx)}{\Sigma(m)}, \quad \ldots\ldots\ldots\ldots(1)$$

if $\bar{x}$ be the projection of OG.

In particular, if O coincide with G, we have

$$\Sigma(mx) = 0. \qquad\qquad (2)$$

If the projections be orthogonal, the sum $\Sigma(mx)$ is called the 'linear moment,' or 'first moment,' of the mass-system with respect to the plane $x = 0$.

In the particular case where the masses are all equal, the formula (1) reduces to

$$\bar{x} = \frac{1}{n}\Sigma(x). \qquad\qquad (3)$$

This formula shews that the distance (positive or negative) of the mass-centre from any plane is the arithmetic mean of the distance of the several particles from that plane. Since the relations are purely geometrical, it is convenient to have a name which shall have no explicit reference to *mass*, and the point G is accordingly called, in the present case, the 'centre of mean position,' or the 'mean centre,' of the points $P_1, P_2, \ldots, P_n$. This term is also naturally extended to the case of points continuously and uniformly distributed over an area or a volume.

If the masses, though unequal, be commensurable, each may be regarded as made up by superposition of a finite number of particles, all of the same mass, and the above interpretation will still apply. And since incommensurable magnitudes may be regarded as the limits of commensurables, the formula (1) may in all cases be interpreted as shewing that the distance of the mass-centre from any plane is in a sense the mean distance of the whole mass from that plane.

It follows from (1) that if the Cartesian coordinates of $P_1, P_2, \ldots, P_n$ relative to any axes, rectangular or oblique, be $(x_1, y_1, z_1), (x_2, y_2, z_2), \ldots, (x_n, y_n, z_n)$, the mass-centre $(\bar{x}, \bar{y}, \bar{z})$ is given by the formulæ

$$\bar{x} = \frac{\Sigma(mx)}{\Sigma(m)}, \quad \bar{y} = \frac{\Sigma(my)}{\Sigma(m)}, \quad \bar{z} = \frac{\Sigma(mz)}{\Sigma(m)} \qquad\qquad (4)$$

We shall have occasion to consider also the case of projection on a plane, by parallel lines drawn in some fixed direction. It is evident at once that the geometric sum of the projections of any series of vectors will be equal to the projection of their sum; and, further, that the projection of a vector $m.\mathrm{OP}$ is m times the projection of OP. Hence if our system of points $P_1, P_2, \ldots, P_n$ be projected by parallel lines on to a plane, the mass-centre of a system of particles $m_1, m_2, \ldots, m_n$ situate at the respective projections will coincide with the projection of the mass-centre G of the original system. This result is, indeed, already contained in (4); for if the projecting lines be supposed parallel to Oz, the symbols x, y in the numerators of the expressions for $\bar{x}$, $\bar{y}$ are the coordinates of the projection of the particle m.

67. Continuous Distributions. Simple Cases.

If we have a continuous distribution of matter instead of a finite number of discrete particles, the summations of Art. 66 are to be replaced by integrations. There are, however, a number of simple cases where the position of the mass-centre can be found by more or less elementary considerations.

Ex. 1. Thus the mean centre of a straight line, of the area of a rectangle, a parallelogram, an equilateral triangle, or a circle, the volume of a parallelepiped, a regular tetrahedron, or a sphere, considered in each case as representing a uniform distribution of matter, is obviously at the geometrical centre.

Ex. 2. Again, if a triangular *area* ABC be divided into infinitely narrow strips parallel to BC, the mean centre of each strip will lie on the median line AA', and the mean centre of the whole is therefore in AA'. We infer that it is at the point of concurrence of the three medians, and so (by Art. 64, Ex. 1) coincides with the mass-centre of three equal particles situate at A, B, C.

Ex. 3. The volume of a uniform tetrahedron $ABCD$ may be divided into thin laminæ by planes parallel to BCD. The mean centre of each of these laminæ is in the line AA' joining A to the mean centre of the triangular area BCD. Hence the mean centre of the whole lies in AA'. Similarly it lies in BB', CC', and DD', where B', C', D' are the mean centres of the triangular faces opposite B, C, D, respectively. The mean centre G of the volume therefore coincides with that of the four vertices A, B, C, D (Art. 64, Ex. 2); in particular it divides AA' so that $AG = \frac{3}{4}AA'$.

From this we can derive the general case of a pyramid, or a cone, on a plane base. The sections of the solid by planes parallel to the base will be geometrically similar, and the mean centre of the whole will therefore lie in the line OH drawn from the vertex O to the mean centre H of the area of the base. Again, in the case of a polygonal base, the pyramid can be divided into triangular pyramids having a common vertex O. The mean centre of each of these is in a plane drawn parallel to the base at a distance from O equal to three-fourths the altitude of the figure. Since a curve of any shape may be regarded as the limit of a polygon, the statement applies also to a cone on any plane base. Hence in all such cases the mean centre of the whole is a point in OH such that $OG = \frac{3}{4}OH$.

Ex. 4. Again, it is known that the area of any zone of a spherical surface bounded by parallel circles is proportional to the distance between the planes of these circles. Hence if such a zone be divided by equidistant parallel planes into infinitely narrow annuli, the areas of these will be equal, and the mean centre of the whole area of the zone will therefore lie on its axis, half-way between the bounding planes. In particular the mean centre of the curved surface of a hemisphere bisects the axial radius.

Ex. 5. The volume of a hemisphere may be regarded as made up of infinitely acute pyramids, or cones, having their vertices at the centre O of the spherical surface. If we imagine the mass of each of these to be transferred to its centre of mass, we get a uniform hemispherical sheet of radius $\frac{3}{4}a$, where a is the radius of the solid hemisphere. And the distance of the mass-centre of this from O is $\frac{3}{8}a$.

68. Integral Formulæ.

We denote by ρ the 'volume-density' at a point P whose coordinates are (x, y, z). By this is meant that the mass contained in an elementary space δv containing P may be taken as ultimately equal to $\rho\delta v$. If the axes are rectangular, we may put $\delta v = \delta x \delta y \delta z$, and in any case δv is equal to $\delta x \delta y \delta z$ multiplied by some constant factor depending on the mutual inclinations of the axes. Since the position of the mass-centre is unchanged when all the masses are altered in the same ratio, we have, finally,

$$\bar{x} = \frac{\iiint x\rho\, dx\, dy\, dz}{\iiint \rho\, dx\, dy\, dz}, \quad \bar{y} = \frac{\iiint y\rho\, dx\, dy\, dz}{\iiint \rho\, dx\, dy\, dz}, \quad \bar{z} = \frac{\iiint z\rho\, dx\, dy\, dz}{\iiint \rho\, dx\, dy\, dz}, \ldots(1)$$

as the general formulæ applicable to continuous distributions.

If the density be uniform, ρ goes outside the integral signs, and cancels.

The adaptation of the formulæ (1) to special problems furnishes a number of excellent exercises in integration; but except in a few of the simpler cases the results are interesting from a geometrical rather than from a mechanical point of view. For this reason only a few of the leading cases are treated here

If a finite mass be supposed concentrated in a line, and if μ be the 'line-density,' i.e. the mass per unit length, the mass m of a particle is represented by $\mu\,\delta s$, where δs is an element of arc. Thus in the case of a plane curve

$$\bar{x}=\frac{\int x\mu\,ds}{\int \mu\,ds},\quad \bar{y}=\frac{\int y\mu\,ds}{\int \mu\,ds}. \quad \text{.................. (2)}$$

It is convenient here to adopt rectangular axes.

Ex. 1. In the case of a uniform circular arc, if the origin be taken at the centre, and the axis of x along the medial line, we have $\bar{y}=0$. Also, writing $x=a\cos\theta$, $\delta s=a\,\delta\theta$, we have

$$\int x\,ds=\int_{-a}^{a} a\cos\theta\,.\,a\,d\theta=2a^2\sin\alpha,\qquad \int ds=2a\alpha,$$

if 2α be the angle which the whole arc subtends at the centre. Hence

$$\bar{x}=\frac{\sin\alpha}{\alpha}\,.\,a. \quad \text{.................................... (3)}$$

For the semicircle we have $\alpha=\frac{1}{2}\pi$, and $\bar{x}=2a/\pi=\cdot 637a$.

If a finite mass be concentrated in a surface, and if σ be the 'surface-density,' i.e. the mass per unit area, we put $m=\sigma\delta S$, where δS is an element of area. In the case of a plane distribution, referred to coordinate axes in its own plane, δS bears a constant ratio to $\delta x\delta y$, viz. it is equal to $\delta x\delta y\sin\omega$, where ω is the inclination of the axes. Hence

$$\bar{x}=\frac{\iint x\sigma\,dx\,dy}{\iint \sigma\,dx\,dy},\quad \bar{y}=\frac{\iint y\sigma\,dx\,dy}{\iint \sigma\,dx\,dy}, \quad \text{.............. (4)}$$

whether the axes be rectangular or oblique.

If σ be constant, and if the plane area has a line of symmetry, then taking this as axis of x and a perpendicular to it as axis of y, the mass-element may be taken to be proportional to $y\delta x$, where y is the ordinate of the bounding curve. Then

$$\bar{x}=\frac{\int xy\,dx}{\int y\,dx},\quad \bar{y}=0. \quad \text{.................... (5)}$$

Ex. 2. For a segment of the parabola

$$y^2=Cx, \quad \text{.................................... (6)}$$

bounded by the double ordinate $x=h$, we have

$$\int_0^h xy\,dx=\tfrac{2}{5}C^{\frac{1}{2}}h^{\frac{5}{2}},\qquad \int_0^h y\,dx=\tfrac{2}{3}C^{\frac{1}{2}}h^{\frac{3}{2}},$$

and therefore

$$\bar{x}=\tfrac{3}{5}h. \quad \text{.................................... (7)}$$

Ex. 3. For a semicircular area of radius a, we have

$$\int_0^a xy\,dx = \int_0^a x\sqrt{(a^2-x^2)}\,dx = a^3\int_0^{\frac{1}{2}\pi}\sin\theta\cos^2\theta\,d\theta = \tfrac{1}{3}a^3,$$

$$\int_0^a y\,dx = \tfrac{1}{4}\pi a^2,$$

whence $$\bar{x} = 4a/3\pi = \cdot 4244a. \qquad\qquad (8)$$

In the case of a homogeneous solid, if the area of a section by a plane perpendicular to Ox be denoted by $f(x)$, the x-coordinate of the mass-centre of the volume included between two such sections is given by the formula

$$\bar{x} = \frac{\int xf(x)\,dx}{\int f(x)\,dx}, \qquad\qquad (9)$$

taken between the proper limits of x.

In the case of a solid of revolution, taking the axis of x coincident with the axis of symmetry, we have $f(x) = \pi y^2$, if y be the ordinate of the generating curve. Hence

$$\bar{x} = \frac{\int xy^2\,dx}{\int y^2\,dx}. \qquad\qquad (10)$$

Ex. 4. In the case of a right circular cone, the origin being at the vertex, y varies as x, so that

$$\bar{x} = \int_0^h x^3\,dx \div \int_0^h x^2\,dx = \tfrac{3}{4}h, \qquad\qquad (11)$$

as already proved in Art. 67, Ex. 3.

Ex. 5. For a hemisphere of radius a, putting $y^2 = a^2 - x^2$, we have

$$\int_0^a x(a^2-x^2)\,dx = \tfrac{1}{4}a^4, \quad \int_0^a (a^2-x^2)\,dx = \tfrac{2}{3}a^3,$$

and therefore $$\bar{x} = \tfrac{3}{8}a, \qquad\qquad (12)$$

as in Art. 67, Ex. 5.

The same formula gives the mass-centre of the half of the ellipsoid

$$\frac{x^2}{a^2} + \frac{y^2}{b^2} + \frac{z^2}{c^2} = 1, \qquad\qquad (13)$$

which lies on the positive side of the plane yz. For the section by a plane x=const. is an ellipse of semiaxes

$$b\sqrt{(1-x^2/a^2)} \text{ and } c\sqrt{(1-x^2/a^2)},$$

and therefore of area $\pi bc(1-x^2/a^2)$, so that in this case also $f(x)$ varies as a^2-x^2.

Ex. 6. In the paraboloid generated by the revolution of the curve (6) about the axis of x, the area of a circular section varies as x. Hence for the segment cut off by a plane $x=h$, we have

$$\bar{x}=\int_0^h x^2dx \div \int_0^h xdx = \tfrac{2}{3}h. \quad \text{......................(14)}$$

69. Theorems of Pappus.

The following properties of the mean centre are to be noticed, although their interest is mainly geometrical.

1°. If an arc of a plane curve revolve about an axis in its plane, not intersecting it, the *surface* generated is equal to the length of the *arc* multiplied by the length of the path of its mean centre.

Let the axis of x coincide with the axis of rotation, and let y be the ordinate of the generating curve. The surface generated in a complete revolution is, by a formula of the Integral Calculus, equal to $2\pi\int y ds$, the integration extending over the arc. But if $\bar{y}$ refer to the mean centre of the arc, we have

$$\bar{y}=\int y ds \div \int ds,$$

by Art. 68 (2). Hence

$$2\pi\int y ds = 2\pi\bar{y}\times\int ds, \quad \text{....................(1)}$$

which is the theorem.

2°. If a plane area revolve about an axis in its plane, not intersecting it, the *volume* generated is equal to the *area* multiplied by the length of the path of its mean centre.

If δS be an element of the area, the volume generated in a complete revolution is $2\pi\int y dS$. But if $\bar{y}$ refer to the mean centre of the area, we have

$$\bar{y}=\int y dS \div \int dS.$$

Hence
$$2\pi\int y dS = 2\pi\bar{y}\times\int dS, \quad \text{..................(2)}$$

which is the theorem.

The revolutions have been taken to be complete, but this is obviously unessential*.

The theorems may be used conversely to find the mean centre of a plane arc, or of a plane area, when the surface, or the volume, generated by its revolution is known independently.

Ex. 1. The surface of the ring generated by the revolution of a circle of radius b about a line in its own plane at a distance a ($>b$) from its centre is $2\pi b \times 2\pi a$, or $4\pi^2 ab$; and the volume of the ring is $\pi b^2 \times 2\pi a$, or $2\pi^2 ab^2$.

Ex. 2. In the case of a semicircular *arc* revolving about the diameter joining its extremities, we have

$$\pi a \times 2\pi\bar{y} = 4\pi a^2,$$

whence $$\bar{y} = 2a/\pi.$$

Again, for a semicircular *area* revolving about its bounding diameter,

$$\tfrac{1}{2}\pi a^2 \times 2\pi\bar{y} = \tfrac{4}{3}\pi a^3,$$

whence $$\bar{y} = 4a/3\pi.$$

Cf. Art. 68, Exx. 1, 3.

70. Quadratic Moments.

We proceed to the consideration of the 'plane,' 'polar,' and 'axial' 'quadratic moments' of a mass-system. Of these the axial moments alone have a dynamical importance, but the others are useful as subsidiary conceptions.

If $h_1, h_2, \ldots, h_n$ be the perpendicular distances of the particles $m_1, m_2, \ldots, m_n$ from any fixed plane, the sum $\Sigma(mh^2)$ is called the 'quadratic moment with respect to the plane.'

If $p_1, p_2, \ldots, p_n$ be the perpendicular distances of the particles from any given axis, the sum $\Sigma(mp^2)$ is called the 'quadratic moment with respect to the axis.' In Dynamics it is also known as the 'moment of inertia' about the axis.

If $r_1, r_2, \ldots, r_n$ be the distances of the particles from a fixed point, or 'pole,' the sum $\Sigma(mr^2)$ is called the 'quadratic moment with respect to the pole.'

If we divide any one of the above quadratic moments by the total mass $\Sigma(m)$, the result is called the 'mean square' of the

* These theorems are contained in a treatise on Mechanics by Pappus, who flourished at Alexandria about A.D. 300. They were given as new by Guldinus, *de centro gravitatis* (1635–1642). (Ball, *History of Mathematics.*)

distances of the particles from the respective plane, axis, or pole. If in the case of an axial moment we put

$$\frac{\Sigma(mp^2)}{\Sigma(m)} = k^2, \qquad \text{..........................(1)}$$

the linear magnitude k is called (in Dynamics) the 'radius of gyration' of the system about the axis in question.

If we take rectangular axes through any point O, the quadratic moments with respect to the coordinate *planes* are

$$A' = \Sigma(mx^2), \quad B' = \Sigma(my^2), \quad C' = \Sigma(mz^2); \qquad \text{......(2)}$$

those with respect to the coordinate *axes* are

$$A = \Sigma\{m(y^2+z^2)\}, \quad B = \Sigma\{m(z^2+x^2)\}, \quad C = \Sigma\{m(x^2+y^2)\}; \qquad \text{.........(3)}$$

whilst the quadratic moment with respect to the origin is

$$J_0 = \Sigma\{m(x^2+y^2+z^2)\}. \qquad \text{..................(4)}$$

We notice that

$$A = B' + C', \quad B = C' + A', \quad C = A' + B', \qquad \text{......(5)}$$

and

$$J_0 = A' + B' + C' = \tfrac{1}{2}(A + B + C). \qquad \text{...........(6)}$$

Another important type of quadratic moment is that of 'deviation-moments,' or 'products of inertia.' The sum $\Sigma(mxy)$, for example, is called the product of inertia with respect to the (rectangular) planes $x = 0$, $y = 0$.

71. Two-Dimensional Examples.

In the case of bodies whose mass is distributed over lines, surfaces, or volumes, the summations are of course to be replaced by integrations. In some cases the work is simplified by the relations (5) and (6) of the preceding Art.

Ex. 1. To find the radius of gyration (κ) of a uniform thin straight bar about a line through its centre perpendicular to the length.

If $2a$ be the length, we have

$$\kappa^2 = \frac{\int_{-a}^{a} x^2\,dx}{\int_{-a}^{a} dx} = \tfrac{1}{3}a^2. \qquad \text{.............................(1)}$$

The same result evidently holds for the radius of gyration of a rectangular plate about a line through the centre parallel to a pair of edges, if $2a$ be the length of the plate perpendicular to this line.

Ex. 2. The moment of inertia of a uniform thin circular ring of mass M and radius a about its axis is evidently

$$C=Ma^2. \quad \text{.....................................(2)}$$

By Art. 70, (5), we have, on account of the symmetry,

$$C=A'+B'=2A', \quad \text{.....................................(3)}$$

where A' is the moment of inertia about a diameter. Hence

$$A'=\tfrac{1}{2}Ma^2. \quad \text{.....................................(4)}$$

Ex. 3. To find the radii of gyration of a uniform circular plate about its axis, and about a diameter.

If we divide the disk into concentric circular annuli, the area of one of these will be represented by $2\pi r\,\delta r$, and its radius of gyration by r. Hence, for the radius of gyration (κ) about the axis,

$$\kappa^2=\frac{\int_0^a r^2 . 2\pi r dr}{\pi a^2}=\tfrac{1}{2}a^2. \quad \text{..........................(5)}$$

The formula (3) applies here also to shew that for a diameter we have

$$\kappa^2=\tfrac{1}{4}a^2. \quad \text{.....................................(6)}$$

To find the radius of gyration (k), about the axis of x, of the area included between a curve

$$y=\phi(x), \quad \text{..........................(7)}$$

the axis of x, and two bounding ordinates, we may divide the area into elementary strips $y\,\delta x$. The square of the radius of gyration of a strip is $\frac{1}{3}y^2$. Hence

$$k^2=\frac{\int \frac{1}{3}y^2 . y\,dx}{\int y\,dx}=\frac{\frac{1}{3}\int y^3 dx}{\int y\,dx}, \quad \text{..................(8)}$$

the integrals being taken between proper limits of x.

Ex. 4. To find the radius of gyration (k) of a triangular area ABC about the side BC.

The area of a strip parallel to BC, at a distance y from it, may be represented by $(h-y)/h . a\delta y$, where h is the perpendicular from A to BC. Hence

$$k^2=\frac{a}{h}\int_0^h (h-y)\,y^2\,dy \div \tfrac{1}{2}ah=\tfrac{1}{6}h^2. \quad \text{.....................(9)}$$

It is easily seen that this is the same as for three equal particles at the middle points of the sides.

Ex. 5. To find the radius of gyration (κ) of the area bounded by the ellipse

$$\frac{x^2}{a^2}+\frac{y^2}{b^2}=1 \quad \text{.....................................(10)}$$

with respect to the axis of x. We have, by (8),

$$\kappa^2 = \frac{1}{3}\int_{-a}^{a} y^3 dx \div \tfrac{1}{2}\pi ab. \quad \text{.........................(11)}$$

If we put $x = a\cos\phi$, $y = b\sin\phi$, this becomes

$$\kappa^2 = \frac{4b^2}{3\pi}\int_0^{\frac{1}{2}\pi} \sin^4\phi\, d\phi = \tfrac{1}{4}b^2. \quad \text{.......................(12)}$$

Similarly, for the radius of gyration about the minor axis we should find

$$\kappa^2 = \tfrac{1}{4}a^2. \quad \text{..................................(13)}$$

72. Three-Dimensional Problems.

The following problems in three dimensions are of interest.

Ex. 1. The polar quadratic moment of a uniform thin spherical shell with respect to its centre is evidently

$$J_0 = Ma^2, \quad \text{..................................(1)}$$

where a is the radius. The origin being at the centre, we have, in the notation of Art. 70,

$$J_0 = A' + B' + C' = 3A', \quad \text{............................(2)}$$

on account of the symmetry. Hence the moment of inertia about a *diameter* is

$$C = A' + B' = 2A' = \tfrac{2}{3}Ma^2. \quad \text{.........................(3)}$$

Ex. 2. In the case of a uniform solid sphere, we divide the volume into concentric spherical shells. The volume of one of these may be represented by $4\pi r^2 \delta r$. Hence, using the result (3), we find, for the radius of gyration about a diameter,

$$\kappa^2 = \int_0^a \tfrac{2}{3}r^2 . 4\pi r^2 dr \div \tfrac{4}{3}\pi a^3 = \tfrac{2}{5}a^2. \quad \text{....................(4)}$$

A general formula for the radius of gyration, about its axis, of a uniform solid of revolution is obtained as follows. Dividing the solid into circular laminæ by planes perpendicular to the axis, which is taken as axis of x, the volume of any one of these may be represented by $\pi y^2 \delta x$, where y is the ordinate of the generating curve, and the square of its radius of gyration by $\frac{1}{2}y^2$ (see Art. 71, Ex. 3). Hence

$$\kappa^2 = \frac{\int \frac{1}{2}y^2 . \pi y^2 dx}{\int \pi y^2 dx} = \frac{\frac{1}{2}\int y^4 dx}{\int y^2 dx}. \quad \text{.................(5)}$$

Ex. 3. To find the radius of gyration of a right circular cone about its axis, we put

$$y = (a/h) . x,$$

where a is the radius of the base, and h the altitude. Thus

$$\kappa^2 = \frac{1}{2}\frac{a^2}{h^2} . \int_0^h x^4 dx \div \int_0^h x^2 dx = \tfrac{3}{10}a^2. \quad \text{....................(6)}$$

Ex. 4. For a solid sphere of radius a,

$$\kappa^2 = \tfrac{1}{2}\pi \int_{-a}^{a} y^4 dx \div \tfrac{4}{3}\pi a^3 = \frac{3}{4a^3}\int_0^a (a^2-x^2)^2 dx = \tfrac{2}{5}a^2, \quad \text{.........(7)}$$

as in (4), above. A similar result can be obtained for an ellipsoid of revolution.

Ex. 5. In the case of a uniform ellipsoid bounded by the surface

$$\frac{x^2}{a^2}+\frac{y^2}{b^2}+\frac{z^2}{c^2}=1, \quad \text{..................................(8)}$$

since its section by a plane perpendicular to x is an ellipse of area

$$\pi bc\,(1-x^2/a^2),$$

the mean square of the distances from the plane $x=0$ is

$$\pi bc \int_{-a}^{a} x^2\left(1-\frac{x^2}{a^2}\right)dx \div \tfrac{4}{3}\pi abc = \tfrac{1}{5}a^2. \quad \text{..................(9)}$$

The relations (5) of Art. 70 then shew that the squares of the radii of gyration about the principal *axes* of the ellipsoid are

$$\kappa_1^2 = \tfrac{1}{5}(b^2+c^2), \quad \kappa_2^2 = \tfrac{1}{5}(c^2+a^2), \quad \kappa_3^2 = \tfrac{1}{5}(a^2+b^2). \quad \text{.........(10)}$$

73. Comparison of Quadratic Moments with respect to Parallel Planes or Axes.

Let (x, y, z) be the rectangular coordinates of any particle m of the system, and let us write

$$x = \bar{x} + \xi, \quad y = \bar{y} + \eta, \quad z = \bar{z} + \zeta, \quad \text{...............(1)}$$

where $\bar{x}$, $\bar{y}$, $\bar{z}$ refer to the mass-centre, and ξ, η, ζ therefore denote coordinates relative to the mass-centre. By Art. 66 (2) we have

$$\Sigma(m\xi) = 0, \quad \Sigma(m\eta) = 0, \quad \Sigma(m\zeta) = 0. \quad \text{.........(2)}$$

Hence

$$\Sigma(mx^2) = \Sigma\{m(\bar{x}+\xi)^2\} = \Sigma(m).\bar{x}^2 + 2\bar{x}.\Sigma(m\xi) + \Sigma(m\xi^2)$$
$$= \Sigma(m).\bar{x}^2 + \Sigma(m\xi^2), \quad \text{.................................(3)}$$

by (2). If we divide by $\Sigma(m)$ we have the theorem that the mean square of the distances of the particles from any plane exceeds the mean square of the distances from a parallel plane through the mass-centre G, by the square of the distance of G from the former plane.

A similar formula holds for $\Sigma(my^2)$, and by addition we find

$$\Sigma\{m(x^2+y^2)\} = \Sigma(m).(\bar{x}^2+\bar{y}^2) + \Sigma\{m(\xi^2+\eta^2)\}. \quad \text{......(4)}$$

Since x^2+y^2 is the square of the distance of a point from the axis of z, this expresses that the moment of inertia about any axis exceeds the moment of inertia about a parallel axis through G by the product of the mass into the square of the distance between the two axes. Or, if k denote the radius of gyration with respect to any axis, κ that with respect to a parallel axis through G, and h the perpendicular distance between these axes, then on division by $\Sigma(m)$ we have

$$k^2 = \kappa^2 + h^2. \quad\text{.........................(5)}$$

This relation is often useful in Dynamics, e.g. in the theory of the Compound Pendulum.

Again, as regards products of inertia, we have

$$\begin{aligned}\Sigma(mxy) &= \Sigma\{m(\bar{x}+\xi)(\bar{y}+\eta)\}\\ &= \Sigma(m).\bar{x}\bar{y} + \bar{y}\,\Sigma(m\xi) + \bar{x}\,\Sigma(m\eta) + \Sigma(m\xi\eta)\\ &= \Sigma(m).\bar{x}\bar{y} + \Sigma(m\xi\eta), \quad\text{..............................(6)}\end{aligned}$$

with a similar interpretation.

Ex. 1. The radius of gyration of a rectangle about a side is given by

$$k^2 = a^2 + \tfrac{1}{3}a^2 = \tfrac{4}{3}a^2, \quad\text{..............................(7)}$$

if $2a$ be the length perpendicular to that side.

Ex. 2. The radius of gyration of a uniform circular disk of radius a about an axis through a point of the circumference normal to the plane of the disk is given by

$$k^2 = a^2 + \tfrac{1}{2}a^2 = \tfrac{3}{2}a^2. \quad\text{..............................(8)}$$

Similarly, the radius of gyration about a tangent line is given by

$$k^2 = a^2 + \tfrac{1}{4}a^2 = \tfrac{5}{4}a^2. \quad\text{..............................(9)}$$

74. Lagrange's Theorems.

The preceding results may be generalized into the formula

$$\Sigma\{m\phi(x, y, z)\} = \Sigma(m).\phi(\bar{x}, \bar{y}, \bar{z}) + \Sigma\{m\phi(\xi, \eta, \zeta)\}, \text{...(1)}$$

where $\phi(x, y, z)$ denotes any homogeneous quadratic function of x, y, z, with coefficients which are constants, i.e. they are the same for all particles of the system. For the various terms are of the types Ax^2 and Cxy, and the result follows by (3) and (6) of Art. 73.

Thus, taking the case of

$$\phi(x, y, z) = x^2 + y^2 + z^2, \quad\text{.................(2)}$$

we get a theorem relating to *polar* quadratic moments, viz.

$$\Sigma\{m(x^2+y^2+z^2)\} = \Sigma(m).(\bar{x}^2+\bar{y}^2+\bar{z}^2) + \Sigma\{m(\xi^2+\eta^2+\zeta^2)\}, \quad\text{........(3)}$$

or, in our previous notation,

$$\Sigma(m.OP^2) = \Sigma(m).OG^2 + \Sigma(m.GP^2). \quad\text{...........(4)}$$

This formula is due to Lagrange*. When written in the form

$$\frac{\Sigma(m.OP^2)}{\Sigma(m)} = \frac{\Sigma(m.GP^2)}{\Sigma(m)} + OG^2, \qquad \text{.............(5)}$$

it expresses that the mean square of the distances of the particles from any point O exceeds the mean square of their distances from the mass-centre G by OG^2. The mass-centre is therefore the point the mean square of whose distances from the various particles of the system is least†.

Another interesting theorem, also due to Lagrange, may be obtained as follows. If in (4) we make the point O coincide with $P_1, P_2, \ldots, P_n$ in succession, we obtain

$$\left.\begin{aligned} 0 + m_2.P_1P_2^2 + \ldots + m_n.P_1P_n^2 &= \Sigma(m.GP^2) + \Sigma(m).GP_1^2, \\ m_1.P_2P_1^2 + 0 + \ldots + m_n.P_2P_n^2 &= \Sigma(m.GP^2) + \Sigma(m).GP_2^2, \\ \ldots\ldots\ldots\ldots\ldots\ldots\ldots\ldots &\ldots\ldots\ldots\ldots\ldots\ldots\ldots\ldots \\ m_n.P_nP_1^2 + m_n.P_nP_2^2 + \ldots + 0 &= \Sigma(m.GP^2) + \Sigma(m).GP_n^2. \end{aligned}\right\} \qquad \text{........(6)}$$

If we multiply these equations by $m_1, m_2, \ldots, m_n$, respectively, and add, we find, on division by 2,

$$\Sigma(mm'.PP'^2) = \Sigma(m).\Sigma(m.GP^2), \qquad \text{...........(7)}$$

where in the summation on the left hand each pair of particles m, m' is taken *once* only. This result, written in the form

$$\frac{\Sigma(m.GP^2)}{\Sigma(m)} = \frac{\Sigma(mm'.PP'^2)}{\{\Sigma(m)\}^2}, \qquad \text{.............(8)}$$

expresses the mean square of the distances of the particles from the centre of mass in terms of their masses and *mutual* distances.

* J. L. Lagrange (1736–1813). The theorems (4) and (7) were given in a memoir of date 1783.

† This theorem has an interesting application in the Theory of Errors. Suppose that a number of independent measurements of the position of a star are made, and that P_1, P_2, ... are the positions given by the observations, as marked on a chart. The question arises, what point best represents the whole series of observations? If G were the true position, GP_1, GP_2, ... would be the 'errors' of the several observations. On the theory referred to, that point is to be chosen which makes the sum of the squares of the errors least, i.e. (by the theorem in the text) the mean centre of the points P_1, P_2, ... This is on the supposition that the observations are judged to be equally good. If not, different 'weights' m_1, m_2, ... are attached to them by estimation, and the mass-centre is taken.

Ex. Considering the case of four equal particles at the vertices of a regular tetrahedron, we infer that the radius R of the circumscribing sphere is given by

$$R^2 = \tfrac{3}{8}a^2,$$

if a be the length of an edge.

The proof of Lagrange's two theorems can be put in a very concise form if we make use of the notion of the *scalar product* of two vectors (Art. 47).

We denote the position-vector of any one of the points $P_1, P_2, \ldots, P_n$ relative to O by $\mathbf{r}$, with the proper suffix, and that of the same point relative to the mass-centre G by $\boldsymbol{\rho}$. The vector OG is denoted by $\bar{\mathbf{r}}$. The formulæ (1) and (2) of Art. 64 may accordingly be written

$$\Sigma(m\boldsymbol{\rho}) = 0, \quad \ldots\ldots(9)$$

$$\bar{\mathbf{r}} = \frac{\Sigma(m\mathbf{r})}{\Sigma(m)}. \quad \ldots\ldots(10)$$

Then, since $\mathbf{r} = \mathrm{OP} = \mathrm{OG} + \mathrm{GP} = \bar{\mathbf{r}} + \boldsymbol{\rho}$,

we have

$$\Sigma(m.\mathbf{r}^2) = \Sigma\{m(\bar{\mathbf{r}}+\boldsymbol{\rho})^2\} = \Sigma(m).\bar{\mathbf{r}}^2 + 2\bar{\mathbf{r}}\,\Sigma(m\boldsymbol{\rho}) + \Sigma(m\boldsymbol{\rho}^2)$$
$$= \Sigma(m).\bar{\mathbf{r}}^2 + \Sigma(m\boldsymbol{\rho}^2). \quad \ldots\ldots(11)$$

Since the scalar square of a vector OP is simply the square of its length, viz. OP^2, this formula is identical with (4).

Again, consider the expression

$$\Sigma\{mm'(\boldsymbol{\rho}-\boldsymbol{\rho}')^2\}, \quad \ldots\ldots(12)$$

where the summation includes every pair m, m' of particles once only. If we expand this, the coefficient of $\boldsymbol{\rho}_1{}^2$ is seen to be

$$m_1m_2 + m_1m_3 + \ldots + m_1m_n = m_1.\Sigma(m) - m_1{}^2,$$

and so on. Hence the expression (12) is equal to

$$\Sigma(m).(m_1\boldsymbol{\rho}_1{}^2 + m_2\boldsymbol{\rho}_2{}^2 + \ldots + m_n\boldsymbol{\rho}_n{}^2) - (m_1\boldsymbol{\rho}_1 + m_2\boldsymbol{\rho}_2 + \ldots + m_n\boldsymbol{\rho}_n)^2,$$

the latter part of which vanishes, by (9). Hence

$$\Sigma\{mm'(\boldsymbol{\rho}-\boldsymbol{\rho}')^2\} = \Sigma(m).\Sigma(m\boldsymbol{\rho}^2). \quad \ldots\ldots(13)$$

Since $(\boldsymbol{\rho}-\boldsymbol{\rho}')^2 = (\mathrm{OP}-\mathrm{OP}')^2 = \mathrm{PP}'^2 = PP'^2, \quad \ldots\ldots(14)$

we see that (13) is equivalent to (7).

75. Moments of Inertia of a Plane Distribution. Central Ellipse.

Some further theorems relating to plane distributions, and in particular to plane areas, are important not only in Dynamics, but also in Hydrostatics and in the theory of Elasticity (see Arts. 95, 149).

To compare the moments of inertia with respect to different lines through any point O, in the plane, denote by (x, y) the coordinates of any particle m with respect to rectangular axes through O. The perpendicular distance of m from a line OP making an angle θ with Ox is then $x \sin\theta - y\cos\theta$, and the moment of inertia with respect to OP is therefore

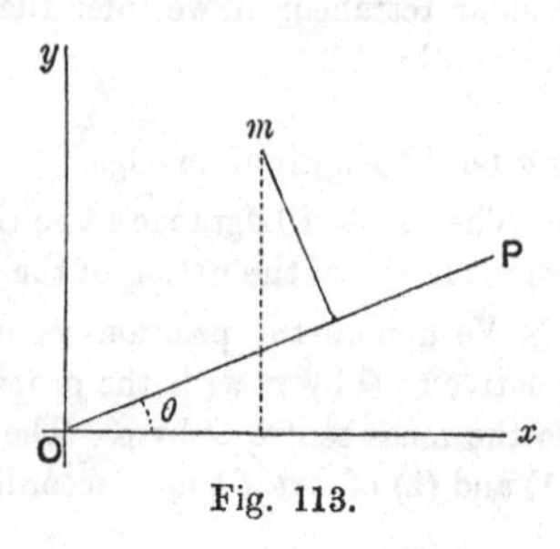

Fig. 113.

$$I = \Sigma\{m(y\cos\theta - x\sin\theta)^2\}$$
$$= \Sigma(my^2).\cos^2\theta - 2\Sigma(mxy)\cos\theta\sin\theta + \Sigma(mx^2).\sin^2\theta, \quad \text{...(1)}$$

or, if we write

$$A = \Sigma(my^2), \quad H = \Sigma(mxy), \quad B = \Sigma(mx^2), \quad \text{......(2)}$$
$$I = A\cos^2\theta - 2H\cos\theta\sin\theta + B\sin^2\theta. \quad \text{.........(3)}$$

Here A, B denote the moments of inertia about Ox, Oy, and H is the product of inertia with respect to these axes.

Now consider the conic

$$Ax^2 - 2Hxy + By^2 = M\epsilon^4, \quad \text{.................(4)}$$

where M, $= \Sigma(m)$, is the total mass, and ϵ is any convenient linear magnitude. The intercept r which this conic makes on the line OP is found by putting $x = r\cos\theta$, $y = r\sin\theta$. Hence, by comparison with (3),

$$I = \frac{M\epsilon^4}{r^2}. \quad \text{..........................(5)}$$

This conic therefore indicates the relative magnitudes of the moments of inertia about different diameters, viz. these moments are inversely proportional to the squares of the respective diameters. Since I is an essentially positive quantity the conic must be an ellipse; it is called the 'momental ellipse' at O.

If this ellipse be referred to its principal diameters as axes, its equation will be of the form

$$Ax^2 + By^2 = M\epsilon^4. \quad \text{......................(6)}$$

Since the coefficient H is now zero, we learn that there is always one position of the coordinate axes at O for which the product of inertia $\Sigma(mxy)$ vanishes. The axes thus characterized are called the 'principal axes of inertia' of the plane distribution at O, and the corresponding moments of inertia are called the 'principal moments.'

In the particular case where the principal moments A, B are equal, the momental ellipse is a circle; the moments of inertia about all diameters of the circle are equal, and each diameter is a principal axis.

Let us now write, further,

$$A = Mb^2, \quad B = Ma^2, \quad \epsilon^4 = a^2b^2, \quad \text{.................(7)}$$

so that a^2 and b^2 denote the mean squares of the abscissæ and ordinates, respectively, of the particles, relatively to the principal axes at O. The special momental ellipse thus obtained has the equation

$$\frac{x^2}{a^2} + \frac{y^2}{b^2} = 1, \quad \text{.........................(8)}$$

and if k denote the radius of gyration about OP, we have from (5),

$$k = \frac{ab}{r} \quad \text{..............................(9)}$$

If p be the perpendicular from O on a tangent parallel to OP, we have $pr = ab$, by a known property of conics, and the relation (9) becomes

$$k = p, \quad \text{........(10)}$$

simply. In words, the radius of gyration about any axis through O is one-half the breadth of the ellipse (8) in the direction perpendicular to that axis.

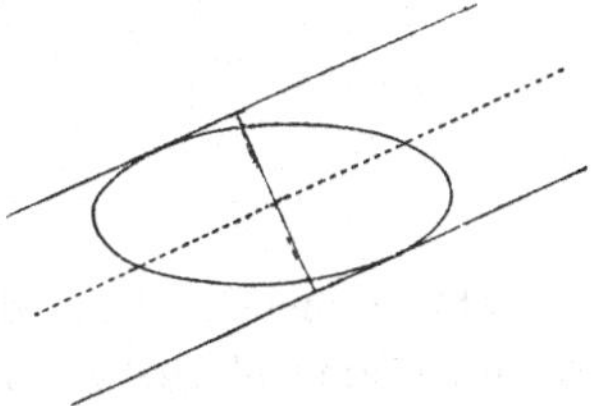

Fig. 114.

The case where O coincides with the mass-centre G is of course the most important. The ellipse (8) is then called the 'central ellipse' of the plane system. With its help we can construct the radius of gyration about any axis through G, and thence, by the theorem of Art. 73 (5), the radius of gyration about any parallel axis

Ex. 1. In the case of a rectangular area, if the coordinate axes be taken parallel to the sides ($2a$, $2b$), the equation of the central ellipse is

$$\frac{x^2}{\frac{1}{3}a^2}+\frac{y^2}{\frac{1}{3}b^2}=1. \qquad\qquad (11)$$

It is therefore similar to the ellipse which touches the sides of the rectangle at their middle points.

Ex. 2. The central ellipse for an area in the shape of an equilateral triangle, a square, or any other regular polygon is necessarily a circle; for an ellipse cannot have more than two axes of symmetry unless it be a circle.

In the case of the equilateral triangle, it is easily proved from the result of Art. 71, Ex. 4, that the mean square of the distances from a line through the centre parallel to a side is $\frac{1}{6}a^2$, if $2a$ be the length of the side. The equation of the circle is therefore

$$x^2+y^2=\tfrac{1}{6}a^2. \qquad\qquad (12)$$

Another geometrical representation of the relations between

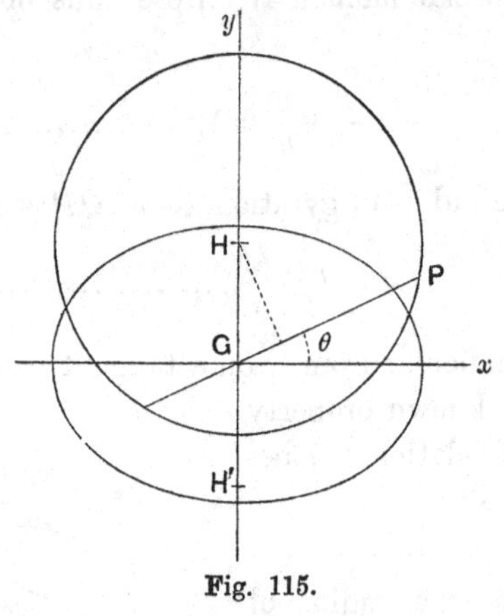

Fig. 115.

moments of inertia about different axes may be noticed. If on the minor axis of the central ellipse we take two points H, H' such that

$$GH = GH' = \sqrt{(a^2-b^2)}, \qquad\qquad (13)$$

where a is supposed $> b$, and if with either of these points (say H) as centre we describe a circle of radius a, the radius of gyration about any axis through G is one-half the chord intercepted by the circle on this axis*. For if θ denote as before the angle which

* This construction is due to O. Mohr (1870).

any line GP makes with the major axis, the square of half the length of the chord is

$$a^2 - GH^2 \cos^2 \theta = a^2 \sin^2 \theta + b^2 \cos^2 \theta,$$

which is, by (3) and (7), the square of the radius of gyration about GP.

It appears from Art. 73 (6) that Hy is a principal axis at H. And since the square of the radius of gyration about an axis through H parallel to Gx is $(a^2 - b^2) + b^2$, the momental ellipse at H (or H') is a circle.

76. Transformation by Parallel Projection.

Reference has already been made (Art. 66) to the method of parallel projection, by which (in particular) a plane mass-system may be transformed into another such system, with the centres of mass corresponding to one another.

Some simple geometrical properties of this kind of projection may be recalled. The projections of parallel lines are themselves parallel, so that parallelograms project into parallelograms. Hence equal and parallel lines project into equal and parallel lines, so that all lines having any given direction have their lengths altered in a constant ratio, this ratio varying however (in general) with the direction.

Again, an ellipse projects into an ellipse, and conjugate diameters into conjugate diameters. For if PCP', DCD' be any two conjugate diameters of the original ellipse, and QV any ordinate to PCP' (parallel to DD'), we have

$$\frac{QV^2}{CD^2} = \frac{PV.P'V}{CP^2} \quad \text{.........................(1)}$$

Now, in the projection, QV and CD, being parallel, are altered in the same ratio, and the same holds with regard to PV, $P'V$, CP. Hence the relation (1) remains true if the letters be supposed to refer to the *projections* of the original points. The locus of the projection of Q is therefore an ellipse of which the projections of PP', DD' are conjugate diameters.

In particular, the projection of a circle is an ellipse, and any two perpendicular diameters of the circle project into conjugate diameters of the ellipse. Hence, since the principal axes of an ellipse are conjugate, there must be one pair of mutually perpendicular directions in one figure which project into mutually perpendicular directions in the other.

If we adopt these as the directions of coordinate axes in the two planes, the origins O, O' being at corresponding points, the relations between the coordinates of any other pair of corresponding points P, P' will be of the form

$$x = \alpha x', \quad y = \beta y', \qquad \text{.......................(2)}$$

where α, β are the constant ratios in which lines parallel to Ox, Oy are respectively altered*. Thus, the circle

$$x'^2 + y'^2 = c'^2 \qquad \text{.........................(3)}$$

corresponds to the ellipse

$$\frac{x^2}{a^2} + \frac{y^2}{b^2} = 1, \qquad \text{..........................(4)}$$

provided

$$a = \alpha c', \quad b = \beta c'. \qquad \text{.......................(5)}$$

It follows that the mean squares of distances from the axes $O'x'$, $O'y'$ will be related to the mean squares of distances from Ox, Oy, respectively, in the same way as the squares of any other lines having the respective directions, since

$$\Sigma (mx^2) = \alpha^2 \Sigma (mx'^2), \quad \Sigma (my^2) = \beta^2 \Sigma (my'^2). \qquad \text{......(6)}$$

Ex. 1. We have seen (Art. 68) that the mean centre of a semicircular area is on the radius perpendicular to the bounding diameter, at a distance $4/3\pi$ of its length from the centre. Since areas are altered by parallel projection in a constant ratio, the mean centre of a semi-ellipse, cut off by any diameter, lies in the conjugate semi-diameter, at a distance of $4/3\pi$ of its length from the centre.

Ex. 2. The mean squares of the distances of points within the circle

$$x'^2 + y'^2 = c'^2 \qquad \text{..................................(7)}$$

from the coordinate axes are $\frac{1}{4}c'^2$, $\frac{1}{4}c'^2$; hence for the elliptic area

$$\frac{x^2}{a^2} + \frac{y^2}{b^2} = 1, \qquad \text{..................................(8)}$$

the mean squares of the distances from Oy, Ox will be $\frac{1}{4}\alpha^2 c'^2$, $\frac{1}{4}\beta^2 c'^2$, or $\frac{1}{4}a^2$, $\frac{1}{4}b^2$, respectively. (Cf. Art. 71, Ex. 5.)

* It is to be observed that the formulæ (2) represent a transformation somewhat more general than that of parallel projection in that they include the case where a uniform magnification or diminution of scale, of any amount, is superposed.

In the particular case of orthogonal projection, if the axis of x be parallel to the common section of the planes, we have $\alpha = 1, \beta = \cos\theta$, where θ is the angle between the planes.

77. Properties of the Central Ellipse.

Now considering any mass-system, let it be projected (e.g. by orthogonal projection) so that its central ellipse, having its centre at G, becomes a circle. Its principal axes Gx, Gy, say, will project into a pair of perpendicular diameters $G'x'$, $G'y'$ of the circle. Moreover, since

$$\Sigma(mx'y') = \frac{1}{\alpha\beta}\Sigma(mxy) = 0, \quad \text{.................(1)}$$

$G'x'$, $G'y'$ will be principal axes of inertia of the projected system at G'. Again, in this latter distribution, the mean squares of the distances from $G'x'$, $G'y'$ will be equal, by Art. 76. The circle in question will therefore be the central ellipse of the projected system, and any diameter will be a principal axis at G'.

Let us now, changing the notation, take coordinate axes Gx, Gy coincident with *any* two conjugate diameters of the original central ellipse, and let $G'x'$, $G'y'$ be the corresponding axes in the projection. These latter lines will, as we have seen, be mutually perpendicular, and since they are principal axes we shall have

$$\Sigma(mx'y') = 0. \quad \text{......................(2)}$$

The relations between corresponding points in the two planes will still be of the form

$$x = \alpha x', \quad y = \beta y'. \quad \text{......................(3)}$$

Hence if the lengths of the semi-diameters of the original ellipse in the directions Gx, Gy be a, b, respectively, we shall have

$$a = \alpha c', \quad b = \beta c', \quad \text{......................(4)}$$

where c' is the radius of the circle in the plane $x'y'$. Hence, if M denote the total mass of either system,

$$\left.\begin{aligned} \Sigma(mx^2) &= \alpha^2\Sigma(mx'^2) = M\alpha^2c'^2 = Ma^2, \\ \Sigma(my^2) &= \beta^2\Sigma(my'^2) = M\beta^2c'^2 = Mb^2; \end{aligned}\right\} \quad \text{........(5)}$$

whilst, from (2),

$$\Sigma(mxy) = 0. \quad \text{.........................(6)}$$

The properties (5) and (6), which are thus proved to hold with respect to any pair of conjugate diameters of the central ellipse of a plane distribution, have an application in Hydrostatics (Art. 95).

In the above discussion we might have started from the formulæ (2) of Art. 76 as defining the relation between the two figures to be compared, without any reference to projections. As already stated, these formulæ represent a transformation more general than that of parallel projection.

A similar method of transformation may be used in three dimensions, the formulæ being

$$x = \alpha x', \quad y = \beta y', \quad z = \gamma z'. \quad \text{.........................(7)}$$

It is easily proved that mass-centres transform into mass-centres, and there are obvious relations between the mean squares of distances from the co-ordinate planes in the two systems. In this way the principal moments of inertia of a uniform solid ellipsoid can be derived from those of a sphere.

78. Equimomental Systems.

Two distributions of matter are said to be 'equimomental' when the moment of inertia of one system about any axis whatever is equal to that of the other. In the case of plane distributions the conditions for this are that the two systems shall have the same total mass, the same mass-centre, and the same central ellipse.

Ex. The mass-centre G of a uniform triangular plate coincides with that of a system of three particles, each of one-third the mass of the plate, situate at the middle points of the sides. Again, we have seen (Art. 71, Ex. 4) that the plate and the three particles have the same moment of inertia about any side. By Art. 73 (5), the same statement holds with regard to the lines drawn through G parallel to the sides. Hence the central ellipses of the two systems have three diameters, and therefore six points, in common. They are accordingly identical, and the two systems are equimomental.

79. Graphical Determination of Linear and Quadratic Moments.

The construction of Art. 36 may be applied to find the moments of a plane system of particles with respect to a line in their plane, the masses being represented by forces parallel to the given line.

If we denote by x the distance (positive or negative) of a particle m from the given line, the quadratic moment of the system, viz. $\Sigma(mx^2)$, or $\Sigma\{(mx).x\}$, may be found by attributing to each particle a (positive or negative) mass equal to the moment mx of the original mass as above found, and repeating the process. The construction of a second force-diagram and funicular polygon may however be avoided by the use of a planimeter.

Thus, suppose we have four particles numbered 1, 2, 3, 4, whose masses are indicated by the corresponding lines in the force-polygon (Fig. 116), these being drawn parallel to the line p with respect to which the moments are required. Take a pole O, and construct the corresponding funicular $ZABCD\ldots$; and let the sides meet p in the points $H, K, L, M, \ldots$. The moment of the first particle with respect to p is represented on a certain scale by HK; to obtain its quadratic moment we multiply by the perpendicular distance of A from p. The result is represented by twice the area of the triangle AHK; and so on. The quadratic

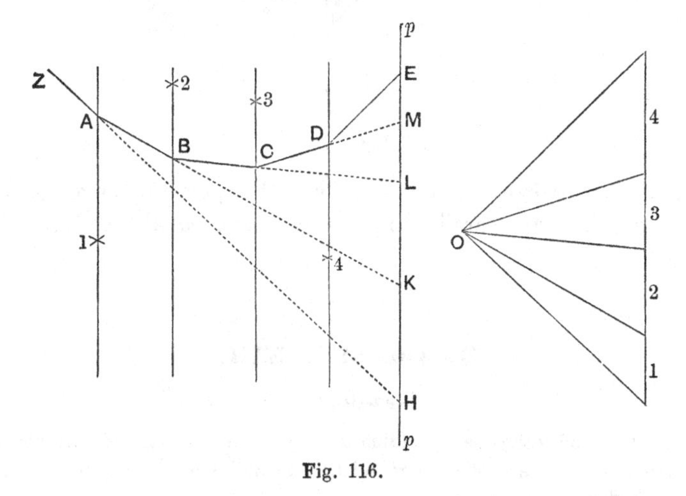

Fig. 116.

moment of the system is accordingly represented by twice the area $AHEDCBA$. The areas are of course to be taken positively in all cases.

If some of the masses lie on one side of p and some on the other, the quadratic moment of each set is to be taken, and the results added. This is illustrated by Fig. 117, where the quadratic moment in question is represented by the sum of the shaded areas. It appears that the quadratic moment is least, for different parallel positions of p, when p passes through the point X in the figure, which is the intersection of the extreme sides of the funicular, i.e. when p goes through the mass-centre of the given system of particles. Cf. Art. 73.

The graphical methods of finding linear and quadratic moments are used as modes of approximation in cases of plane figures of

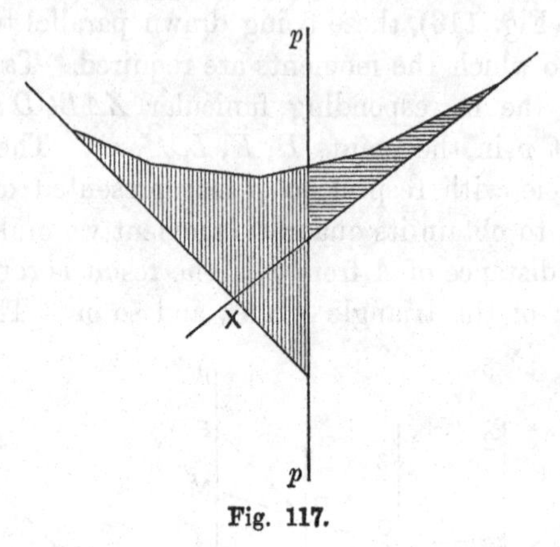

Fig. 117.

complicated or irregular outline. For this purpose the areas are divided into strips parallel to p, preferably of equal breadth. Cf. Art. 37.

EXAMPLES. XIII.

(Mass-Centres.)

1. A cylindrical vessel of radius a and length l is made of thin sheet-metal; it is closed except for a round hole of radius b at the centre of one end. Find its mass-centre.

2. Three rods of the same material and cross-section, of lengths 3, 4, 5 ft., respectively, are put together in the form of a right-angled triangle. Find the distances of the centre of gravity from the two shorter sides. [$1\frac{1}{3}$ ft., 1 ft.]

3. A uniform wire, bent into the form of a triangle ABC, hangs over a smooth peg at A; find where a plumb-line suspended from the peg will cross the side BC, having given $BC=7$, $CA=6$, $AB=5$. [Distance from $B=3{\cdot}64$.]

4. Prove that the mean centre of a trapezium divides the line joining the middle points of the parallel sides in the ratio $2a+b : a+2b$, where a, b are the lengths of the parallel sides.

5. $ABCD$ is a uniform quadrilateral lamina; O is the intersection of the diagonals, and P, Q are points in BD, AC such that $QA=OC$, $PB=OD$. Prove that the centre of gravity of the plate coincides with that of the triangle OPQ.

6. Prove that the mean centre of the area of a plane quadrilateral coincides with the mass-centre of four equal masses m at the corners and a mass $-m$ at the intersection of the diagonals.

7. The upper surface of a trough is a rectangle of sides a, b; the sides are equal trapeziums meeting in an edge c, parallel to a; and the ends are equal triangles. Prove that when the trough is filled with water the centre of gravity of the water is at a depth

$$\frac{a+c}{4a+2c} \cdot h,$$

where h is the depth of the trough.

8. If a tetrahedron be bisected by a plane parallel to a pair of opposite edges, the distance of the mean centre of either half from this plane is $\frac{3}{16}h$, where h is the shortest distance between the edges in question.

9. Find the mass-centre of a homogeneous sphere containing a spherical cavity whose surface touches that of the sphere; and deduce the position of the mass-centre of the thin shell which remains when the radii are nearly equal.

10. The mass-centre of a homogeneous hemispherical shell whose inner and outer radii are a and b is at a distance

$$\frac{3}{8} \frac{(a+b)(a^2+b^2)}{a^2+ab+b^2}$$

from the centre of the curved surfaces.

11. A sphere of radius a is divided into two segments by a plane at a distance c from the centre. Prove that the distances of the mass-centres of the two segments from the geometrical centre are

$$\frac{3(a \pm c)^2}{4(2a \pm c)}.$$

12. A uniform solid hemisphere is perforated by a cylindrical hole whose axis passes through the centre and is at right angles to the base. Prove that the distance of the mass-centre from the base is $\frac{3}{8}h$, where h is the length of the hole.

13. Find the mass-centre of a solid spherical sector, i.e. the portion cut from a sphere (of radius a) by a right cone (of semi-angle α) having its vertex at the centre. [Distance from vertex $= \frac{3}{4}a \cos^2 \frac{1}{2}\alpha$.]

14. Find the mass-centre of a thin wedge cut from a solid sphere by two planes meeting in a diameter.

Also the mean centre of the curved surface of the wedge.

[The distances from the straight edge are $\frac{3}{16}\pi a$ and $\frac{1}{4}\pi a$, where a is the radius.]

15. Prove that the mass-centre of the segment of the elliptic paraboloid

$$2x = \frac{y^2}{p} + \frac{z^2}{q}$$

cut off by a plane $x = h$ is at a distance $\frac{2}{3}h$ from the vertex.

What is the corresponding result when the section is made by an oblique plane?

16. If a circular disk of radius a, whose thickness varies as $\sqrt{(1 - r^2/a^2)}$, where r denotes distance from the centre, be bisected by a diameter, find the centre of gravity of either half. [Distance from centre $= \frac{3}{8}a$.]

17. Find the mass-centre of a circular disk whose (small) thickness varies as the distance from a given tangent line. [Distance from centre $= \frac{1}{4}a$.]

18. Prove that the distance $\bar{x}$ of the mean centre of any area on the surface of a sphere from a plane through the centre is given by the formula

$$\frac{\bar{x}}{a} = \frac{\Sigma}{S},$$

where S is the area, and Σ its orthogonal projection on the plane in question.

Apply this to find the mean centre (1) of a hemispherical area, (2) of a spherical lune.

19. A thin uniform rod of length l is bent into the form of a circular arc whose radius a is large compared with l. Prove that the displacement of the mass-centre is $\frac{1}{24}l^2/a$, approximately.

20. Water is poured into a vessel of any shape. Prove that at the instant when the centre of gravity of the vessel and the contained water is lowest it is at the level of the water surface.

21. A groove of semicircular section, of radius b, is cut round a cylinder of radius a; prove that the volume removed is

$$\pi^2 ab^2 - \tfrac{4}{3}\pi b^3.$$

Also that the surface of the groove is $2\pi^2 ab - 4\pi b^2$.

22. If a lamina receive infinitesimal rotations $\omega_1, \omega_2, \ldots$, in its own plane, about points $P_1, P_2, \ldots$, respectively, prove that the result is equivalent to a rotation $\Sigma(\omega)$ about the mass-centre of a system of particles whose masses are proportional to $\omega_1, \omega_2, \ldots$, situate at $P_1, P_2, \ldots$.

EXAMPLES. XIV.

(Quadratic Moments.)

1. The square of the radius of gyration, about the axis, of a solid ring whose section is a rectangle with the sides parallel and perpendicular to the axis, is

$$\tfrac{1}{2}(a^2 + b^2),$$

where a, b are the inner and outer radii.

2. Find the radius of gyration of a truncated solid cone about its axis of symmetry.
$$\left[\kappa^2=\frac{3}{10}\frac{a^5-b^5}{a^3-b^3}.\right]$$

3. A disk is cut from a solid sphere by two parallel planes at equal distances c from the centre. If κ be its radius of gyration about the axis of symmetry,
$$\kappa^2=\frac{15a^4-10a^2c^2+3c^4}{10(3a^2-c^2)}.$$

4. Find the radius of gyration of a segment of a uniform thin spherical shell, with respect to the axis of symmetry.

[$\kappa^2=\frac{1}{3}a^2(1-\cos\alpha)(2+\cos\alpha)$, where a is the radius of the sphere, and α the angular radius of the segment.]

5. A uniform solid sphere has a cylindrical hole bored axially through it. Prove that the square of the radius of gyration about the axis of the hole is $a^2-\frac{3}{5}h^2$, if a be the radius of the hemisphere, and $2h$ the length of the hole.

6. The density of a globe of radius a at a distance r from the centre is
$$\rho=\rho_0\left(1-\beta\frac{r^2}{a^2}\right);$$
prove that the square of the radius of gyration (κ) about a diameter is
$$\frac{2}{5}a^2\cdot\frac{1-\frac{5}{7}\beta}{1-\frac{3}{5}\beta}.$$

If the mean density be twice the surface density, prove that $\kappa^2=\frac{1}{3}\frac{2}{5}a^2$.

7. Find the radius of gyration (κ) of a solid circular cylinder of radius a, and length $2h$, about an axis through its centre at right angles to the length.

If a cylindrical bar be 20 cm. long, and the diameter of its (circular) section be 1 cm., prove that κ^2 is greater than if the thickness had been neglected in the ratio 1·001875.

8. The square of the radius of gyration of a solid anchor ring about its axis is
$$a^2+\tfrac{3}{4}b^2,$$
where b is the radius of the circular section, and a that of the locus of its centre.

9. If k be the radius of gyration of an anchor ring about a diameter of the circle through the centres of the cross-sections, prove that
$$k^2=\tfrac{1}{2}a^2+\tfrac{5}{8}b^2,$$
where a is the radius of the aforesaid circle, and b is the radius of the cross-section.

10. The radius (κ) of gyration of a uniform circular arc of radius a and angle 2α about an axis through its mass-centre, perpendicular to the plane of the arc, is given by

$$\kappa^2 = a^2\left(1 - \frac{\sin^2 \alpha}{\alpha^2}\right).$$

And the radius of gyration (k) about a parallel axis through the middle point of the arc is given by

$$k^2 = 2a^2\left(1 - \frac{\sin \alpha}{\alpha}\right).$$

11. Two regular polygons of p and q sides, respectively, are inscribed in two circles of radii a and b, whose centres are at a distance c apart; find the sum of the squares of all the straight lines which can be drawn from a vertex of one polygon to a vertex of the other. [$pq(a^2+b^2+c^2)$.]

12. Two regular tetrahedra are inscribed in a sphere of radius a. Prove that the sum of the squares of all the straight lines which can be drawn from a vertex of the one to a vertex of the other is $32a^2$.

13. Apply Lagrange's second theorem (Art. 74) to prove that the mean square of the mutual distances of the points inside a spherical surface of radius a is $\frac{6}{5}a^2$.

EXAMPLES. XV.

(Plane Distributions.)

1. Prove that if the coordinate axes be the principal axes at any point O of a plane system, the product of inertia with respect to other rectangular axes Ox', Oy' is

$$(A - B)\sin\theta\cos\theta,$$

where θ is the angle xOx', and A, B are the principal moments at O.

2. If as in Art. 75 A, B, H denote the moments and product of inertia of a plane system relative to axes Ox, Oy, prove that the product of inertia with respect to any other rectangular axes Ox', Oy' having the same origin is

$$\tfrac{1}{2}(A - B)\sin 2\theta + H\cos 2\theta.$$

Hence find the directions of the principal axes of inertia at O.

3. ABC is a uniform triangular plate, C being a right angle; and $CA = a$, $CB = b$. Find the angles which the principal axes of inertia at C make with CA. $\left[\tfrac{1}{2}\tan^{-1}\dfrac{ab}{b^2 - a^2}.\right]$

4. Find the eccentricity of the momental ellipse at the corner (1) of a square, (2) of an equilateral triangle. [$e^2 = \frac{6}{7}$; $e^2 = \frac{8}{9}$.]

5. Find the radius of gyration of a rectangle, whose sides are a, b, about a diagonal. $\left[k^2 = \frac{1}{6}\frac{a^2b^2}{a^2+b^2}.\right]$

6. Find the radius of gyration of a triangular plate about an axis through its centre of mass, normal to its plane. $[\kappa^2 = \frac{1}{36}(a^2+b^2+c^2).]$

7. Find the squares of the radii of gyration of a regular hexagonal plate of side a, (1) about a diameter, (2) about a side, (3) about an axis through a vertex normal to the plate. $[\frac{5}{24}a^2;\ \frac{23}{24}a^2;\ \frac{17}{12}a^2.]$

8. Prove that the radius of gyration of a regular polygon of n sides about an axis through its centre perpendicular to its plane is given by

$$\kappa^2 = \tfrac{1}{6}(R^2 + 2r^2),$$

where R, r are the radii of the circumscribed and inscribed circles.

9. The central ellipse of a rhombus is similar to the ellipse having the diagonals as its principal axes; and the ratio of linear dimensions is $1/\sqrt{6}$.

10. Prove that the central ellipse of a parallelogram is similar and similarly situated to the ellipse which touches the sides at their middle points.

11. Prove that an ellipse can be described to touch the sides of any given triangle at their middle points. Shew that the central ellipse of the triangular area is concentric, similar, and similarly situated to this ellipse; and find the ratio of the linear dimensions. $[\frac{1}{2}\sqrt{2}.]$

12. Prove that there are in general two and only two points in a plane mass-system for which the momental ellipse is a circle.

13. If S, S' be the foci of the momental ellipse at any point O of a plane lamina, the radius of gyration about any diameter varies as the chord of the auxiliary circle drawn through S (or S') perpendicular to this diameter.

14. Prove that the radius of gyration of a plane system about any line in the plane is given by

$$k^2 = a^2 + p_1 p_2,$$

where p_1, p_2 are the perpendiculars on the line from the points H, H' of Fig. 115, these perpendiculars having the same or opposite signs according as H, H' are on the same or on opposite sides of the given line.

15. Prove that in a parallel projection of a plane mass-system the central ellipse projects into the central ellipse, although the directions of the principal axes do not in general correspond.

16. Prove that a plane distribution of matter is equimomental with a system of four equal particles situate at the extremities of any pair of conjugate diameters of the ellipse which is obtained by magnifying the central ellipse in the ratio $\sqrt{2}:1$.

CHAPTER IX

FLEXIBLE CHAINS

80. Tangential and Normal Resolution.

The perfectly flexible string, or chain, of theory is conceived as a line of matter such that the mutual action between any two adjacent portions reduces to a stress in the direction of the tangent. This stress is of the nature of a *tension* resisting separation of the two parts; we shall denote its amount by T. The forms assumed by the curve on these somewhat ideal suppositions will give a good representation of the case of a chain of loose links, or even of a wire rope, if the curvature be not excessive.

The physical assumption which we make is that for equilibrium the forces acting on an infinitesimal element δs of the length must fulfil the same conditions as in the case of a rigid body. Among these forces must of course be included the forces exerted by the adjacent portions on the ends of δs. As a necessary consequence, the ordinary conditions of equilibrium will be satisfied for any finite portion of the string.

We shall consider only cases where the form assumed by the string is that of a *plane* curve. It is often useful to resolve the forces acting on an element δs in the directions of the tangent and normal, respectively. If PQ be the element in question, and if ψ denote the angle which the tangent (drawn in the direction of s increasing) makes with some fixed direction, the inclination of the tangents at P, Q is $\delta\psi$. The tangential forces T and $T + \delta T$ at P and Q, respectively, when resolved

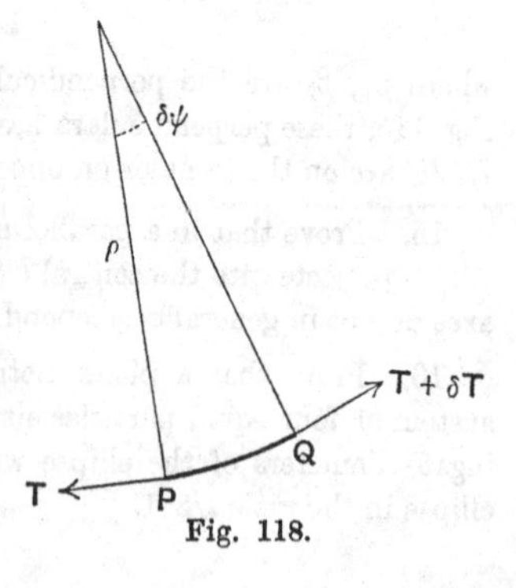

Fig. 118.

along the tangent at P give a component $(T + \delta T) \cos \delta\psi - T$, whilst the component along the normal at P is $(T + \delta T) \sin \delta\psi$. Hence, to the first order of small quantities, the tangential and normal components are

$$\delta T \quad \text{and} \quad T\delta\psi, \qquad\qquad (1)$$

respectively.

Ex. The simplest application of this process is a dynamical one. If we have an endless chain of uniform line-density m, in the form of a circle of radius a, revolving about its centre in its own plane with circumferential velocity v, the mass of an element is $ma\,\delta\psi$, and its acceleration towards the centre is v^2/a, whence

$$T\delta\psi = ma\,\delta\psi \,.\, v^2/a, \quad \text{or} \quad T = mv^2. \qquad\qquad (2)$$

It will be understood that T is here given in dynamical measure.

81. String under Constraint. Friction.

We may apply the method to the case of a string stretched over a smooth curve, and subject to no external forces except the pressure of the curve. The normal pressure on an element δs will ultimately be proportional to δs; we denote it by $R\,\delta s$. Hence, resolving along the tangent and normal, we have

$$\delta T = 0, \quad T\delta\psi - R\,\delta s = 0, \qquad\qquad (1)$$

or

$$\frac{dT}{ds} = 0, \quad R = \frac{T}{\rho}, \qquad\qquad (2)$$

where ρ, $= ds/d\psi$, is the radius of curvature. The former of these equations shews that the tension T is the same at all points, and the latter gives the normal pressure (per unit length) when T is known.

When friction is operative, we have in addition a tangential force which we may denote by $F\delta s$. We will suppose, for definiteness, that this acts in the direction opposite to that of s increasing. We have then

$$\delta T - F\delta s = 0, \quad T\delta\psi - R\,\delta s = 0, \qquad\qquad (3)$$

or

$$F = \frac{dT}{ds}, \quad R = T\frac{d\psi}{ds}. \qquad\qquad (4)$$

If the string is in limiting equilibrium we shall have $F = \mu R$ at all points, where μ is the coefficient of friction. Thus

$$\frac{dT}{ds} = \mu T\frac{d\psi}{ds}, \quad \text{or} \quad \frac{1}{T}\frac{dT}{d\psi} = \mu, \qquad\qquad (5)$$

whence $\log T = \mu\psi + \text{const.}, \quad T = T_0 e^{\mu\psi}$, ...(6)

if T_0 be the tension corresponding to $\psi = 0$.

Thus if a rope be wrapped n times round a post, a tension at one end can balance a tension $e^{2n\pi\mu}$ times as great at the other. For instance, if $\mu = \cdot 2$, $n = 5$, the limiting ratio of the tensions is 535.

82. Constrained Chain subject to Gravity.

We next take the case of a chain* in contact with a smooth curve in a vertical plane, the weight being now taken into account.

The weight of an element δs may be denoted by $w\delta s$, where w, the weight per unit length, is not for the moment assumed to be constant. If ψ now denote the inclination of the tangent line to the horizontal, the tangential and normal components of the downward force $w\delta s$ will be $-w\delta s \sin\psi$ and $w\delta s \cos\psi$, respectively. Hence the conditions of equilibrium are

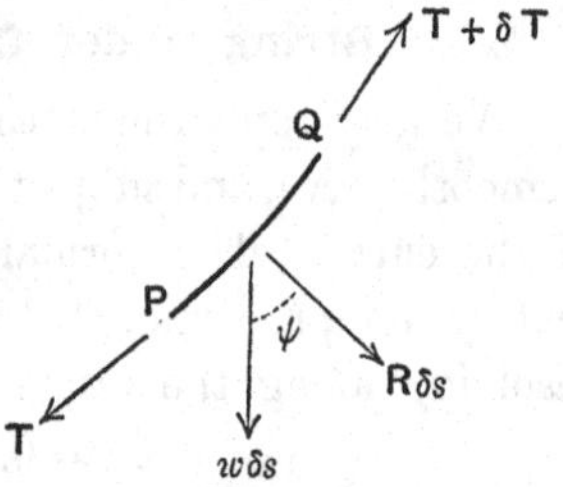

Fig. 119.

$$\left.\begin{aligned} \delta T - w\delta s \sin\psi &= 0 \\ T\delta\psi - R\delta s - w\delta s \cos\psi &= 0 \end{aligned}\right\}, \ldots(1)$$

or $$\frac{dT}{ds} = w \sin\psi, \quad R = \frac{T}{\rho} - w\cos\psi. \quad \text{...........(2)}$$

The form of the curve being supposed given, the former of these equations, combined with the terminal conditions, enables us to find T, and the second equation then gives R.

If the chain be *uniform* and inextensible w is a constant. Moreover, if y denote vertical altitude above some fixed level, we have $\sin\psi = dy/ds$, and therefore, by (2),

$$\frac{dT}{ds} = w\frac{dy}{ds}, \text{..........................(3)}$$

whence $$T = wy + \text{const.} \quad \text{.......................(4)}$$

The difference of tension at any two points is therefore proportional to the difference of level. For instance, if a length of

* It is convenient to use the word 'chain' when the weight or the mass is to be taken into account, and the word 'string' when this is neglected.

uniform chain is in equilibrium in a vertical plane, part of it in contact with a smooth curve, and part hanging vertically, the free ends must be at the same level.

Ex. A uniform chain passes one and a half times round a smooth horizontal circular cylinder of radius a, and the ends hang freely; to find what must be the lengths of the straight portions in order that the portion of the chain underneath the cylinder may be everywhere in contact with it.

At the lowest point of this portion we must have $T > wa$, by (2), and therefore at the level of the axis of the cylinder $T > 2wa$, by (4). Hence the length of each vertical portion of the chain must exceed $2a$.

The above formulæ cover of course the case of a chain hanging freely between any two points, the only difference being that the pressure R is now absent. Thus we have

$$\frac{dT}{ds} = w \sin \psi, \quad \frac{T}{\rho} = w \cos \psi, \quad \text{.................(5)}$$

and, in the case of a uniform chain,

$$T = w(y - y_0), \quad \text{.......................(6)}$$

where y_0 refers to some fixed level.

83. Chain hanging freely.

We proceed to consider more particularly the case of a chain hanging freely under gravity. It is usually most convenient to begin by forming the conditions of equilibrium of a finite portion.

Consider the portion extending from the lowest point A to any other point P, and let ψ denote as before the inclination to the horizontal of the tangent at P. The forces acting on the portion

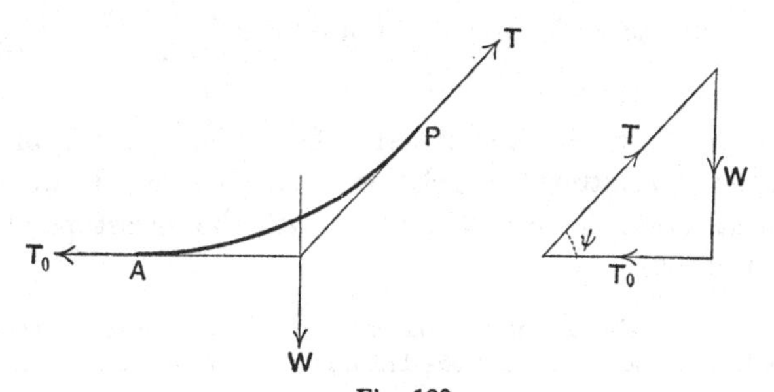

Fig. 120.

AP may be reduced to three, viz. a tangential force T at P, a horizontal force T_0 at A, and the weight W. For equilibrium these forces must be concurrent, and their ratios must be those given by a triangle of forces. Thus

$$T \cos \psi = T_0, \quad W = T_0 \tan \psi. \quad \text{..............(1)}$$

The former of these expresses the obvious fact that the horizontal component of the tension is constant.

In the case of a *uniform* chain* we may put $W = ws$, where w is the weight per unit length, and s denotes the arc AP. We write also

$$T_0 = wc, \quad \text{............................(2)}$$

i.e. c is the length of a portion of chain whose weight would equal the horizontal tension. Hence, from (1),

$$s = c \tan \psi, \quad \text{..........................(3)}$$

which is the 'intrinsic' equation of the curve in which the chain hangs.

From this the Cartesian equation may be derived as follows. If the axes of x and y be drawn horizontally, and vertically upwards, respectively, we have

$$\frac{dx}{ds} = \cos \psi, \quad \frac{dy}{ds} = \sin \psi. \quad \text{..................(4)}$$

Hence
$$\left.\begin{aligned} \frac{dx}{d\psi} &= \frac{dx}{ds}\frac{ds}{d\psi} = \cos \psi . c \sec^2 \psi = c \sec \psi, \\ \frac{dy}{d\psi} &= \frac{dy}{ds}\frac{ds}{d\psi} = \sin \psi . c \sec^2 \psi = c \tan \psi \sec \psi. \end{aligned}\right\} \quad \text{......(5)}$$

Integrating, we have

$$\left.\begin{aligned} x &= c \log (\sec \psi + \tan \psi) + \alpha, \\ y &= c \sec \psi + \beta. \end{aligned}\right\} \quad \text{..............(6)}$$

The constants of integration merely affect the position of the origin (hitherto arbitrary) in relation to the curve. If we take the origin at a distance c vertically beneath the lowest point, we have $\alpha = 0$, $\beta = 0$.

* The form of a freely hanging uniform chain appears to have been first ascertained in 1690 by James Bernoulli (1654–1705), professor of mathematics at Bâle 1687–1705.

On this understanding we have, from (6),

$$\sec\psi + \tan\psi = e^{x/c}, \quad \text{.....................(7)}$$

and therefore

$$\sec\psi - \tan\psi = e^{-x/c}, \quad \text{.....................(8)}$$

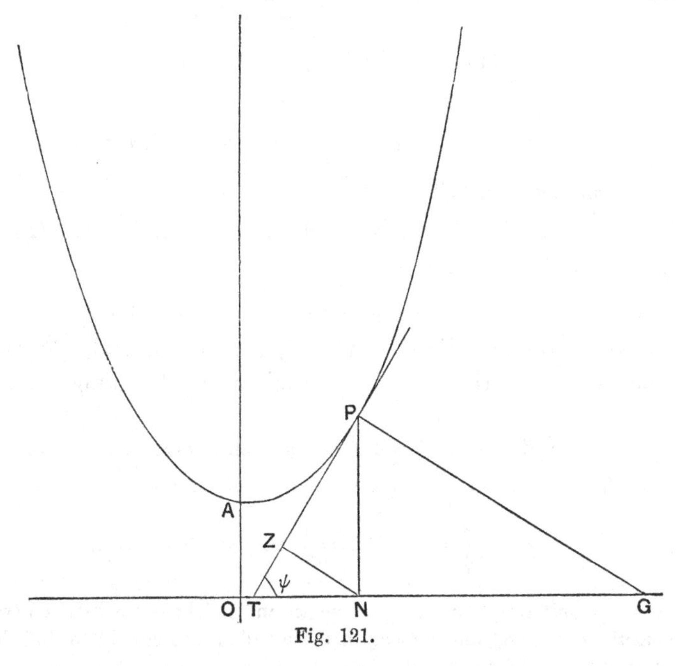

Fig. 121.

since the quantities on the left-hand of these equations are reciprocals. Hence

$$y = c\sec\psi = \tfrac{1}{2}c(e^{x/c} + e^{-x/c}) = c\cosh\frac{x}{c}, \quad \text{.........(9)}$$

and

$$s = c\tan\psi = \tfrac{1}{2}c\,(e^{x/c} - e^{-x/c}) = c\sinh\frac{x}{c}. \quad \text{......(10)}$$

The curve defined by (9) is called the 'uniform catenary.' Since $\cosh(x/c)$ is an even function of x, the curve is symmetrical with respect to the vertical through the lowest point, as was to be expected. The linear magnitude c is called the 'parameter' of the catenary, and determines its scale, all uniform catenaries being geometrically similar*. The horizontal line at a depth c below

* It is obvious from (9) that if x, y, c be altered in the same ratio the equation is still satisfied.

the lowest point (i.e. the axis of x of our present formulæ) is called the 'directrix.'

84. Properties of the Uniform Catenary.

From equations (1) and (9) of Art. 83 we have

$$T = wc \sec \psi = wy, \quad \text{.....................(1)}$$

so that the tension varies as the height above the directrix

Also, from (9) and (10),

$$y^2 = s^2 + c^2, \quad \text{..........................(2)}$$

a relation which is often useful.

Some geometrical properties follow easily from the figure. If PN be the ordinate, PT the tangent, PG the normal, NZ the perpendicular from the foot of the ordinate on the tangent, we have

$$NZ = y \cos \psi = c, \quad PZ = c \tan \psi = s. \quad \text{...........(3)}$$

Also, for the radius of curvature,

$$\rho = \frac{ds}{d\psi} = c \sec^2 \psi = y \sec \psi = PG. \quad \text{............(4)}$$

Ex. 1. A uniform chain of given length and weight hangs between two points at the same level, and the sag in the middle is measured; to find the pull on the points of support.

If $2s$ be the length, y the height of the ends above the directrix, k the sag, we have

$$(k+c)^2 = y^2 = c^2 + s^2,$$

whence $$c = (s^2 - k^2)/2k. \quad \text{..............................(5)}$$

This gives the horizontal tension wc; the vertical tension at the ends is ws, and the resultant tension is wy, where

$$y = c + k = (s^2 + k^2)/2k. \quad \text{...........................(6)}$$

Thus if the length be 10 ft., the weight 15 lbs., and the sag 3 ft., the horizontal pull is 4 lbs., the vertical pull 7·5 lbs., and the resultant pull 8·5 lbs.

Ex. 2. A uniform chain of given length is stretched between two given points at the same level; to find the parameter of the catenary in which it hangs.

In the equation $$s = c \sinh \frac{x}{c}, \quad \text{...............................(7)}$$

s and x are now given, and the question is to find c. This can only be effected by means of a table of hyperbolic functions*. If we put

$$x/c = u, \qquad\qquad (8)$$

we have

$$\frac{\sinh u}{u} = \frac{s}{x}. \qquad\qquad (9)$$

With the help of the tables we can trace the curve whose abscissa is u and ordinate $(\sinh u)/u$.

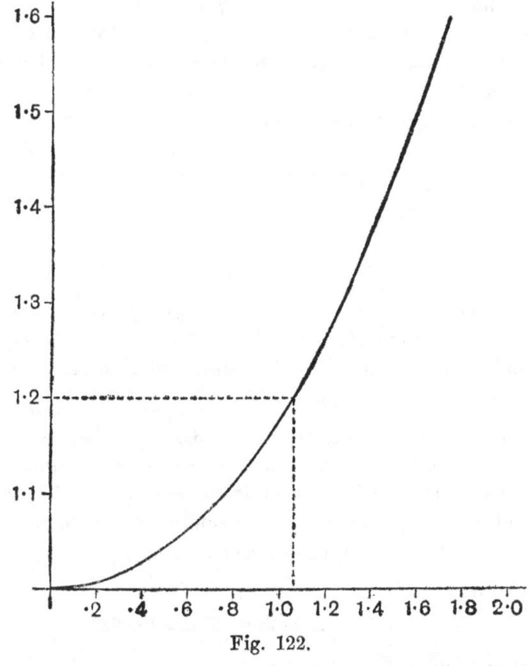

Fig. 122.

The horizontal line whose ordinate is s/x (which is necessarily > 1) will meet this curve in two points whose abscissæ are, say, $\pm u_1$. We have then from (8)

$$c = x/u_1. \qquad\qquad (10)$$

If greater accuracy is required than can be attained by a drawing, we may find u by interpolation from the tables. For instance, if the span be 100 ft., and the length 120 ft., we have $s/x = 1{\cdot}2$. On reference to the tables we find, as corresponding values,

$$u = 1{\cdot}06, \quad (\sinh u)/u = 1{\cdot}1981,$$
$$u = 1{\cdot}07, \quad (\sinh u)/u = 1{\cdot}2021,$$

* Such as is given, for example, in J. B. Dale's *Five-Figure Tables of Mathematical Functions*, London, 1903.

whence, by interpolation, $u_1 = 1{\cdot}065$. Hence $c = 46{\cdot}95$ ft., $y = \sqrt{(c^2+s^2)} = 76{\cdot}18$ ft., and the droop in the middle is $29{\cdot}32$ ft.

The figure shews however that the method is not susceptible of any great accuracy when the length of the chain only slightly exceeds the span. It appears from (9) that u is then small, and c accordingly great compared with x. A slight alteration in the length may then cause a considerable proportional change in the value of c.

Ex. 3. To find the greatest possible horizontal span for a uniform wire of given material. The limitation is supposed supplied by the condition that the tension at the points of support is not to exceed the weight of a certain length λ of the wire.

At the ends we have $y=\lambda$, and therefore

$$\lambda = c \cosh \frac{x}{c}, \qquad (11)$$

or, if we put

$$u = x/c, \qquad (12)$$

$$\frac{x}{\lambda} = \frac{u}{\cosh u}. \qquad (13)$$

As u increases from 0 to ∞, the function $u/(\cosh u)$ at first increases from 0, and finally tends asymptotically to 0. It must therefore have at least one maximum; and the condition for this is found on differentiation to be

$$\tanh u = 1/u. \qquad (14)$$

Now as u increases from 0 to ∞, $\tanh u$ steadily increases from 0 to 1, whilst $1/u$ steadily decreases from ∞ to 0. Hence there is one and only one (positive) value of u for which the expressions are equal. From the tables we find that this is $u = 1{\cdot}200$, nearly. The required span is therefore

$$2x = 2\lambda u/\cosh u = 2\lambda/\sinh u = 1{\cdot}325\lambda\,; \qquad (15)$$

the length of wire is

$$2s = 2c \sinh u = 2\lambda \tanh u = 2\lambda/u = 1{\cdot}667\lambda\,; \qquad (16)$$

and the sag in the middle is

$$\lambda - c = \lambda\,(1 - \operatorname{sech} u) = {\cdot}447\lambda. \qquad (17)$$

Also, since

$$\tan\psi = \sinh u = (\cosh u)/u = y/x, \qquad (18)$$

it appears that the tangents at the points of support meet at the origin. Their inclination to the horizontal is found to be about 56°.

85. Wire stretched nearly horizontal.

The relation between the sag, the tension, and the span of a wire, e.g. a telegraph wire, stretched nearly straight between two points B, C at the same level, is found most easily from first principles.

Let l be the length, W the total weight, T_0 the horizontal tension, k the sag AN at the lowest point A. Taking moments about C for the portion AC of the wire, we have

$$T_0.k = \tfrac{1}{2}W.\tfrac{1}{4}l$$

approximately, or $$T_0 = \frac{1}{8}\frac{l}{k}.W. \quad \text{.........................(1)}$$

The same result follows from Art. 84 (5), which reduces to

$$c = s^2/2k, \quad \text{..........................(2)}$$

when k/s is small.

The tension at B or C will exceed T_0 by wk, or Wk/l, or $8k^2/l^2.T_0$. This is by hypothesis a small fraction of T_0.

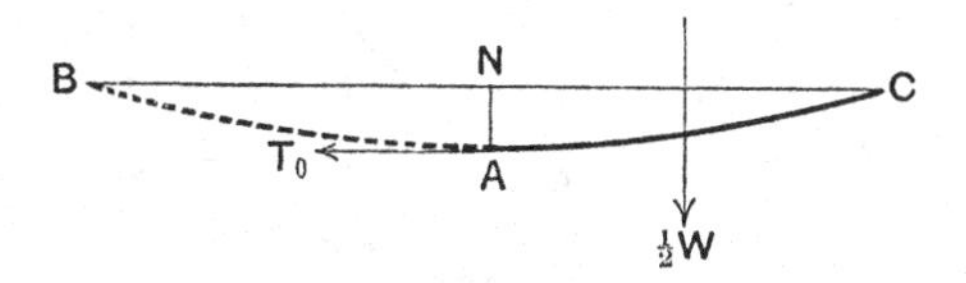

Fig. 123.

The calculation of the excess of the length over the span is a matter of Geometry. If R be the radius of curvature, and 2ψ the angle which BC subtends at the centre of curvature, we have

$$2Rk = \tfrac{1}{4}l^2, \quad R\psi = \tfrac{1}{2}l, \quad \text{..................(3)}$$

nearly. Hence

$$\text{arc } BC - \text{chord } BC = 2R(\psi - \sin\psi) = \tfrac{1}{3}R\psi^3 = \tfrac{8}{3}k^2/l, \text{...(4)}$$

approximately.

Ex. A telegraph wire has a span of 88 ft.; to find the sag in the middle if the tension is not to exceed 150 lbs., assuming that 20 ft. of the wire weighs 1 lb.

Here $T_0=150$, $W=4{\cdot}4$, $l=88$. Hence, from (1), $k={\cdot}32$ ft.

86. Parabolic Catenary.

Of problems relating to chains of variable line-density the most interesting is that of the parabolic catenary.

We have seen already (Art. 12) that if the load on any part of a chain varies as the horizontal projection of this part, the form assumed is that of a parabola with vertical axis*. The same thing may be proved more directly as follows.

Take the lowest point A as origin, and the tangent there as axis of x, and let x, y be the rectangular coordinates of any other point P on the curve. The weight of any element of the chain

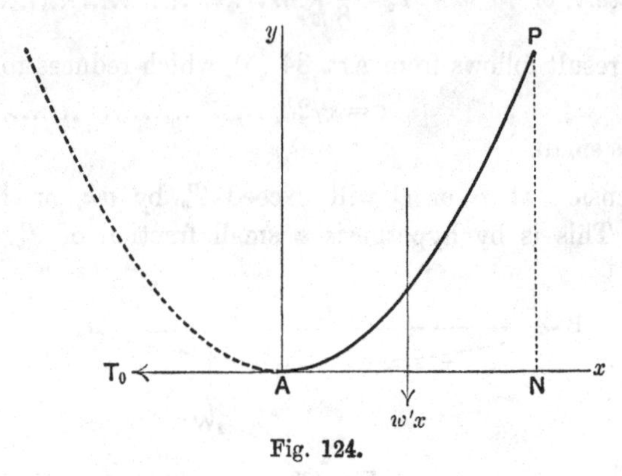

Fig. 124.

will be $w'\delta x$, where w' is the given constant load per unit length of the horizontal projection. The horizontal coordinate of the centre of gravity of the arc AP is therefore $\frac{1}{2}x$, being the same as for a uniform line of matter in the position AN. Hence, taking moments about P for the portion AP of the chain, we have

$$T_0 . y = w'x . \tfrac{1}{2}x,$$

or

$$y = \frac{w'}{2T_0} . x^2, \quad \text{..........................(1)}$$

which is the equation of the parabola in question.

87. Catenary of Uniform Strength.

In a uniform chain a limit is set to the possible span by the consideration that the chain would break (or be dangerously strained) at the points of attachment if the tension there exceeded a certain value. This limit could obviously be enlarged by making the chain lighter in the lower portions and stronger near the supports. We are thus led to the notion of the 'catenary of uniform strength,' where the cross-section is supposed to be

* This result also is attributed to James Bernoulli.

adjusted so as to be everywhere proportional to the tension; in other words, the stress-intensity, i.e. the tension per unit area of the cross-section, is assumed to be the same at all points*.

We assume, then, that the weight w per unit length varies as the tension T, so that

$$T = w\lambda, \qquad \text{.........................(1)}$$

where λ is a certain constant length. Resolving along the normal, we have

$$T\delta\psi = w\delta s \cos\psi, \qquad \text{....................(2)}$$

whence

$$\frac{ds}{d\psi} = \lambda \sec\psi. \qquad \text{.......................(3)}$$

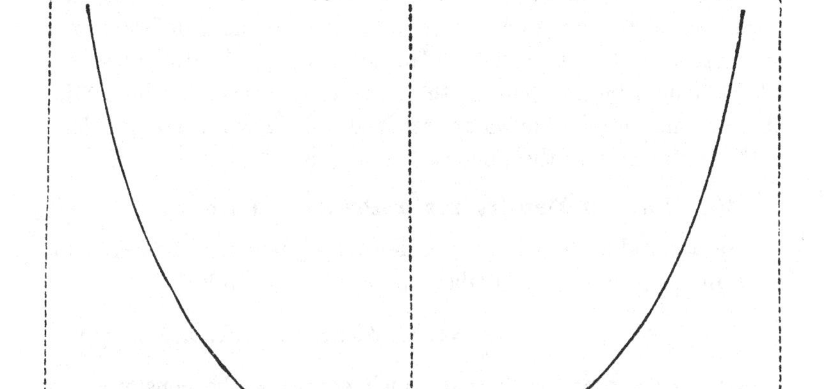

Fig. 125.

Integrating, we have the intrinsic equation

$$s = \lambda \log(\sec\psi + \tan\psi), \qquad \text{.................(4)}$$

no additive constant being necessary if the origin of s be at the lowest point ($\psi = 0$).

* This problem seems to have been first investigated by Davies Gilbert (1826).

To deduce the Cartesian equation, we have

$$\left.\begin{aligned}\frac{dx}{d\psi}&=\frac{dx}{ds}\frac{ds}{d\psi}=\cos\psi\,.\,\lambda\sec\psi=\lambda,\\ \frac{dy}{d\psi}&=\frac{dy}{ds}\frac{ds}{d\psi}=\sin\psi\,.\,\lambda\sec\psi=\lambda\tan\psi,\end{aligned}\right\}\quad\text{......(5)}$$

by (3). Hence

$$x=\lambda\psi,\quad y=\lambda\log\sec\psi,\quad\text{..................(6)}$$

provided the origin be taken at the lowest point. The required equation is therefore

$$y=\lambda\log\sec\frac{x}{\lambda}.\quad\text{........................(7)}$$

We note that for $x=\pm\frac{1}{2}\pi\lambda$ we have $y=\infty$, so that the curve is included between two vertical asymptotes. For a given material the linear magnitude λ, which determines the stress-intensity, cannot exceed a certain value if permanent deformation, or rupture, is to be avoided. Hence, however the thickness be adjusted, no cable can hang with a span so great as $\pi\lambda$, where λ is the extreme admissible value referred to. We have seen that with a uniform wire the limit to the span is $1{\cdot}325\,\lambda$.

88. Law of Density for a Prescribed Form.

By a suitable distribution of density a chain may be made to hang in the form of a prescribed curve. The formula

$$\frac{T}{\rho}=w\cos\psi,\quad\text{..........................(1)}$$

of Art. 82, combined with that which expresses the constancy of the horizontal tension, viz.

$$T=T_0\sec\psi,\quad\text{.........................(2)}$$

gives

$$w=\frac{T_0}{\rho}\sec^2\psi.\quad\text{..........................(3)}$$

When the intrinsic equation of the curve is given, this determines w as a function of ψ.

Ex. 1. In a parabola with axis vertical we have

$$\rho=2a\sec^3\psi,\quad\text{..................................(4)}$$

if $2a$ be the latus-rectum. Hence

$$w=\frac{T_0}{2a}\cos\psi,\quad\text{...................................(5)}$$

in agreement with Art. 86.

Ex. 2. For a circular arc we have

$$w = \frac{T_0}{a}\sec^2\psi, \qquad\qquad (6)$$

where a is the radius. This makes w infinite for $\psi = \frac{1}{2}\pi$, as we should expect, since with a finite tension the tangent can never be exactly vertical, there being necessarily a horizontal component T_0.

If we put $w'\delta x = w\delta s$, so that w' denotes, as in Art. 86, the weight per unit length of the horizontal projection, then, considering the equilibrium of a portion AP of the chain extending from the lowest point A, we have

$$\int_0^x w'\,dx = T_0 \tan\psi = T_0\frac{dy}{dx}, \qquad\qquad (7)$$

by Art. 83 (1). Hence, differentiating with respect to x, we have the equation

$$T_0\frac{d^2y}{dx^2} = w'. \qquad\qquad (8)$$

This equation, which may be obtained otherwise from (3), putting $w = w'\cos\psi$, is of some importance.

Ex. 3. If w' be constant we find on integration

$$y = \frac{w'}{2T_0}x^2 + Ax + B, \qquad\qquad (9)$$

which is the equation of a parabola with vertical axis.

EXAMPLES. XVI.

1. Prove *from first principles* that when a cord wrapped round a rough circular cylinder is in limiting equilibrium the tension varies in geometrical progression as the angular coordinate increases in arithmetical progression.

2. Two weights P, Q hang in limiting equilibrium from a string which passes over a rough circular cylinder in a plane perpendicular to the axis, which is horizontal. If P be on the point of descending, what weight may be added to Q without causing it to descend? [$P^2/Q - Q$.]

3. A smooth elliptic cylinder, having the minor axis of the cross-section vertical, is surrounded by an endless chain of length equal to the perimeter. What must be the tension at the highest point in order that the chain may be in contact everywhere? [$w(a^2+2b^2)/b$.]

4. A chain hangs beneath a smooth cycloidal arc whose base is horizontal and vertex downwards; what must be the tension at the cusps in order that the chain may be everywhere in contact?

[$6wa$, if a be the radius of the generating circle.]

5. A parabola whose axis is vertical and concavity upwards is described so as to touch a catenary at the vertex, and to have the same curvature there. Does the parabola lie above or below the catenary?

If a uniform chain be enclosed in a smooth parabolic tube whose axis is vertical and concavity upwards, and if it be just free at the vertex, find the pressure on the tube at a point where the tangent makes an angle ψ with the vertical. [$R=\frac{1}{2}w\sin\psi\cos^2\psi$.]

6. A length of uniform chain hangs over two or more smooth pegs in a vertical plane. Prove that the catenaries formed by the various portions have the same directrix.

7. A uniform chain hangs between two fixed points, and various weights are attached to it at intermediate points. Prove that the intervening portions of the chain form arcs of equal catenaries.

8. A freely hanging chain is made up of two parts of different densities. Prove that the curvature is discontinuous at the junction, the curvatures on the two sides being proportional to the respective densities.

9. A uniform chain AB of length l hangs vertically from A. If the end B be pulled horizontally by a force equal to the weight of a length a of the chain, prove that in equilibrium the horizontal and vertical projections of AB will be

$$a\log\frac{l+\sqrt{(l^2+a^2)}}{a}, \qquad \sqrt{(l^2+a^2)}-a.$$

10. The end links of a uniform chain of length l can slide on two smooth rods in the same vertical plane which are inclined in opposite ways at equal angles α to the vertical. Prove that the sag in the middle is $\frac{1}{2}l\tan\frac{1}{2}\alpha$.

11. The end links of a uniform chain can slide on a fixed rough horizontal rod. Prove that the ratio of the extreme span to the length of the chain is

$$\mu\log\frac{1+\sqrt{(1+\mu^2)}}{\mu},$$

where μ is the coefficient of friction.

12. One end of a uniform chain ABC of length l is attached to a fixed point A at a height h above a rough table. The portion BC is straight and rests on the table in a vertical plane through A. If the end C be free, prove that in limiting equilibrium the length s of the hanging portion is given by the equation

$$s^2+2\mu hs=h^2+2\mu hl.$$

13. A uniform chain 20 ft. long hangs between two points at the same level, and the sag in the middle is measured and found to be 5 ft. Find the parameter of the catenary in which it hangs. Also if the total weight be 50 lbs., find the pull on either point of support. [7·5 ft.; 31·25 lbs.]

14. A uniform chain of length l and weight W hangs between two fixed points at the same level, and a weight W' is attached at the middle point. If k be the sag in the middle, prove that the pull on either point of support is

$$\frac{k}{2l}W + \frac{l}{4k}W' + \frac{l}{8k}W.$$

15. A chain ABC is fixed at A, and passes over a smooth peg at B, and the portion BC hangs vertically. The length of each of the portions AB, BC is 20 ft., and the depth of B below the horizontal through A is 15 ft. Find the horizontal pull on A, having given that the weight per foot of the chain is 1 lb. [16·9 lbs.]

16. A uniform chain has a horizontal span of 100 ft., and the droop in the middle is 50 ft. Find (with the help of tables) the length of the wire. [149·6 ft.]

17. A uniform wire 150 ft. long has a horizontal span of 100 ft.; find (with the help of tables) the droop in the middle. [50·26 ft.]

18. A chain of length l is stretched nearly straight between two points at *different* levels. If W be the weight, T the tension, prove that the sag, measured vertically from the middle point of the chord, is $\frac{1}{8}Wl/T$.

19. A wire rope weighs 2 lbs. per foot; it has a horizontal span of 150 ft.; and the droop in the middle is 5 ft. Find approximately the length of the rope, and the tension. [150·44 ft.; 1125 lbs.]

20. A telegraph wire weighing 1 oz. per foot is stretched nearly horizontal between two points at a distance of 100 yds.; what must be the sag in the middle in order that the tension may not exceed 150 lbs.? [4·68 ft.]

21. A uniform chain hangs between two fixed points A, B at the same level. Prove that the horizontal pull on the supports will be increased by shortening the wire.

22. A uniform chain of length $2l$ hangs symmetrically over two smooth pegs at the same level, at a distance $2a$ apart. Prove that l cannot be less than $2{\cdot}718a$.

Also prove that if l exceed this value there will be two possible positions of (unstable) equilibrium.

23. A uniform chain of length l hangs between two points A, B at the same level, and the depth of the lowest point below AB is k. If the distance AB $(=a)$ be increased by the small quantity δa, the vertex of the catenary will be raised through a height

$$\frac{k\cos\psi}{a - l\cos\psi}\delta a,$$

where ψ denotes the inclination of the chain to the horizontal at A or B.

24. Prove that in the preceding Example the centre of gravity of the chain is raised through a height

$$\tfrac{1}{2}\cot\psi\,.\,\delta a.$$

25. A uniform chain of length l hangs between two points whose horizontal and vertical distances apart are h, k, respectively; prove that the parameter c of the catenary is determined by

$$\sqrt{(l^2-k^2)}=2c\sinh\frac{h}{2c}.$$

Discuss the solution of this equation (graphically or otherwise).

26. A solid of revolution has an axis of given length, and the meridian has a given length; what must be the shape of the meridian curve in order that the surface may be a maximum?

27. Shew geometrically that if the weight of any portion of a chain be proportional to its projection on the horizontal, the subnormal (measured along the vertical through the lowest point) is constant, and thence that the curve is a parabola.

28. Prove that in the parabolic catenary the tension varies as the square root of the height above the directrix.

29. The coordinates (horizontal and vertical) of the centre of gravity of an arc of a parabolic catenary are

$$\bar{x}=\tfrac{1}{2}(x_1+x_2),\quad \bar{y}=\tfrac{1}{6}(y_1+4y'+y_2),$$

where (x_1, y_1), (x_2, y_2) refer to the extremities of the arc, and y' is the ordinate half-way between y_1 and y_2.

30. Prove that in the catenary of uniform strength:

(i) The projection of the radius of curvature on the vertical is constant;

(ii) The tension varies as the radius of curvature;

(iii) $$T=T_0\cosh(s/k).$$

31. Find the law of density in order that a chain may hang in the form of a cycloid with horizontal base. [$w\propto\sec^3\psi$.]

32. A uniform chain of length l rotates about the line AB joining its extremities with constant angular velocity ω. Prove that if l be only slightly greater than $AB\,(=a)$, the form assumed by the chain in steady motion is

$$y=\frac{2}{\pi}\sqrt{\{(l-a)\,a\}}\sin\frac{\pi x}{a};$$

and that the tension is $M\omega^2 a/\pi^2$ in *dynamical* measure, where M is the total mass of the chain.

33. A chain whose ends are fixed at A and B revolves about AB with constant angular velocity. Prove that if it has the form of a curve of sines having zero curvature at A and B, the mass of any portion must be proportional to the orthogonal projection of that portion on AB.

CHAPTER X

LAWS OF FLUID PRESSURE

89. Density, and Specific Gravity.

The 'mean density' of any portion of a substance is the ratio of the mass to the volume. It is therefore expressed in pounds per cubic foot, or grammes per cubic centimetre, or in some other similar way, according to the fundamental units of mass and length adopted.

A substance is said to be of 'uniform' density if the mean density of all its parts is the same. When the density is not uniform, the 'density at a point' is defined as the limit to which the mean density in a small region containing that point tends as the linear dimensions of the region are indefinitely diminished.

The 'specific gravity' of a substance is the ratio of its density to some standard density, e.g. that of pure water when (under ordinary conditions of pressure) its density is greatest, i.e. at a temperature of about 4° Centigrade.

In the metric system, the kilogramme and the metre are defined practically by independent material standards, viz. by a particular piece of platinum, and by the length of a particular bar at a specified temperature. But the kilogramme was adjusted, in the first instance, so that its mass (or weight) should be as nearly as possible equal to that of a cubic decimeter of water under the conditions above stated; and the agreement is in fact so extremely close that for almost all purposes the terms 'density' and 'specific gravity' may on the metric system be taken as synonymous. On the British foot-pound system, on the other hand, the density of water at its maximum is about 62·426, whilst its specific gravity is of course unity.

For ordinary purposes, the variations of density with change of pressure or temperature, and in different specimens of the same substance, may in the case of solids and liquids be neglected. The following table gives, on this understanding, the specific gravities of a few substances.

Copper	8·9	Sea-water	1·026
Gold	19·3	Alcohol	·8
Iron	7·8	Ether	·73
Lead	11·3	Mercury	13·60
Platinum	21·5	Sulphuric acid	1·85
		Oil of Turpentine	·87

90. Stress in a Fluid.

The mutual action of the two portions A, B of a body, whether solid or fluid, which lie on the two sides of an ideal surface S drawn across it, consist partly of actions at a distance such as gravitational attraction, or (it may be) electric or magnetic forces, and partly of molecular forces exerted between the portions of the substance which are close to S on either side. It is these latter actions that are comprehended in the term 'stress.' Owing to the excessively small range of molecular forces, the portions of matter immediately concerned are confined to two strata bounded by S and by parallel surfaces drawn very close to it on the two sides. The thickness of these strata is in reality far below the limits even of microscopic vision.

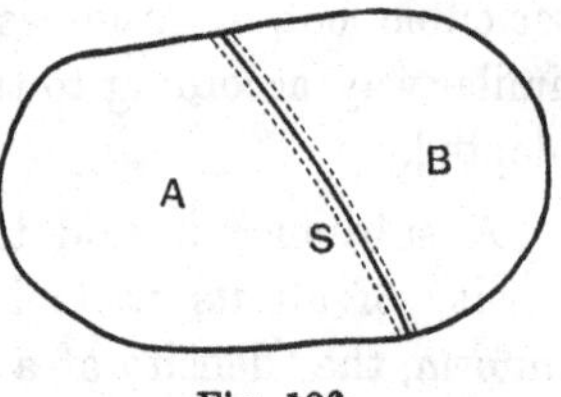

Fig. 126.

Now consider the portion of one of these strata which corresponds to a small area drawn on S. It is assumed that the dimensions of this area may be very large compared with the thickness of the stratum, whilst still small compared with ordinary linear magnitudes, and in particular with the radii of curvature of the surface S. Since adjacent equal areas of this description are then under practically the same conditions, we infer that the action of A on a small area of the stratum in B is ultimately proportional to the area. We therefore specify the intensity of the stress at any point of S by its amount *per unit area* in the neighbourhood of that point. To make the specification of the stress complete we must also know its direction. It may be, and in solid bodies generally is, oblique to the area.

The characteristic property of a *fluid*, however, is that when it is *in equilibrium* the mutual action between adjacent portions is everywhere normal to the common surface, or 'interface.' In other words, a fluid in equilibrium does not exert tangential stress. This is of course a physical assumption; it cannot be verified directly, but is to be justified by the agreement of the theoretical results derived from it with observed facts.

It is to be remarked, that tangential stresses do occur in fluids in motion; they constitute in fact the phenomenon known as 'viscosity.' But as a fluid approaches a state of equilibrium the tangential component of the stress decays with more or less rapidity, until finally, when the permanent condition is attained, it vanishes altogether.

Under ordinary conditions the mutual action referred to is of the nature of a *pressure*, i.e. its tendency is to resist approach of the parts.

The distinction between a 'liquid' and a 'gas' hardly needs to be formally stated. A liquid occupies a definite volume, independent of the dimensions of the vessel in which it is contained, and its density is nearly constant. We may, in fact, as already stated, for most purposes neglect the compressibility of actual liquids. A gas, on the other hand, spreads throughout any region to which it has access, and is readily susceptible of very great variations of density.

91. Uniformity of Pressure-Intensity about a Point.

We now fix our attention on a particular plane drawn through a fluid in equilibrium. The 'mean pressure-intensity'* over any area of this plane is the ratio of the force exerted across this area, by the fluid on one side upon the fluid on the other, to the area.

As throughout the preceding pages, we shall assume for the most part that the forces with which we are concerned are

* The *pressure-intensity* at a point is to be distinguished from the *total pressure* on an area. It is true that the word 'pressure' is often used indifferently in Hydrostatics in either of these senses, but at first it is well to accentuate the distinction.

expressed in *gravitational* units (Art. 6). The pressure-intensity will accordingly be supposed to be expressed in pounds per square foot, or grammes per square centimetre, or in some such way.

If the mean pressure-intensity is the same for all parts of the plane, the pressure-intensity is said to be 'uniform' over that plane. If it is not uniform, the pressure-intensity 'at a point' of the plane is defined as the limit to which the mean pressure-intensity over a small area, containing the point in question, tends as the dimensions of the area are indefinitely diminished.

We assume that the forces acting on any particular portion of fluid, including the pressures exerted on its surface by surrounding fluid or by the walls of a containing vessel, must fulfil the same conditions of equilibrium as in the case of a solid.

It is an important consequence of the hypothesis of purely normal stress that the pressure-intensity at a point P is the same for all planes through P.

Suppose, in the first place, that the fluid is free from external forces such as gravity, so that the only forces with which we are concerned are the mutual pressures between adjacent parts, and the pressures exerted on it at the boundaries, as, for instance, by the walls of a containing vessel. Consider a portion of fluid in the form of a prism whose ends are perpendicular to its length, and whose section is a triangle ABC. The pressures on the various elements of a plane area constitute a system of parallel forces, and have therefore a single resultant equal to their sum. Resolving parallel to the length of the prism, we see that the resultant pressures on the two ends must be equal, and opposite, and must therefore cancel in the geometric sum of the forces. It follows that the geometric sum of the resultant pressures on the three lateral faces of the prism must also vanish. These resultants, being parallel to the plane ABC, and normal to the sides BC, CA, AB, respectively, must therefore be proportional to these sides (Art. 23, Ex. 1), and so proportional to the areas of the faces on which they act. The mean pressure-*intensities* on these faces are accordingly equal. If we now suppose the dimensions of the prism to be infinitely small, we learn that the pressure-intensities at a point P are the same for all planes through that point.

This latter statement holds even when the influence of external forces is allowed for, provided that these forces vary (as in the case of gravity) ultimately as the *volumes* of the fluid elements on which they act. For when the dimensions of the prism are indefinitely diminished, these 'body-forces,' as they are sometimes called, are ultimately of the third order of small quantities, whilst the pressures on the faces are of the second. The former class of forces therefore ultimately disappear from the equations.

In the case of gravity this fundamental argument may be put more explicitly as follows. Let the two sides AB, BC of the triangular section be vertical and horizontal, respectively. Let ρ be the mean density of the fluid forming the prism, and let p_1, p_2, p_3 be the mean pressure-intensities on the faces represented by BC, CA, AB, respectively. The actual pressures will be $p_1.BC.l$, $p_2.CA.l$, $p_3.AB.l$, where l is the length of the prism; and the weight of the fluid is in gravitation measure $\frac{1}{2}\rho.AB.BC.l$. Then, resolving parallel to BC,

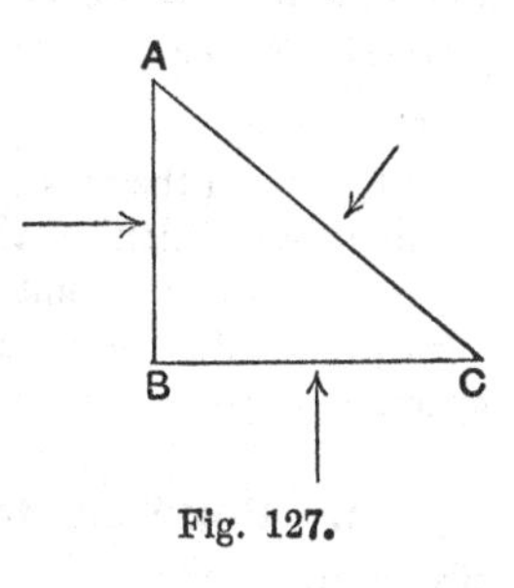

Fig. 127.

$$p_3.AB.l = p_2.AC.l.\cos A = p_2.AB.l;$$

and, resolving vertically,

$$p_1.BC.l = p_2.AC.l.\cos C + \tfrac{1}{2}\rho.AB.BC.l$$
$$= p_2.BC.l + \tfrac{1}{2}\rho.AB.BC.l.$$

Hence $$p_3 = p_2, \quad p_1 = p_2 + \tfrac{1}{2}\rho.AB. \quad \text{.................. (1)}$$

With the above definitions of the symbols, these results are exact, whatever the dimensions of the prism. If we now suppose the dimensions to become infinitely small, we get

$$p_1 = p_2 = p_3. \quad \text{.......................... (2)}$$

Since the plane face represented by AC may have *any* direction, the theorem in question is verified.

The state of stress at any point P of a fluid in equilibrium is therefore completely specified by a single symbol p, which denotes the pressure-intensity on *any* small area through P.

Again, in a fluid free from body-force, the pressure-intensity is not only the same for all planes through a point, but is the same at all points. For consider the equilibrium of a cylindrical or prismatic portion of the fluid, of small section, whose ends (perpendicular to the length) are at any two points P, Q. Resolving parallel to the length we see that the pressures on the two ends must be equal, and since the areas on which they act are equal, the intensities at P and Q must be the same.

This suggests various theorems of pure Statics. Thus we learn that a system of uniformly distributed pressures over any closed surface will be in equilibrium. If the surface be that of a polyhedron with plane faces, we infer that forces acting at right angles to the faces, at the respective mean centres, and proportional to the respective areas, will be in equilibrium provided they act all inwards (or all outwards).

Again, if a liquid completely enclosed in a vessel is in equilibrium under given forces and the pressure of the walls, the equilibrium will not be affected by the superposition of a uniform distribution of pressure. Conversely, if the pressure at any point be increased by any means, the pressure at every other point will be increased to an equal extent. This is the principle of the hydraulic press.

92. Fluid Subject to Gravity. Conditions for Equilibrium.

From this point onwards we consider more especially the case of fluids subject to gravity. Since the pressures on the two ends of a cylindrical column of small section, whose length is *horizontal*, must still be equal, we learn that the pressure-intensity is uniform over any connected horizontal area in the fluid.

Again, along any vertical line drawn in the fluid, the pressure-intensity increases downwards with a gradient equal to the density. For let z denote depth below some fixed horizontal plane of reference; and consider a small cylindrical portion of fluid whose length δz is vertical, and whose ends are horizontal, of small sectional area ω. The forces which act on this are gravity, the pressures on the ends, and the horizontal pressures on the sides. Resolving vertically, we see that the upward pressure on the base must exceed the downward pressure on the top by the weight of the column. Hence if p and

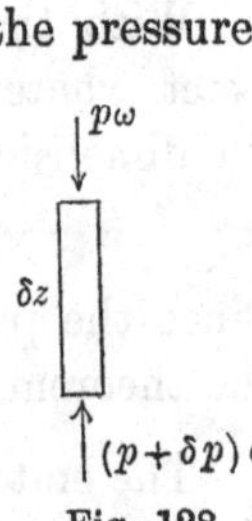

$p\omega$

δz

$(p+\delta p)\,\omega$

Fig. 128.

$p + \delta p$ be the pressure-intensities at the top and bottom, and ρ the density, we have

$$(p + \delta p)\,\omega = p\omega + \rho\omega\delta z,$$

whence $\delta p = \rho\delta z$, or, ultimately,

$$\frac{dp}{dz} = \rho. \qquad (1)$$

It may be well to repeat that a gravitational unit of force is here implied. In absolute (dynamical) units the weight of the cylinder would be denoted by $g\rho\omega dz$, and the factor g would be required on the right-hand side of (1).

So far, nothing is assumed with respect to ρ except that it denotes the *local* density. We have seen, however, that p is a function of the depth z only; the equation (1) shews therefore that the same must be true of ρ. In other words, in a fluid (whether liquid or gaseous) in equilibrium under gravity, the surfaces of equal density, as well as the surfaces of equal pressure-intensity, must be horizontal planes.

Moreover, the density of any fluid is determined by the pressure and the temperature. It follows that for equilibrium it is also necessary that the temperature should be uniform over any horizontal area. If in consequence of one-sided heating this condition is violated, convection currents at once set in.

Another important consequence is that the free surface of a liquid (where it is in contact with the atmosphere), or the common boundary of two liquids of different densities, must be a horizontal plane. For it would otherwise be possible to draw a horizontal plane over which the density was not uniform. The fact that the free surface of a liquid, such as mercury, furnishes an optically perfect plane mirror is a striking confirmation of the hypothesis of normal pressure, on which the above proposition rests.

As already stated, the variations of density in a liquid may as a rule be disregarded. Hence, treating ρ as a constant, we have from (1)

$$p = \rho z + C, \qquad (2)$$

where the arbitrary constant C is determined by the value of the pressure-intensity at some particular level. Thus if the pressure-intensity at the plane $z = 0$ be p_0, we have

$$p = p_0 + \rho z. \quad \text{.........................(3)}$$

If the origin be at a free surface, p_0 is the atmospheric pressure.

Since p has a uniform vertical gradient, it easily follows that the different parts of the free surface of a liquid, in communicating vessels, are all at the same level.

Ex. If the height of the barometer be 76 cm., the temperature being 0° C., the atmospheric pressure is

$$76 \times 13{\cdot}60 = 1034 \text{ gms. per sq. cm.}$$

If the height be 30 in., the pressure is

$$2{\cdot}5 \times 13{\cdot}60 \times 62{\cdot}43 = 2123 \text{ lbs. per sq. ft.,}$$

or 14·74 lbs. per sq. in.

A pressure-intensity of about this value is called an 'atmosphere.' The height of a water column which would produce the same pressure-intensity is called the 'height of the water-barometer.' With the above data this would be about 10·3 metres, or 34 feet, respectively.

93. Resultant Pressure on a Plane Area. Centre of Pressure.

The statical effect of the fluid pressures on any plane area is known when we know the magnitude and line of action of the resultant.

The magnitude of the resultant pressure is given, in the case of liquids under gravity, by the following rule: The mean pressure-intensity over any plane area is equal to the pressure-intensity at the mean centre of the area. For let δS be an element of the area (S), at a depth z, and let $\bar{z}$ refer to the mean centre of the area, so that

$$\Sigma (z . \delta S) = \bar{z} . S. \quad \text{.......................(1)}$$

The mean intensity in question is therefore

$$\frac{\Sigma (p . \delta S)}{S} = \frac{\Sigma \{(p_0 + \rho z)\, \delta S\}}{S} = p_0 + \rho \bar{z}, \quad \text{...........(2)}$$

which is the pressure-intensity at the depth $\bar{z}$.

The line of action of the resultant is determined when we know the point in which it intersects the plane of the area; this point is called the 'centre of pressure.' If the pressure-intensity were uniform, as in the case of a horizontal area, the centre of pressure would of course coincide with the mean centre; but in other cases it occupies a lower position, in consequence of the increase of pressure-intensity with depth.

In determining resultant pressures, and centres of pressure, it is generally proper to omit the term p_0 in the formula (3) of Art. 92, since what we are really concerned with is usually not so much the absolute value of the pressure-intensity, as its *excess* over the atmospheric value. For instance, the pressure of the water on the side of a tank is in part compensated by the atmospheric pressure on the outside.

The position of the centre of pressure of a plane area can be assigned at once in a number of cases from the consideration that the problem is the same as that of finding the mass-centre of a thin lamina of the same shape, whose thickness varies as the depth below the free surface, and therefore as the distance from the surface-line in the plane of the area.

Exx. In the case of a rectangle having one side in the surface, the thin lamina has the form of a very acute wedge. The centre of pressure is therefore in the median line at the point of trisection furthest from the surface.

In the case of a triangle having one side in the surface, the lamina is a very flat tetrahedron, of which the surface-line forms one edge; and the mass-centre is the middle point of the line joining this to the middle point of the opposite (infinitely short) edge. Hence the centre of pressure of the triangle bisects the median line.

Similarly, the centre of pressure of a triangular area having one vertex in the surface and the opposite side horizontal is in the median line, at a distance of three-fourths its length from the vertex.

Other cases can be deduced from these by composition. Thus, in the case of a trapezium whose parallel sides are a, b, the side a being in the surface, we divide the figure into three triangles whose areas are proportional to $\frac{1}{2}a$, $\frac{1}{2}a$, b. The mean pressure-intensities over these areas are proportional to the depths of their mean centres, and are therefore as $1:1:2$. Hence the resultant pressures on the three triangles are as $a:a:4b$. The distances of the respective centres of pressure from the upper side have

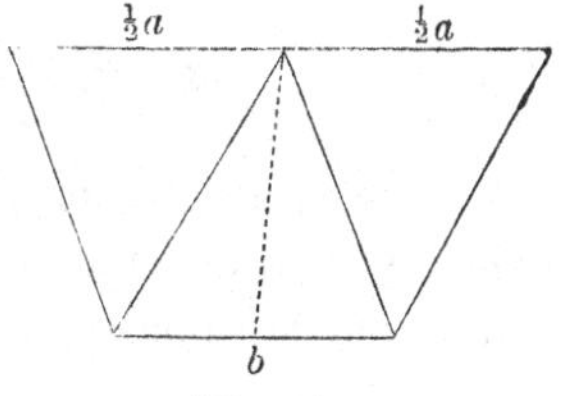

Fig. 129.

been shewn to be $\frac{1}{3}h$, $\frac{1}{2}h$, $\frac{3}{4}h$, where h is the breadth of the trapezium. Hence, taking moments, we find for the distance of the centre of pressure from the upper side

$$\frac{a \cdot \frac{1}{3}h + a \cdot \frac{1}{2}h + 4b \cdot \frac{3}{4}h}{a+a+4b} = \frac{a+3b}{2a+4b} \cdot h. \quad \text{.....................(3)}$$

The point lies of course in the line bisecting the parallel sides.

94. Formulæ for Centre of Pressure.

General formulæ for the position of the centre of pressure can be written down from the theory of parallel forces. If we take rectangular axes in the plane of the area, the pressure on an element $\delta x \, \delta y$ will be denoted by $p \, \delta x \, \delta y$, and its moment about the axis of y by $xp \, \delta x \, \delta y$. Since the moment of the total pressure is equal to the sum of the moments of the elementary pressures, we find, for the coordinates of the required point,

$$\xi = \frac{\iint xp \, dx \, dy}{\iint p \, dx dy}, \quad \eta = \frac{\iint yp dx \, dy}{\iint p dx dy} \quad \text{..............(1)}$$

These formulæ hold also if the axes are oblique, for if ω be their inclination, the element of area, now a parallelogram, will be $\delta x \, \delta y \sin \omega$, whilst the distances of an element, and of the centre of pressure, from the axis of y will be $x \sin \omega$ and $\xi \sin \omega$. On substitution the factor $\sin \omega$ will divide out.

In the case of a liquid subject to gravity it is convenient to make the axis of y coincide with the line in which the plane of the area meets the free surface. If the axis of x be taken perpendicular to this, we have, ignoring the atmospheric pressure for the reason stated,

$$p = \rho z = \rho x \sin \alpha, \quad \text{......................(2)}$$

where α is the inclination of the plane to the horizontal. The factor $\sin \alpha$ disappears on substitution in (1), so that

$$\xi = \frac{\iint x^2 dx \, dy}{\iint x dx dy}, \quad \eta = \frac{\iint xy \, dx \, dy}{\iint x dx dy}. \quad \text{.............(3)}$$

The integrals are of the types which we have met with in Chapter VIII as linear and quadratic moments, and their values in a number of cases have been computed.

If the axis of x be a line of symmetry of the area, the numerator in the expression for η will vanish, owing to the cancelling of terms due to elements in the positions (x, y) and $(x, -y)$. Also if k be the radius of gyration of the area about the surface-line, and h the distance of the mean centre of the area from this line, we have

$$\iint x^2 dx\, dy = Sk^2, \quad \iint x\, dx\, dy = Sh, \quad \text{..........(4)}$$

where S is the total area. Hence

$$\xi = k^2/h, \quad \eta = 0. \quad \text{......................(5)}$$

If κ be the radius of gyration about a line through the mean centre parallel to the surface-line, we have $k^2 = \kappa^2 + h^2$, by Art. 73, and therefore

$$\xi = h + \kappa^2/h. \quad \text{.........................(6)}$$

Exx. In the case of a rectangle of sides a, b, with the side b in the surface, we have

$$h = \tfrac{1}{2}a, \quad k^2 = \tfrac{1}{3}a^2, \quad \xi = \tfrac{2}{3}a, \quad \text{..........................(7)}$$

as before.

For a semicircular area of radius a, with the diameter in the surface,

$$h = 4a/3\pi, \quad k^2 = \tfrac{1}{4}a^2, \quad \xi = \tfrac{3}{16}\pi a = \cdot 585a. \quad \text{.................(8)}$$

For a circular area at any depth

$$k^2 = h^2 + \tfrac{1}{4}a^2, \quad \xi = h + \tfrac{1}{4}a^2/h. \quad \text{.......................(9)}$$

The formula (6) shews that with increasing depth of immersion the centre of pressure approximates more and more to the mean centre of the area, as is otherwise evident, since the distribution of pressure-intensity becomes more and more nearly uniform.

95. Centre of Pressure and Central Ellipse.

The formulæ (3) of Art. 94 will apply even if the axis of x be oblique to the surface-line, since the depth of any point of the area below the free surface will still be in a constant ratio to the abscissa x, and the element of area will be in a constant ratio to $\delta x \delta y$. We will suppose that the axis of x passes through the mean centre G, whose abscissa is, say, h. If we now transfer the origin to this point, writing $x + h$ for x and $\xi + h$ for ξ, the axis of y remaining parallel to the surface-line, we have

$$\iint x\, dx\, dy = 0, \quad \iint y\, dx\, dy = 0, \quad \text{................(1)}$$

and therefore

$$\left.\begin{aligned}\iint(x+h)^2dxdy=\iint x^2dxdy+h^2\iint dxdy,\\ \iint(x+h)\,ydxdy=\iint xydxdy,\qquad \iint(x+h)\,dxdy=h\iint dxdy.\end{aligned}\right\}\quad(2)$$

Hence $$\xi=\frac{\iint x^2dxdy}{h\iint dxdy},\quad \eta=\frac{\iint xydxdy}{h\iint dxdy}.\ \ldots\ldots\ldots\ldots(3)$$

The direction of the axis of x is still arbitrary. It is convenient so to choose it that the integral $\iint xydxdy$ shall vanish. This will be the case, by Art. 77, if the axes of x, y are conjugate diameters

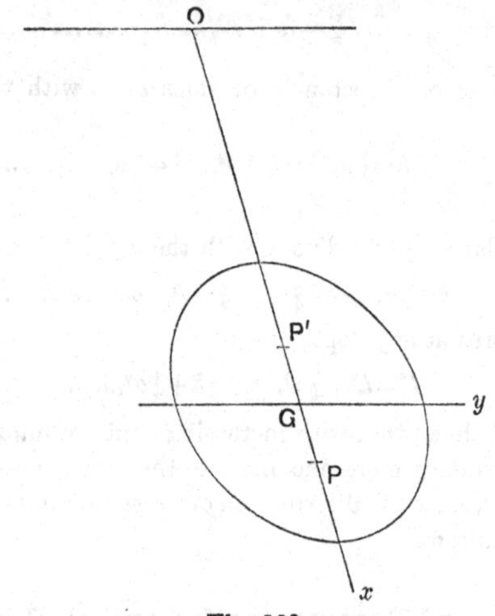

Fig. 130.

of the central ellipse of the area. The equation of this ellipse is then of the form

$$\frac{x^2}{a^2}+\frac{y^2}{b^2}=1,\ \ldots\ldots\ldots\ldots\ldots\ldots(4)$$

where $$a^2=\frac{\iint x^2dxdy}{\iint dxdy},\quad b^2=\frac{\iint y^2dxdy}{\iint dxdy}\ \ldots\ldots\ldots\ldots(5)$$

Hence, relatively to axes Gx, Gy through the mean centre of the area, of which Gy is parallel to the surface-line, and Gx is conjugate to it with respect to the central ellipse, we have

$$\xi=a^2/h,\quad \eta=0,\ \ldots\ldots\ldots\ldots\ldots\ldots(6)$$

if a^2 denote the mean square of the abscissæ of the points of the area, and $-h$ the abscissa of the surface-line.

The point (P) thus determined is on the same diameter of the central ellipse, and at the same distance from its centre, as the *pole* (P') of the surface-line, but on the opposite side. It is therefore called the 'anti-pole' of the surface-line*.

Ex. 1. In the case of any regular polygon the central ellipse is a circle. Thus for a square, the centre of pressure is in the line through the centre of the square perpendicular to the surface-line, at a distance of $\frac{1}{3}a^2/h$ from this centre, if $2a$ be the length of the side.

For an equilateral triangle of side $2a$ the distance is $\frac{1}{6}a^2/h$.

Ex. 2. The centre of pressure of any triangle which is wholly immersed coincides with the mass-centre of three particles at the middle points of the sides whose masses are proportional to the depths of these points below the free surface.

If (x_1, y_1), (x_2, y_2), (x_3, y_3) be the positions of these three particles relative to the pair of conjugate diameters of the central ellipse (of the area) above referred to, we have

$$\Sigma(x)=0, \quad \Sigma(y)=0, \quad \Sigma(xy)=0, \quad \tfrac{1}{3}\Sigma(x^2)=a^2, \qquad \text{.........(7)}$$

where a has the same meaning as in (5). For a system of three *equal* particles in the same positions would have the same mass-centre, and the same central ellipse, as the triangular area, by Art. 78. Hence the coordinates of the mass-centre of three particles at the middle points of the sides, whose masses are proportional to $h+x_1$, $h+x_2$, $h+x_3$, are

$$\left.\begin{aligned}\bar{x}&=\frac{\Sigma\{(h+x)x\}}{\Sigma(h+x)}=\frac{\Sigma(x^2)}{3h}=\frac{a^2}{h},\\ \bar{y}&=\frac{\Sigma\{(h+x)y\}}{\Sigma(h+x)}=0,\end{aligned}\right\} \qquad \text{.........(8)}$$

which are, by (6), the coordinates of the centre of pressure.

96. Pressures on Curved Surfaces.

The pressures on the various elements of a *curved* area constitute a three-dimensional system of forces, and are not necessarily equivalent to a single resultant. The simplest mode of reduction is to resolve the pressure on each element into its components in three directions, one of which may conveniently be

* It is the pole of the surface-line with respect to the imaginary conic

$$\frac{x^2}{a^2}+\frac{y^2}{b^2}=-1.$$

taken to be vertical, whilst the other two are horizontal and at right angles to one another. In this way we obtain three systems of parallel forces, each of which may be replaced by a single force. The three mutually perpendicular forces to which the original system is thus reduced may be called the 'resultant pressures' in the directions specified. But their lines of action do not necessarily intersect, and no further simplification is as a rule possible.

If, however, the area in question has a vertical plane of symmetry, it is evident that the resultant pressure in a direction normal to this plane will vanish. There remain a resultant vertical pressure, and a resultant horizontal pressure, in the plane of symmetry, and these can of course be replaced by a single resultant.

The rule for finding the resultant pressure in any given horizontal direction is very simple. If we project the contour of the area orthogonally on a vertical plane perpendicular to this direction, we obtain a curve bounding a plane area S'. The required resultant is equal in magnitude to the pressure on S', and acts in a line through its centre of pressure. For let δS be an element of the given curved area, and $\delta S'$ its projection. A horizontal column of fluid whose ends are formed by the elements δS, $\delta S'$ would be in equilibrium under gravity and the normal pressures on its surface. Resolving parallel to the length we see that the component of the pressure on δS in the direction of the length must be equal as well as opposite to the pressure on $\delta S'$. Cases may arise where there is more than one surface-element having the same projection $\delta S'$, but if we take account of the *sense* of the horizontal components of the pressures on them, we find that the above rule holds in all cases.

As regards the vertical resultant pressure, we project orthogonally on a horizontal plane at the level of zero pressure (or at the free surface, if we leave the atmospheric pressure out of account). Let $\delta S''$ be the projection of an element δS of the given area. Considering the forces which would act on a vertical column of fluid whose ends are δS and $\delta S''$, we see that the vertical component of the pressure on δS is equal to the weight of the column. The resultant vertical pressure is therefore equal in magnitude to the weight of a mass of fluid bounded (i) by the curved area, (ii) by a cylindrical surface whose generators are vertical lines through the edge of the area, and (iii) by the plane of zero pressure; and it acts in a vertical line through the centre of gravity of this mass. It will be seen that the case where more than one element δS has the same projection $\delta S''$ is allowed for in the rule*.

The resultant horizontal pressure on a *closed* surface, in any direction, will vanish. The resultant vertical pressure therefore represents the whole effect of the fluid pressures on the surface; it is equal to the weight of fluid which would fill the included space, and acts through the centre of gravity of this fluid.

Ex. To find the resultant thrust on the (outer) curved surface of an immersed hemisphere.

Let a be the radius, h the vertical depth of the centre, θ the inclination of the plane of the rim to the horizontal. For definiteness we will suppose that the hemispherical surface considered lies *below* this plane. The projection of the rim on a vertical plane perpendicular to the vertical plane of symmetry is an ellipse of semi-axes a and $a \sin \theta$, and the resultant *horizontal* thrust is therefore

$$\rho h . \pi a^2 \sin \theta. \quad \text{.....................................(1)}$$

The volume of liquid whose weight is equal to the resultant *vertical* thrust consists of a hemispherical portion $\frac{2}{3}\pi a^3$ and a cylindrical portion whose upper end is an ellipse of semi-axes a and $a \cos \theta$, and whose lower end is a circle of radius a, the mean height being h. The resultant in question is therefore

$$\rho \left(\tfrac{2}{3}\pi a^3 + \pi a^2 h \cos \theta\right). \quad \text{..............................(2)}$$

Since the pressures on the curved surface act in lines through the geometrical centre, the resultant of (1) and (2) must pass through this point.

97. Work of Fluid Pressure. Potential Energy of a Liquid.

When a fluid mass is slightly deformed without change of volume, the work done by a *uniform* pressure over its surface vanishes.

For consider, in the first place, a fluid filling a rigid vessel which has a number of apertures occupied by moveable pistons of

* In each of the preceding questions, the case where the contour of the projected area intersects itself would require examination, but the matter is hardly important enough to call for further discussion.

areas $S_1, S_2, \ldots, S_n$, and let these be pushed in through small spaces (positive or negative) $\nu_1, \nu_2, \ldots, \nu_n$, respectively. If p be the uniform pressure-intensity, the work done is

$$pS_1 . \nu_1 + pS_2 . \nu_2 + \ldots + pS_n . \nu_n,$$

which vanishes, since

$$\nu_1 S_1 + \nu_2 S_2 + \ldots + \nu_n S_n = 0, \quad \ldots\ldots\ldots\ldots\ldots\ldots(1)$$

by the condition of constancy of volume.

The extension to the general case is merely a matter of notation. If δS be an element of area of the boundary, ν the small extent to which it is pushed inwards, the work is

$$\Sigma (p\,\delta S . \nu) = p\Sigma (\nu\,\delta S) = 0. \quad \ldots\ldots\ldots\ldots\ldots\ldots(2)$$

This is the principle of virtual velocities as now proved on the particular hypothesis of uniform pressure.

Conversely, if we were to assume as a physical axiom that the internal forces of a fluid mass do no work when the deformation involves no change of volume, we could infer the uniformity of pressure-intensity. Thus if, in the case first considered, we suppose only two of the pistons to be displaced, we should have

$$p_1 S_1 . \nu_1 + p_2 S_2 . \nu_2 = 0, \quad \text{with} \quad \nu_1 S_1 + \nu_2 S_2 = 0, \quad \ldots\ldots\ldots\ldots\ldots(3)$$

whence $p_1 = p_2$.

We go on to the case of a liquid subject to gravity. With our former notation, the work of the fluid pressures on the boundary of the mass considered will be

$$\Sigma (p\,\delta S . \nu) = \Sigma \{(p_0 + \rho z)\,\delta S . \nu\} = \Sigma (\rho\nu\,\delta S . z), \quad \ldots\ldots(4)$$

since $\Sigma (\nu\,\delta S) = 0$, as before. This expression may be identified with the increment of the potential energy, with respect to gravity, of the mass considered. For the change in the potential energy may be regarded as due to the transfer of small amounts of matter from one part of the boundary to another; and if we consider two elements δS_1, δS_2 so chosen that

$$\nu_1\,\delta S_1 = -\,\nu_2\,\delta S_2, \quad \ldots\ldots\ldots\ldots\ldots\ldots\ldots\ldots(5)$$

both sides being supposed positive, the removal of a weight $\rho\nu_1\,\delta S_1$ from the first position to the second involves a gain of potential energy of amount

$$\rho\nu_1\,\delta S_1 (z_1 - z_2) = \rho\nu_1\,\delta S_1 . z_1 + \rho\nu_2\,\delta S_2 . z_2. \quad \ldots\ldots\ldots(6)$$

Since the whole alteration of the boundary may be supposed

brought about by the interchange of pairs of elements related as in (5), the interpretation of the last member in (4) follows.

Hence the work done by the fluid pressures on the boundary of a liquid mass in equilibrium, in any infinitesimal deformation, is equal to the increment of the potential energy. This verifies the principle of Virtual Velocities for the present case, in the form given in Art. 57 (2).

In particular, if the work of the external pressures vanishes, the potential energy must be stationary, for equilibrium.

This statement implies various results already proved, as well as others. For instance, the free surface of a liquid contained in an open vessel must be such that any small deformation involves a change of the *second order* only in the value of the potential energy. It is easily seen that a horizontal plane is the only form of surface fulfilling this condition. For if the surface had any other form, a slight lowering of the elevated portions, accompanied by an equivalent raising of the depressed portions, would involve a diminution of the potential energy by a small quantity of the *first* order. Again, if a solid float in the liquid, the depth of the centre of mass of the whole system must be stationary for all displacements.

We may anticipate, further, that the equilibrium will be stable if, and only if, the potential energy be a minimum. For instance, in the case last mentioned the depth of the centre of gravity of the whole system must be a maximum, for stability*.

EXAMPLES. XVII.

[The weight of a cubic foot of pure water is taken to be 62·43 lbs.]

1. If equal weights of n liquids of specific gravities $s_1, s_2, \ldots, s_n$ are taken and mixed together, without change of volume, the specific gravity of the mixture is

$$n \bigg/ \left(\frac{1}{s_1} + \frac{1}{s_2} + \ldots + \frac{1}{s_n}\right).$$

2. Pure water is added, drop by drop, to a vessel of volume V filled with a salt solution of sp. gr. s, which is allowed to overflow; find the sp. gr. of the solution when a volume v of water has been poured in. [$1+(s-1)\,e^{-v/V}$.]

* Huygens, 'De iis quæ liquido supernatant' (1650), *Oeuvres complètes*, vol. II, p. 93.

3. Prove from first principles that in a fluid free from external force the resultant pressure on any plane circular area acts through the centre, and thence that the pressure-intensity is uniform over any plane.

4. Find the pressure in tons per sq. yd. at a depth of 10 fathoms in the sea, assuming that the sp. gr. of sea-water is 1·026. [15·44.]

5. A triangular area is immersed with one side in the surface of a liquid. Divide it into two parts by a horizontal line so that the pressures on the two parts shall be equal.

6. A cylindrical jar is partly filled with water. When a solid is placed in it, floating freely, the level of the water rises m millimetres, and when the solid is pushed down so as to be just immersed, the level rises n millimetres more. Find the specific gravity of the solid.

7. A hollow sphere is just filled with water; prove that the resultant vertical pressures on the upper and lower halves of the internal surface are $\frac{1}{4}W$ and $\frac{5}{4}W$, respectively, where W is the weight of the water.

8. A cylindrical boiler of circular section, with flat ends, whose length is horizontal, is filled with water; find the ratio of the vertical pressures on the upper and lower halves of the curved surface. [·1203.]

9. A rectangular block is completely immersed in water, one set of edges being horizontal. One pair of opposite faces makes an angle a with the horizontal; prove that the resultant of the pressures on these two faces is $W \cos a$, where W is the weight of water displaced by the block. Prove also that this resultant passes through the centre of the block.

10. A circular cylinder, closed at both ends and filled with water, is held with its axis at any given inclination to the vertical. Prove that the resultant pressure on the *curved* surface acts at right angles to the axis, through its middle point.

11. An open vessel containing water has a flat base. A piece of metal hanging by a string is dipped into the water so as to be totally immersed. How is the pressure on the base affected, (1) when the sides of the vessel are vertical, (2) when they are not? If the vessel is suspended from a spring balance, what will be the effect on the balance in each case?

12. A cylindrical boiler having hemispherical ends is filled with water. The diameter is 6 ft., and the total length is 15 ft. Find the total thrust lengthways on each end, (1) when the length is horizontal, and (2) when it is vertical. [(1) 5296 lbs.; (2) 1765 lbs., 24720 lbs.]

13. An open cylindrical vessel of any form of section contains water to a depth h. If it be tilted through an angle θ, find the pressure on the base (area S), supposing it to remain covered. [$\rho h S \cos \theta$.]

14. An open hemispherical shell 6 inches in diameter is filled with water and closed by a glass plate. It is then inverted and placed on a table. Find, in lbs., the least weight of the shell which will prevent it being lifted by the pressure of the contained water. [1·02 lbs.]

15. A cylindrical vessel with a horizontal base of area A contains two liquids of densities ρ, ρ' and depths h, h', the former above the latter. Prove that if the liquids be thoroughly mixed without change of volume the potential energy is increased by

$$\tfrac{1}{2}(\rho'-\rho)\,hh'A.$$

16. Prove that the resultant of a uniform pressure over *any* area S of a spherical surface of radius a is a force

$$p_0 S \,.\, OG/a,$$

along the line OG joining the centre of curvature to the mean centre of the area S.

EXAMPLES. XVIII.

(Centres of Pressure.)

1. If a plane area consist of two portions A_1, A_2, whose centres of pressure are at depths z_1, z_2, whilst their mean centres are at depths h_1, h_2, respectively; prove that the depth of the centre of pressure of the whole is

$$\frac{h_1 z_1 A_1 + h_2 z_2 A_2}{h_1 A_1 + h_2 A_2}.$$

2. A right-angled isosceles triangle is immersed with one of the shorter sides (a) in the surface of a liquid. Find the distance, from this side, of the centre of pressure of each of the two portions into which the triangle is divided by the perpendicular from the right angle to the hypotenuse. [$\tfrac{1}{4}a$, $\tfrac{7}{12}a$.]

Shew their positions accurately on a figure.

3. A trapezium $ABCD$ is immersed with the side AB in the surface of water, and the sides $AD\,(=a)$, $BC\,(=b)$ are vertical. Prove that the vertical line through the centre of pressure divides AB in the ratio

$$a^2+2ab+3b^2 : 3a^2+2ab+b^2,$$

and that the depth of the centre of pressure is

$$\frac{1}{2}\,\frac{(a+b)(a^2+b^2)}{a^2+ab+b^2}.$$

4. A lamina in the form of a regular hexagon is half immersed in liquid, a diagonal being in the surface. Prove that the centre of pressure of the immersed half is at a depth $\tfrac{5}{8}r$, where r is the radius of the inscribed circle.

5. A cube is immersed with one edge in the surface of water, and the opposite edge vertically beneath this. Prove that the distance of the centre of pressure of either of the lower faces from the centre of the face is $\frac{1}{18}a$, where a is the length of an edge.

6. A cube is totally immersed in a liquid with one corner in the surface, and a diagonal vertical; prove that the depths of the centres of pressure of the faces are $\frac{7}{18}$ and $\frac{25}{36}$ of the length of the diagonal.

7. A segment of a parabola cut off by a double ordinate at a distance h from the vertex is immersed with this ordinate in the surface of a liquid. Prove that the distance of the centre of pressure from this ordinate is $\frac{4}{7}h$.

8. A horizontal boiler has a flat bottom, and its ends are plane and semicircular. Prove that if it be just full of water, the depth of the centre of pressure of either end is seven-tenths of the total depth, very nearly.

9. Shew from a consideration of the forces acting on a hemisphere immersed in a liquid that the centre of pressure of a circular area of radius a is at a distance $(\frac{1}{4}a^2/h).\sin\theta$ from the geometrical centre, h being the depth of this latter point, and θ the inclination of the plane of the circle to the horizontal. (See Art. 96, *ad fin.*)

10. If a plane area, wholly immersed, rotate about its mean centre in its own plane, the locus of the centre of pressure relative to the area is an ellipse.

11. If an area of any shape turn in its own plane about a fixed point in the surface of a liquid, being wholly immersed, the locus of the centre of pressure relatively to the area is a straight line.

12. A square plate is just immersed with its plane vertical. If it turn about the upper corner, the locus of the centre of pressure relatively to the area is a straight line cutting the sides at distances of seven-sixths of their length from the corner.

13. An ellipse is just immersed with its plane vertical. Find the locus of the centre of pressure relatively to the ellipse.

CHAPTER XI

EQUILIBRIUM OF FLOATING BODIES

98. Principle of Archimedes. Buoyancy.

We have seen in Art. 96 that when a body is surrounded by a fluid which is in equilibrium under gravity, the pressures exerted on it by the latter are equivalent to a single force equal to the weight of the portion of fluid displaced by the solid, and acting vertically upwards through the centre of gravity of the displaced fluid. It is otherwise evident that if the body were removed, and its place filled by fluid with the same vertical distribution of density as the surrounding medium, the portion of fluid thus introduced would be in equilibrium under its own gravity and the same system of surface pressures as before.

This is known as the 'principle of Archimedes*,' who enunciated it for the case of liquids. As above indicated, the argument applies to fluids of all kinds, and is not restricted to the case of uniform density.

The resultant upward pressure is called the 'buoyancy' of the body; and the centre of gravity of the displaced fluid is called the 'centre of buoyancy.'

Ex. If a body whose true weight is W and density σ be surrounded by fluid of density ρ, its volume is W/σ, and its buoyancy is therefore $W\rho/\sigma$. Hence its 'apparent weight,' i.e. the true weight diminished by the buoyancy, is

$$W' = (1 - \rho/\sigma)\, W. \qquad \qquad (1)$$

Hence

$$\frac{\sigma}{\rho} = \frac{W}{W - W'}. \qquad \qquad (2)$$

* Born B.C. 287, died B.C. 212, at Syracuse. He wrote a treatise on floating bodies.

This is of course a common method of determining the density of solids heavier than water.

In very delicate weighings the buoyancy of the air has to be allowed for. If σ' be the density of the material of the standard weights in the scale-pan, we have, equating the apparent weights,

$$\left(1-\frac{\rho}{\sigma}\right)W=\left(1-\frac{\rho}{\sigma'}\right)W', \quad\text{.........................(3)}$$

where ρ now denotes the density of the air and W, W' are the true weights of the bodies in the scale-pans. Since ρ/σ is usually a very minute fraction, this is practically equivalent to

$$W=\left(1+\frac{\rho}{\sigma}-\frac{\rho}{\sigma'}\right)W'. \quad\text{..............................(4)}$$

Thus if a quantity of water be weighed against platinum weights, we have, putting $\rho=\cdot00129$, $\sigma=1$, $\sigma'=21$,

$$W/W'=1\cdot00123.$$

99. Conditions of Equilibrium of a Floating Body.

When a body is floating freely, its gravity is exactly balanced by the fluid pressures on its surface. Hence the buoyancy must be equal to the weight of the body, and the centre of buoyancy must be in the same vertical with the centre of gravity. These are necessary and sufficient conditions of equilibrium; the question of stability will be discussed presently.

When a body is immersed partly in a liquid and partly in air, as in the ordinary case of a body floating 'on the surface' of water, we ought in strictness to include the weight of the displaced air in estimating the buoyancy, but this is so small as to be practically unimportant, the density of air as compared with water being only about $\frac{1}{775}$. To neglect this is equivalent to assuming that the atmospheric pressure-intensity has the same uniform value over the upper portion of the surface of the body as over the horizontal surface of the liquid.

Ex. 1. To find the limiting ratio of thickness to radius in order that a hollow spherical shell of specific gravity s may float in water.

If a, b be the external and internal radii, the condition is

$$(a^3-b^3)\,s<a^3,$$

since the volume of a sphere varies as the cube of the radius. Hence

$$\frac{b}{a}>\left(1-\frac{1}{s}\right)^{\frac{1}{3}}, \quad \frac{a-b}{a}<1-\left(1-\frac{1}{s}\right)^{\frac{1}{3}}. \quad\text{......................(1)}$$

Thus for iron of sp. gr. 7·8, the thickness must not exceed ·045 of the outer radius.

Ex. 2. A uniform elliptic cylinder floats with its axis horizontal.

The locus of the mean centre of a segment of an ellipse, of constant area, is a similar ellipse; and the line joining this point to the centre bisects the chord. These statements, being obviously true for the circle, can be inferred for the ellipse by the method of parallel projection (Art. 76). The line in question cannot therefore be perpendicular to the chord unless it coincides with a principal axis. Hence there are four distinct positions of equilibrium of the floating cylinder, viz. those in which a principal axis of the elliptic section is vertical. It will be seen later that two of these positions are stable, and the other two unstable.

Ex. 3. If a solid of *uniform* specific gravity s floats in water, a solid of the same size and shape but of uniform specific gravity $1-s$ can float in the inverted position, with the same plane section in the free surface.

Let H, H' be the mean centres of the two portions V, V' into which the volume of the solid is divided by the plane of the free surface; and let G be the centre of gravity of the whole. Then G will lie in the straight line HH', which is therefore vertical in both cases. Also we have

$$V=(V+V')s, \qquad (2)$$

by hypothesis, and therefore

$$sV'=(1-s)V. \qquad (3)$$

Ex. 4. To find the possible positions of equilibrium of a beam of square section, of uniform specific gravity s, floating with its length horizontal.

The result of Ex. 3 shews that we need only consider the cases where $s<\frac{1}{2}$.

We take rectangular axes through the centre of the middle section, parallel to the sides. We begin with the case where the water-line PQ is parallel

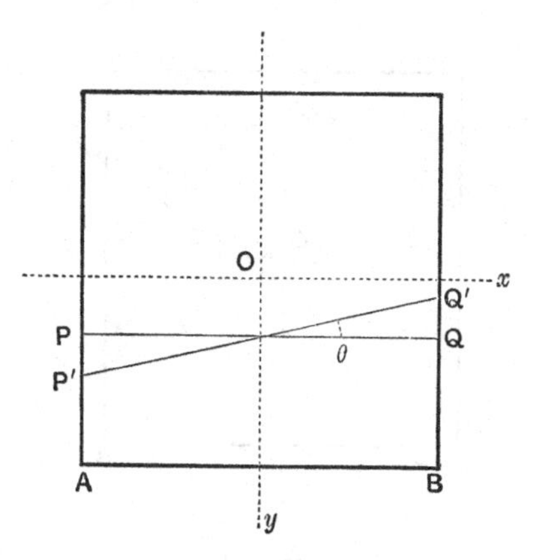

Fig. 131.

to a side. Let $AP = BQ = h$, so that h is the depth immersed. Hence if $2a$ be the length of a side, we have

$$h = 2as. \qquad (4)$$

If the prism be now tilted through an angle θ, so that the water-line assumes the position $P'Q'$ relative to the section, but still intersects two opposite sides, the coordinates of the centre of buoyancy H, i.e. of the mean centre of the trapezium $P'ABQ'$, are found to be

$$\left. \begin{aligned} x &= \frac{1}{3}\frac{a^2}{h}\tan\theta = \frac{1}{6}\frac{a}{s}\tan\theta, \\ y &= a - \frac{1}{2}h - \frac{1}{6}\frac{a^2}{h}\tan^2\theta = a(1-s) - \frac{1}{12}\frac{a}{s}\tan^2\theta. \end{aligned} \right\} \qquad (5)$$

In order that OH may be perpendicular to $P'Q'$, we must have

$$x = y\tan\theta. \qquad (6)$$

This gives the obvious solution $\tan\theta = 0$. The remaining positions of the present type, if any, are given by

$$\tan^2\theta = 12s(1-s) - 2. \qquad (7)$$

Since s is by hypothesis $< \frac{1}{2}$, it appears that the expression on the right hand is positive only if

$$s > \tfrac{1}{2} - \tfrac{1}{6}\sqrt{3}, \text{ or } \cdot 2113.... \qquad (8)$$

We have also the geometrical condition that P' must lie in AP, viz. $\tan\theta \ngtr h/a$, or $2s$. This requires that

$$8s^2 - 6s + 1 > 0, \qquad (9)$$

or $s < \frac{1}{4}$.

We next examine the cases where $\tan\theta > 2s$, but < 1, so that the immersed portion of the section is triangular (Fig. 132).

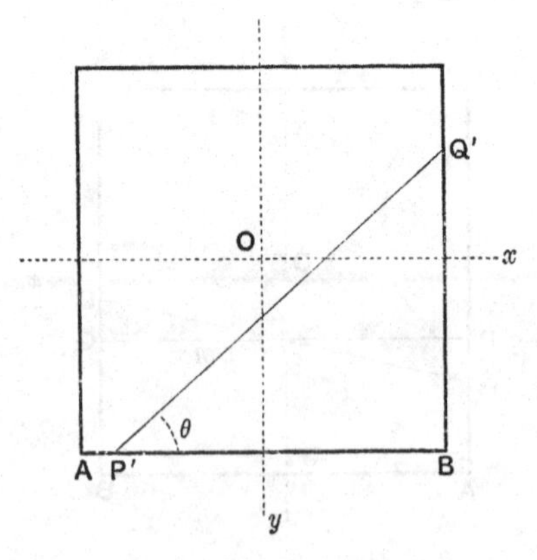

Fig. 132.

We find, in terms of h,

$$P'B=\sqrt{(4ah\cot\theta)},\quad BQ'=\sqrt{(4ah\tan\theta)},\quad \text{.............(10)}$$

or, if we write

$$\tan\theta=z^2,\quad s=\tfrac{9}{8}\omega^2,\quad \text{.............(11)}$$

$$P'B=3\omega z^{-1}a,\quad BQ'=3\omega za.\quad \text{.............(12)}$$

The coordinates of H are then found to be

$$x=a(1-\omega z^{-1}),\quad y=a(1-\omega z).\quad \text{.............(13)}$$

The condition (6) then leads to

$$\omega(z^4-1)=z^3-z.\quad \text{.............(14)}$$

Hence, either $z^2=1$, corresponding to positions in which a diagonal is vertical, or

$$z^2-\frac{z}{\omega}+1=0.\quad \text{.............(15)}$$

The roots of this quadratic are real if $\omega<\frac{1}{2}$, i.e.

$$s<\tfrac{9}{32}\text{ or }\cdot 28125.\quad \text{.............(16)}$$

The further condition that $\tan\theta$ must exceed $2s$ requires that the smaller root of (15) should be greater than $\frac{3}{2}\omega$. It is easily proved that this will be the case if, and only if, $s>\frac{1}{4}$.

To every position which is possible for a given value of s there corresponds an inverted position for a solid having the complementary density $1-s$. Hence, collecting our results, we learn that whilst the symmetrical positions in which a pair of sides, or a diagonal, are vertical are always possible, there are other inclined positions with two angles immersed when s or $1-s$ lies between ·2113... and ·25, and inclined positions with one (or three) angles immersed when s or $1-s$ lies between ·25 and ·28125.

It will appear presently (Art. 107) that the unsymmetrical positions of equilibrium, when they exist, are the only stable ones. When the symmetrical positions are the only ones, one type is stable and the other unstable.

The argument of Art. 96 can evidently be applied to find not only the resultant of the fluid pressures on the whole immersed surface of a solid, but also that of the *vertical* pressures on the strip of the surface included between any two vertical planes. This resultant will be equal to the weight of the fluid displaced by the corresponding portion of the solid. In this way we arrive, in the case of a beam, or other elongated structure, at the notion of the 'buoyancy per unit length.' The formulæ of Art. 27, relating to shearing-stress and bending-moment, will hold in such a case, so far as the horizontal pressures can be left out of account, provided w be now taken to mean the excess of the weight over the buoyancy, both being estimated per unit length. We shall therefore have, in the case of a beam floating freely,

$$\int w\,dx = 0, \quad\text{.........................(17)}$$

if the integral be taken over the whole length. The shearing-stress and bending-moment will not vanish unless $w = 0$ everywhere, i.e. unless the buoyancy and the weight are adjusted to equality over every portion of the length. This matter has an important application in Naval Architecture.

100. Body floating under Constraint.

The application of the principle of Archimedes to bodies floating under partial constraint is simple. Thus in the case of a body free to turn about a fixed horizontal axis, the forces reduce to three, viz. the weight of the body, the buoyancy, and the reaction of the fixed axis. Since the first two of these are vertical, the third must be vertical also, and the conditions of equilibrium are those of three parallel forces. Resolving vertically we find the magnitude of the reaction; and by taking moments we determine the angular coordinate.

Ex. A uniform rod of sp. gr. s is free to turn about its lower end, which is fixed at a depth h in water.

If l be the length of the rod, θ its inclination to the vertical, the weight and the buoyancy are to one another as sl and $h \sec\theta$. Hence, taking moments about the lower end, we have

$$sl \, . \, \tfrac{1}{2} l \sin\theta = h \sec\theta \, . \, \tfrac{1}{2} h \tan\theta. \quad\text{.........................(1)}$$

The solution $\theta = 0$ gives the vertical position of equilibrium. There is also an inclined position determined by

$$\cos\theta = h / l \sqrt{s}, \quad\text{....................................(2)}$$

provided $sl^2 > h^2$. It is easily seen that in this case the vertical position is unstable.

101. Stability of a Floating Body. Vertical Displacements.

The equilibrium of a floating body is stable for vertical displacements. For if the body be depressed, without rotation, through a small space z, the buoyancy is increased by $\rho A z$, where A is the area of the section by the plane of the free surface. Hence a downward force of this amount, acting through the mean centre of the area A, is necessary to maintain the body in its new position. If z be negative, the body is raised, and the requisite force is upward.

In either case, the work required to produce the displacement is $\frac{1}{2}\rho Az.z$, since the mean value of the force is half the final value. The potential energy of the system is therefore increased by $\frac{1}{2}\rho Az^2$.

The equilibrium is evidently neutral for horizontal displacements.

It is assumed, above, that the area of the free surface is large compared with A. If the fluid be contained in a vessel whose area at the free surface is B, the depression z produces an elevation $Az/(B-A)$ of the free surface; the increase of buoyancy is therefore

$$\rho A\left(z+\frac{Az}{B-A}\right)=\frac{\rho ABz}{B-A},$$

and the increment of the potential energy is $\frac{1}{2}\rho ABz^2/(B-A)$.

102. Angular Displacements. Metacentre.

The question of stability for angular displacements is more difficult. We shall consider only cases where the body has a vertical plane of symmetry as regards both geometrical form and distribution of weight, and shall contemplate only small displacements parallel to this plane. Any such displacement may be resolved into a rotation about an arbitrary axis perpendicular to this plane, together with a translation. We shall take the axis to be in the plane of the free surface, and to be so chosen that the volume V of the displaced fluid is unaltered. Since the effect of the translation has been found, we consider now that of the rotation alone.

Let GH be that line in the body which was originally vertical through the centre of gravity G, and so contains the original centre of buoyancy H. Owing to the altered shape of the mass of displaced fluid, the centre of buoyancy is shifted, relatively to the body, to a new position H'. Let the vertical through H' meet HG in M. The weight ρV acting downwards through G, and the buoyancy ρV acting upwards through M, form a couple of moment $\rho V.GM.\sin\theta$, where θ is the angle of rotation. This couple will tend to diminish θ, or to increase it,

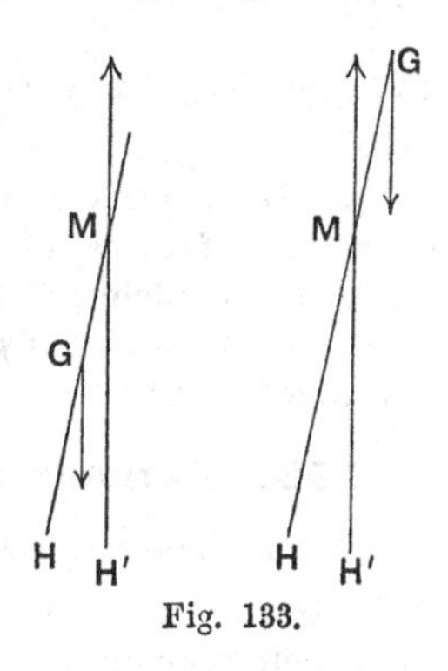

Fig. 133.

according as M is above or below G. In the former case it is called the 'righting moment.' A couple of the same amount, but of the opposite tendency, would be required to maintain the body in its new position. The limiting position of the point M when the angle θ is infinitely small is called the 'metacentre,' and the height (GM in the first figure) of the metacentre above the centre of gravity is called the 'metacentric height.' It is necessary and sufficient for *initial* stability that the metacentric height should be positive. The case of *finite* displacements will be referred to later (Art. 109).

In the case of stability, the work required to turn the body through a small angle θ from the equilibrium position will be equal to θ multiplied by the mean moment of the couple, or $\frac{1}{2}\rho V.GM.\theta^2$. The potential energy of the system composed of the body and the fluid is increased by this amount.

If the immersed surface is spherical, the metacentre coincides with the geometrical centre (O); for the buoyancy, being the resultant of a system of elementary pressures acting in lines through O, must also act through O in all positions of the body.

In the case of a body wholly immersed, e.g. a submarine boat, the metacentre obviously coincides with the centre of buoyancy, and the condition of stability is therefore that the centre of gravity must be below this latter point.

The metacentric height of a ship (for rolling displacements) is found practically by observing the angle through which the ship heels when known weights are shifted across the deck. Thus if a weight W be shifted through a space a, this is equivalent to applying a couple Wa, and we have the equation

$$\rho V.GM.\theta = Wa. \qquad (1)$$

The angle θ is given by the deflection of a long plumb-line suspended from a mast.

Ex. It was found in the case of a ship of 9000 tons displacement that the shifting of a mass of 20 tons laterally through a space of 42 feet caused the bob of a pendulum 20 ft. long to move through 10 in. Here

$$W/\rho V = \tfrac{1}{450}, \quad a = 42, \quad \theta = \tfrac{1}{24},$$

whence $GM = 2{\cdot}24$ ft.

103. Formula for Metacentre.

To calculate the position of the metacentre, we take rectangular axes Ox, Oy in the plane of the water-line section, the axis of x being along the assumed line of symmetry, and that of y

coincident with the axis about which the body is turned. The lower portion of Fig. 134 represents the solid in its equilibrium position, and the lines AOB, $A'OB'$ indicate the original and the displaced positions of the free surface, relative to the solid. The two positions of the centre of buoyancy are denoted by H, H', and HN is drawn perpendicular to AB. We write also, for a moment, $ON = b$, $NH = c$.

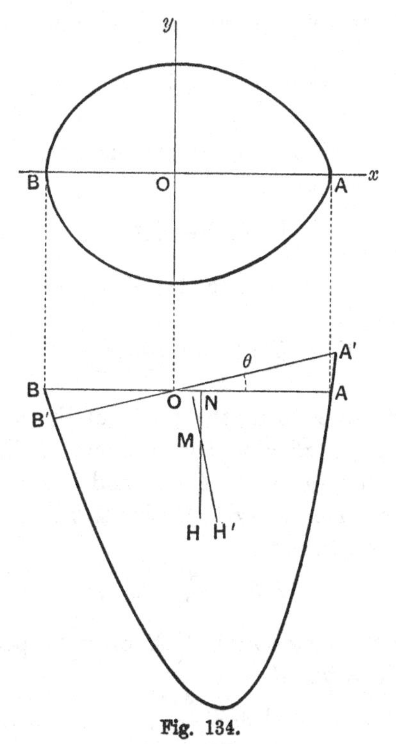

Fig. 134.

If θ be the small angle AOA' through which the body is turned, the difference between the immersed volumes consists of a stratum of variable thickness $x\theta$, positive on one side and negative on the other. This stratum may be considered as made up of prisms of height $x\theta$, standing on bases $\delta x \delta y$, and the total change of volume is therefore

$$\iint x\theta \, dx dy.$$

Since we assume the volume to be unaltered, we must have

$$\iint x\,dx\,dy = 0, \quad \text{.......................(1)}$$

i.e. the origin O must be at the mean centre of the water-line section.

Again, since H' is the mean centre of a volume which is made up of the original immersed volume V, having its mean centre at H, and of the thin stratum referred to, the horizontal projection of HH' is

$$\frac{V.0 + \iint (x-b)\,x\theta\,dx\,dy}{V} = \frac{\theta \iint x^2\,dx\,dy}{V}, \quad \text{........(2)}$$

by (1). If A be the area of the water-line section, κ its radius of gyration about Oy, we have

$$\iint x^2\,dx\,dy = A\kappa^2, \quad \text{.......................(3)}$$

and the result may be written $\theta . A\kappa^2/V$. Similarly, the vertical projection of HH' is

$$\frac{V.0 + \iint (c + \tfrac{1}{2}x\theta)\,x\theta\,dx\,dy}{V}, \quad \text{.................(4)}$$

the factor $c + \frac{1}{2}x\theta$ representing the height of the centre of gravity of an elementary prism $x\theta\,\delta x\,\delta y$ above the level of H. In virtue of (1) and (3), this reduces to $\frac{1}{2}\theta^2 . A\kappa^2/V$, and is accordingly of the second order in θ. Hence HH' is ultimately parallel to AB, and its length is, to the first order,

$$HH' = \frac{A\kappa^2}{V} . \theta. \quad \text{.......................(5)}$$

But if M be the metacentre, $H'M$ must be perpendicular to $A'B'$; hence $HH' = HM . \theta$, and

$$HM = \frac{A\kappa^2}{V}, \quad \text{..........................(6)}$$

which is the formula required.

If we write $V/A = h$, h may be called the 'mean depth' of immersion; in terms of it we have

$$HM = \kappa^2/h. \quad \text{..........................(7)}$$

In the particular case of a solid of revolution floating with its axis vertical we have $\kappa^2 = \frac{1}{4}a^2$, where a is the radius of the circular water-line section, and therefore

$$HM = \tfrac{1}{4}a^2/h. \quad \text{.......................(8)}$$

Ex. 1. A square slab, of uniform density, of side a and thickness h, floats with its square faces horizontal.

If O be the centre of the lower face, and x the depth immersed, we have

$$OH = \tfrac{1}{2}x, \quad HM = \tfrac{1}{12}a^2/x, \quad OG = \tfrac{1}{2}h. \quad \text{.................(9)}$$

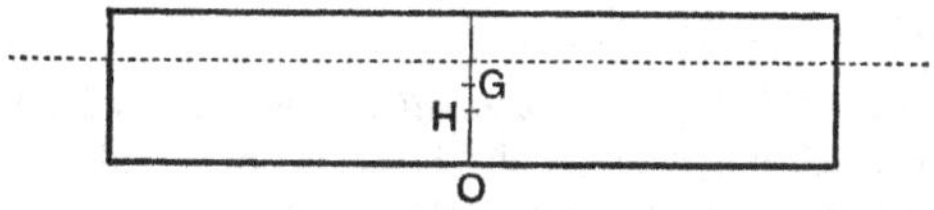

Fig. 135.

Hence for stability we must have

$$\frac{1}{2}x + \frac{1}{12}\frac{a^2}{x} > \frac{1}{2}h, \quad \text{............................(10)}$$

or

$$\frac{h}{a} < \frac{x}{a} + \frac{1}{6}\frac{a}{x}. \quad \text{...............................(11)}$$

The right-hand side is least when $x/a = \frac{1}{6}a/x$, and the above position is therefore stable for all depths of immersion provided

$$h/a < \tfrac{1}{3}\sqrt{6} \text{ or } \cdot 8165. \quad \text{..........................(12)}$$

If $h = a$, so that the solid is a cube, the condition of stability is

$$\frac{x^2}{a^2} - \frac{x}{a} + \frac{1}{6} > 0. \quad \text{...............................(13)}$$

The factors of the left-hand side are $x/a - \alpha$, $x/a - \beta$, where

$$\left.\begin{matrix}\alpha \\ \beta\end{matrix}\right\} = \tfrac{1}{2} \pm \tfrac{1}{6}\sqrt{3} = \left\{\begin{matrix}\cdot 211 \\ \cdot 789\end{matrix}\right. .$$

Hence the specific gravity of the solid, which is equal to x/a, must either be less than the smaller, or exceed the greater, of these two numbers. For intermediate densities the position in question is unstable. Cf. Art. 107, *ad fin.*

Ex. 2. We have seen (Art. 99, Ex. 3) that if a solid of uniform specific gravity s float in a certain position, then a solid of the same shape, but of specific gravity $1 - s$, can float in the *inverted* position. We can now prove that both positions will be stable, or both unstable*.

Let V_1, V_2 be the volumes which are below and above the water-line, respectively, in the first position; let H_1, H_2 be the mean centres of these two volumes, G that of the whole. Then

$$V_1 . H_1G = V_2 . H_2G = \frac{V_1 V_2}{V_1 + V_2} . H_1H_2. \quad \text{.................(14)}$$

Also, if M_1, M_2 be the two metacentres,

$$V_1 . H_1M_1 = V_2 . H_2M_2 = A\kappa^2. \quad \text{.......................(15)}$$

Hence both cases will be stable, or both unstable, according as

$$(V_1 + V_2)\, A\kappa^2 \gtrless V_1 V_2 . H_1H_2. \quad \text{.....................(16)}$$

* This theorem is due to the late Prof. F. Elgar.

104. Factor of Stability. Variation with Draught.

The degree of stability of the floating body is measured by the quantity $\rho V.GM$, which is the initial righting moment per unit angle of inclination. This will of course vary with the extent to which the body is immersed.

Ex. To illustrate the effect of a *small* increase of load on the stability we will suppose for simplicity that the body has a second vertical plane of symmetry at right angles to the former. This will contain the points H, G, and M, and we will suppose that the additional load ρv is applied at a point P in the line HG. We will further assume that the surface of the body near the water-line is vertical, as in the case of a 'wall-sided' ship, so that the quadratic moment $A\kappa^2$ is unaltered by a small vertical displacement. We denote the new positions of the above-named points by H', G', M'. Then, C being any convenient point of reference in the line HG, below H, we have, if O be the point where HG meets the water-line,

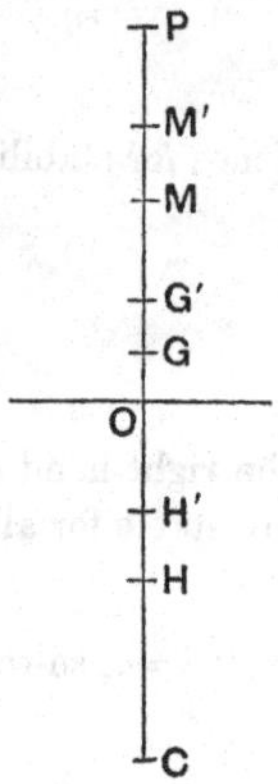

$$\left.\begin{aligned} CH' &= \frac{V.CH+v.CO}{V+v}, \quad CG' = \frac{V.CG+v.CP}{V+v}, \\ H'M' &= \frac{A\kappa^2}{V+v}. \end{aligned}\right\} \qquad \text{......(1)}$$

Fig. 136.

Hence

$$G'M' = CH' + H'M' - CG' = \frac{A\kappa^2 - V.HG - v.OP}{V+v}$$

$$= \frac{V.GM - v.OP}{V+v}, \qquad \text{......(2)}$$

or

$$(V+v)\,G'M' - V.GM = -v.OP. \qquad \text{......(3)}$$

The stability is therefore increased or diminished, according as the added load is below or above the water-line. It will be noticed that the calculation does not really require v to be small, provided P denotes the centre of gravity of the load, and O the point midway between the two water-plane sections.

If the surface near the water-line is not vertical, the formula (3) is replaced by

$$(V+v)\,G'M' - V.GM = -v.OP + A'\kappa'^2 - A\kappa^2. \qquad \text{......(4)}$$

105. Resultant Pressure of a Liquid in a Tank.

The formula (6) of Art. 103 may also be applied to find the line of action of the resultant pressure of a liquid contained in a vessel of any shape, when the latter is slightly tilted. The only difference is that the forces with which we are now concerned act downwards instead of upwards.

Ex. Take the case of a cylindrical tank of circular section, the axis being vertical. If in consequence of a small inclination the centre of gravity of the fluid is shifted from H to H', and if the vertical through H' meet the axis in M, we have

$$HM = A\kappa^2/V = \tfrac{1}{4}a^2/h, \quad \text{...........(1)}$$

where a denotes the radius of the cylinder, h the depth of the fluid.

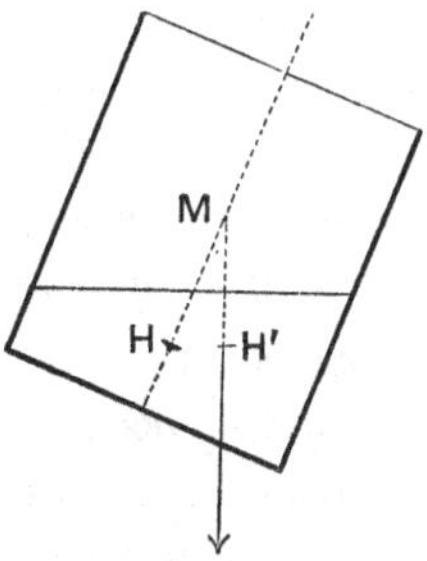

For example, if the tank be pivoted about a horizontal axis through its own centre of gravity, which is at a height c (say) above the base, we find that the upright position will be stable only if

$$\tfrac{1}{2}h + \tfrac{1}{4}a^2/h < c. \quad \text{..................(2)}$$

Fig. 137.

The position is therefore unstable for very small as well as very large values of h. The critical values of h at which the change from instability to stability, or *vice versâ*, takes place are given by

$$h^2 - 2ch + \tfrac{1}{2}a^2 = 0, \quad \text{................................(3)}$$

or

$$h = c \mp \sqrt{(c^2 - \tfrac{1}{2}a^2)}. \quad \text{..............................(4)}$$

If $c^2 < \frac{1}{2}a^2$, there is instability for all depths.

106. Curves of Floatation and Buoyancy. Dupin's Theorems.

A more general view of the problem of the equilibrium and stability of a floating body is afforded by some theorems due to Dupin*. We shall consider these mainly in their two-dimensional forms, the displacements considered being parallel to a vertical plane of symmetry.

A variable plane which cuts off a constant volume from a given solid envelopes a certain surface. In the present application the plane in question is that of the water surface in the various positions which the body may occupy, and is therefore called the 'plane of floatation.' We have seen in Art. 103 that the mean centre O of the water-line section is the point of ultimate intersection of consecutive planes of floatation; it is accordingly the point of contact of the plane of floatation with its envelope. In our special case we are mainly concerned with the section of the surface by the vertical plane of symmetry; this is called the 'curve of floatation.'

Again, the centre of buoyancy (H) has a certain locus in the

* Ch. Dupin (1784–1873), 'De la stabilité des corps flottants' (1814).

body; this is called the 'surface of buoyancy.' In our case, its section by the vertical plane of symmetry is called the 'curve of buoyancy.' We have seen in Art. 103 that the line joining two consecutive positions H, H' of the centre of buoyancy is ultimately parallel to the line of floatation; i.e. the tangent to the curve of buoyancy at any point is parallel to the corresponding position of the line of floatation.

Since, for equilibrium, the line joining the centre of gravity (G) of the body to the centre of buoyancy must be vertical, the problem of determining the possible positions of equilibrium, stable or unstable, of a solid floating in a liquid of given density reduces to that of drawing normals from a given point (G) to the curve of buoyancy. It appears also that if H be the foot of one of these normals, the corresponding metacentre, being the point of intersection of consecutive normals, is the centre of curvature of the curve of buoyancy at H.

Ex. 1. In the case of a cylinder of elliptic section, with its length horizontal, the curve of floatation and the curve of buoyancy are similar and similarly situated ellipses. Hence if the cylinder be of uniform density the normal from the centre must coincide with one or other of the principal axes. (Cf. Art. 99, Ex. 2.)

Ex. 2. In the case of a rectangular parallelepiped with one set of edges horizontal, the line of floatation, so long as it meets two parallel faces, passes through a fixed point, which is accordingly the degenerate form of the curve of floatation. The corresponding part of the curve of buoyancy is a parabola; thus with the notation of Art. 99 (5) we have

$$y = a - \tfrac{1}{2}h - \tfrac{3}{2}hx^2/a^2. \qquad \text{.............................(1)}$$

From any point on the axis of a parabola three real normals or one can be drawn, according as the point does or does not lie beyond the centre of curvature at the vertex.

Ex. 3. Again, if the section of the immersed portion of the solid by the plane of symmetry consists of two straight lines making an angle, the envelope of the line of floatation, which forms with these a triangle of constant area, is a hyperbola having the lines in question as asymptotes. The curve of buoyancy is a hyperbola with the same asymptotes, of two-thirds the linear dimensions. The problem of drawing normals to a hyperbola from an arbitrary point involves in general the solution of a biquadratic equation, which however reduces when the point is on a principal axis. In the particular case of Fig. 132, we have, with our previous notation,

$$(x-a)(y-a) = \omega^2 a^2, \qquad \text{.............................(2)}$$

as the equation of the curve of buoyancy.

Again, it was shewn in Art. 103 that the projection of HH' on the upward normal at H is $\frac{1}{2}\theta^2.A\kappa^2/V$, which is always positive. The curve of buoyancy is therefore always concave upwards. Moreover, since the normals at H, H' are inclined at an angle θ, the same projection must be equal to $\frac{1}{2}R\theta^2$, where R is the radius of curvature. Hence

$$R = \frac{A\kappa^2}{V}, \quad \text{..........................(3)}$$

which shews (again) that the metacentre coincides with the centre of curvature.

107. The various Positions of Equilibrium of a Floating Body. Stability.

We retain the supposition of a vertical plane of symmetry. If G be any point on the normal at a point H of a curve, H will be a point of maximum or minimum distance from G according as G does or does not lie beyond the centre of curvature (M). Hence, comparing with the rule of Art. 102, we learn that the positions of equilibrium of a floating body, which correspond as we have seen to the points on the curve of buoyancy whose distances from the centre of gravity G are stationary, will be stable or unstable according as the distance from G is a minimum or a maximum, respectively.

Since the curve of buoyancy* is necessarily closed there is (if we exclude the case of a circle with G as centre) always one absolute minimum and one absolute maximum distance from G. Hence there is at least one stable and one unstable position of equilibrium. In any case the stable and unstable positions will alternate.

It will be noticed that the conditions of equilibrium and stability are exactly the same as for a cylinder, whose section has the form of the curve of buoyancy, which is free to roll, with its length horizontal, on a horizontal plane, provided the centre of gravity has the same position relative to this curve (Art. 58).

* It does not necessarily consist, of course, of one analytical curve. Thus in the case of a log of rectangular section it is made up of arcs of parabolas and rectangular hyperbolas.

Ex. To ascertain the nature of the various positions of equilibrium of a log of square section floating with its length horizontal. For reasons given it is sufficient to consider cases where the specific gravity (s) is less than $\frac{1}{2}$.

Denote by a^2u the square of the distance GH. Then in the case where the line of floatation meets two opposite sides as in Fig. 131, we have, by Art. 99 (5),

$$u = \frac{x^2+y^2}{a^2} = \frac{t^2}{36s^2} + \left\{(1-s) - \frac{t^2}{12s}\right\}^2, \quad \text{.....................(1)}$$

where t is written for $\tan\theta$. Hence

$$\frac{du}{dt} = \frac{t}{36s^2}\{t^2 - 12s(1-s) + 2\}, \quad \text{.....................(2)}$$

$$\frac{d^2u}{dt^2} = \frac{1}{36s^2}\{3t^2 - 12s(1-s) + 2\}. \quad \text{.....................(3)}$$

The value of du/dt changes sign for $t=0$, and also for

$$t^2 = 12s(1-s) - 2, \quad \text{.....................(4)}$$

in agreement with Art. 99 (7).

In the case $t=0$, u is a minimum only if

$$s^2 - s + \tfrac{1}{6} > 0\,; \quad \text{.....................(5)}$$

i.e. s must be less than ·2113. This gives the range for which the symmetrical position is stable. The values of t given by (4) are then imaginary.

When the symmetrical position becomes unstable, the values of t given by (4) become real; and substituting in (3) we find that they make d^2u/dt^2 positive. Hence the unsymmetrical positions of the present type, when they exist, are stable.

When one edge only is immersed, as in Fig. 132, we have from Art. 99 (13)

$$u = \frac{x^2+y^2}{a^2} = \left(1 - \frac{\omega}{z}\right)^2 + (1-\omega z)^2, \quad \text{.....................(6)}$$

where z is written for $\sqrt{(\tan\theta)}$, and ω for $\sqrt{(\frac{8}{9}s)}$. Hence

$$\frac{du}{dz} = \frac{2\omega^2}{z^3}(z^2-1)\left(z^2 - \frac{z}{\omega} + 1\right), \quad \text{.....................(7)}$$

$$\frac{d^2u}{dz^2} = \frac{2\omega^2}{z^4}\left(z^4 - \frac{2z}{\omega} + 3\right). \quad \text{.....................(8)}$$

The unsymmetrical positions, determined by

$$z^2 - \frac{z}{\omega} + 1 = 0, \quad \text{.....................(9)}$$

make d^2u/dz^2 positive, and are therefore stable when they exist, i.e. when s lies between $\frac{1}{4}$ and $\frac{9}{32}$. The symmetrical position with a diagonal vertical, corresponding to $z=1$, is stable if $\omega > \frac{1}{2}$, or $s > \frac{9}{32}$*.

* It has been thought worth while to give the complete solution of one problem of this kind, notwithstanding the verdict which Huygens (*l.c. ante*) pronounces on his own researches on the subject: 'In his autem utilitas nulla, vel perquam exigua!' The somewhat intricate solution for the case of a log of *any* rectangular section, given in part by Huygens, has been completed by Prof. D. J. Korteweg as editor of the tenth volume of Huygens' *Oeuvres Complètes*.

108. Curvature of Curves of Floatation and Buoyancy.

A formula for the curvature of the curve of floatation was given by Dupin.

The plane of the paper in Fig. 138 is supposed to be a plane of floatation, and the full curve represents the contour of the section, the origin being taken at the mean centre O, and the axis of x along the line of symmetry. A consecutive plane of floatation, parallel to Oy, will meet the surface of the solid in a slightly different curve; this is represented by its orthogonal projection (shewn by the dotted line) on the plane of the figure. Let O' be the projection of the mean centre of the section made by this consecutive plane; by a known theorem (Art. 76) O' will also be the mean centre of the projected area.

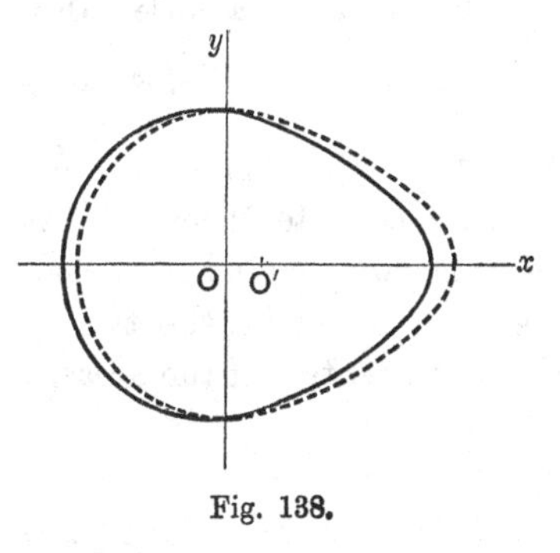

Fig. 138.

Taking moments about Oy and denoting the area of the section by A, we see that $A.OO'$ will be the moment of the *difference* of the areas enclosed by the continuous and the dotted lines. Now if δs be an element of the original contour, ν the normal distance between the two curves, reckoned positive when the dotted curve is external to the other, the element of area included between the two curves will be $\nu\delta s$, whence

$$A.OO' = \int x\nu\, ds, \quad \text{.........................(1)}$$

taken round the contour. If θ denote as usual the small angle between the planes, the perpendicular distance between these at any point will be $x\theta$, so that $\nu = x\theta \tan\psi$, where ψ is the inclination of the surface of the solid (i.e. of its tangent plane) to the vertical. Hence

$$A.OO' = \theta \int x^2 \tan\psi\, ds. \quad \text{..................(2)}$$

But if R_1 be the radius of curvature of the curve of floatation we have $R_1 = OO'/\theta$, ultimately, since θ is the angle between the normals at the two points of contact. Hence

$$R_1 = \frac{\int x^2 \tan\psi\, ds}{A}. \quad \text{.........................(3)}$$

Ex. 1. In the case of a wall-sided vessel we have, in the symmetrical position $\psi = 0$, and therefore $R_1 = 0$; cf. Art. 106, Ex. 2.

Ex. 2. In the case of a solid of revolution with its axis vertical, the water-line section is a circle. Denoting its radius by r, we have

$$\int x^2 ds = 2\pi r \,.\, \tfrac{1}{2} r^2, \quad A = \pi r^2,$$

whence

$$R_1 = r \tan\psi. \quad \text{..................................(4)}$$

In a cognate theorem*, we compare the quadratic moments ($A\kappa^2$) of two *parallel* sections at a short distance δz apart. These are represented by the two curves in Fig. 139, where O, O' are the mean centres of the areas, which we denote by A and $A + \delta A$.

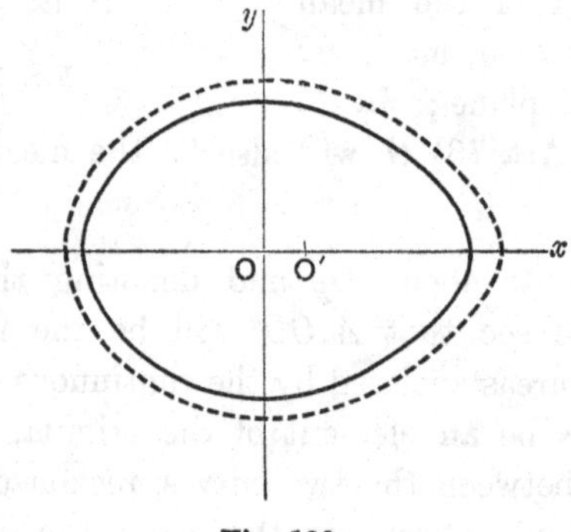

Fig. 139.

If the corresponding moments of inertia, about Oy and a parallel line through O', respectively, be denoted by I and $I + \delta I$, we have, taking quadratic moments with respect to Oy,

$$I + \delta I + (A + \delta A)\, OO'^2 = I + \int x^2 \nu\, ds,$$

or, neglecting the terms of the second order,

$$\delta I = \int x^2 \nu\, ds. \quad \text{.........................(5)}$$

* Due to E. Leclert (1870).

In the present case we have $\nu = \delta z \tan \psi$, whence

$$\frac{dI}{dz} = \int x^2 \tan \psi \, ds. \qquad \text{.....................(6)}$$

Comparing with (3), we have

$$R_1 = \frac{1}{A} \frac{dI}{dz}, \qquad \text{...........................(7)}$$

or, putting $A\delta z = \delta V$, where δV is the change in the volume of immersion,

$$R_1 = \frac{dI}{dV}. \qquad \text{...........................(8)}$$

In the present notation, the expression for the radius of curvature of the curve of buoyancy is, by Art. 106 (3),

$$R = \frac{I}{V}. \qquad \text{...........................(9)}$$

Hence
$$R_1 - R = V \frac{d}{dV}\left(\frac{I}{V}\right) = V \frac{dR}{dV}. \qquad \text{............(10)}$$

109. Stability for Finite Displacements. Metacentric Evolute.

When the body is turned through a *finite* angle, the normal to the curve of buoyancy at the new centre of buoyancy will intersect the line HG in a point M distinct from the metacentre, which we now denote by M_0. The relations are best understood from a consideration of the evolute, to which all the normals are tangents.

If no further condition be imposed, the point on the evolute which corresponds to the equilibrium position will not be a singular point, and M will be above M_0 when the angular displacement is in one sense, and below M_0 when it is in the other.

If however, as in the case of a ship, there is a vertical plane of symmetry at right angles to the plane of the displacement, the curve of buoyancy is symmetrical, and the curvature at the point corresponding to the equilibrium position is a maximum or a minimum. The evolute has then a cusp. In the case of a wall-sided ship, where the curve of buoyancy is parabolic, the cusp points downwards as in Fig. 140, and M is above M_0 for a displacement towards either side.

In any case the righting moment is $\rho V.GM.\sin\theta$, where θ is the angle of displacement. If the cusp of the evolute points *upwards*, GM decreases as the angle in question increases; if it diminishes to zero and changes sign, the stability is of course lost.

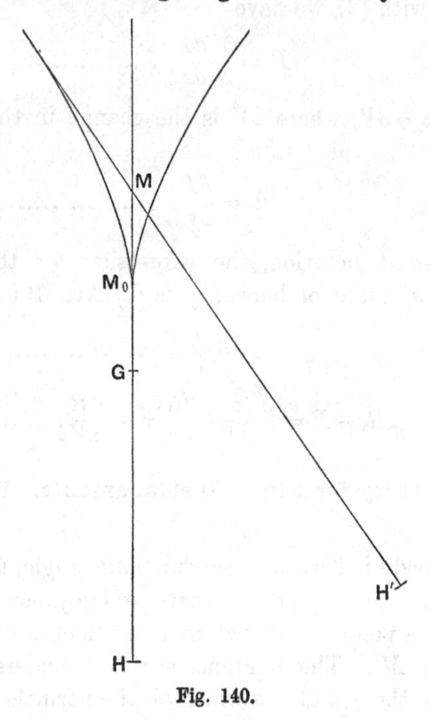

Fig. 140.

110. Energy of a Floating Body.

The conditions of equilibrium and of stability of a floating body may also be deduced from the principle of energy.

For simplicity we will suppose that the area of the water-surface is unlimited, so that no change in its level is produced by any change in the immersion of the solid. The potential energy of the solid alone may be taken as equal to $-W\bar{z}$, where W is its weight, and $\bar{z}$ the depth of its centre of gravity G below the water-surface. If the solid displace a volume V of water, this has the effect that a weight ρV of water whose centre of gravity is at H (the centre of buoyancy) is removed and spread as an infinitely thin film over the plane surface. This involves a gain of potential energy of amount $\rho V . \zeta$, where ζ is the depth of H below the water-surface. The total potential energy of the system may therefore be taken as equal to

$$\rho V\zeta - W\bar{z}. \quad\text{.........................}(1)$$

For equilibrium this must be stationary, and for stability it must be a minimum.

Now when the solid is depressed, without rotation, through a space δz, the volume of displaced fluid is increased by a stratum $A\,\delta z$, whose centre of gravity is at a depth $\frac{1}{2}\delta z$. Since $\delta\bar{z} = \delta z$, the increase of potential energy is

$$\rho V\delta z + \tfrac{1}{2}\rho A(\delta z)^2 - W\delta z, \quad\text{................}(2)$$

and in order that this may vanish to the first order we must have

$$\rho V = W, \quad\text{.........................}(3)$$

in accordance with the principle of Archimedes. The fact that the remaining term in (2) is positive shews that for vertical displacements the equilibrium is stable.

When the condition (3) is fulfilled, the expression (1) for the potential energy takes the form

$$W(\zeta - \bar{z}), \quad\text{.........................}(4)$$

i.e. it is proportional to the difference of level between the centre of gravity G and the centre of buoyancy H. Hence for stability this difference must be a minimum*. It remains to shew that this is equivalent to Dupin's criterion (Art. 107).

If we take G as origin, GH is the radius vector (r) of the curve of buoyancy, and the difference of level between G and H is the perpendicular p from G on the tangent at H. We are therefore concerned with the minimum value of p. Now if R be the radius of curvature of the curve of buoyancy we have

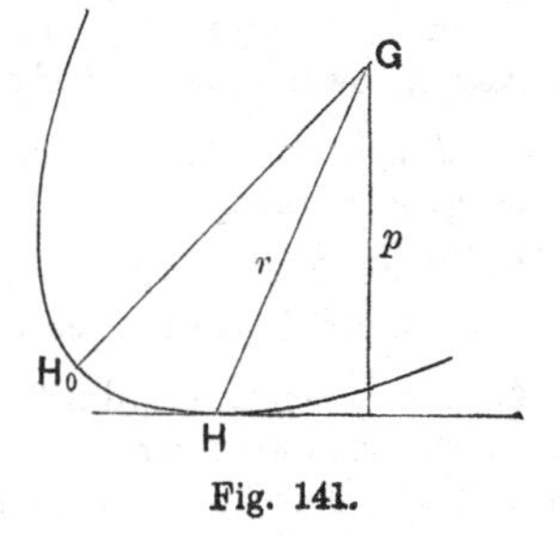

Fig. 141.

$$R = r\frac{dr}{dp}, \quad\text{.........}(5)$$

* Huygens, *l.c.*

by a known formula of the Differential Calculus. Writing this in the form

$$\delta r = \frac{R}{r}\,\delta p, \quad \text{.........................(6)}$$

we see that δp vanishes with δr, i.e. the positions of equilibrium are those for which GH is normal to the curve of buoyancy. Also, denoting by r_0 (or p_0) a stationary value of r, we have, for neighbouring values,

$$r - r_0 = \frac{dr}{dp}(p - p_0), \quad \text{...................(7)}$$

or

$$r - r_0 = \frac{R}{r_0}(p - p_0), \quad \text{...................(8)}$$

approximately, so that $p - p_0$ has the same sign as $r - r_0$. Hence the maxima and minima of r occur simultaneously with those of p. The condition of stability therefore agrees with that formulated by Dupin.

The expression (4) now takes the form Wp. Hence the work required to heel a vessel through a given angle is

$$W(p - p_0). \quad \text{.........................(9)}$$

This is called the 'dynamical stability' corresponding to the given position. It is sometimes represented graphically by a curve with the angle of heel as abscissa. It is easily seen that the gradient of this curve will be proportional to the righting moment.

EXAMPLES. XIX.

1. A hollow vessel floats in a basin; if owing to a leak water flows into the vessel, how will the level of the water in the basin be affected?

2. A sphere floats with one-fourth of its surface above water; find its mean specific gravity. [·844.]

3. Prove that (apart from stability) a floating log of square section, of sp. gr. ·75, can float with one edge in the surface of the water.

4. A rectangular board $ABCD$ floats with the diagonal AC in the surface of water, the lowest corner B being attached to the bottom by a string. Prove that unless the rectangle be a square this position is only possible if the sp. gr. of the board be $\frac{1}{3}$.

5. A uniform rod of length l can turn freely about its upper end, which is fixed at a height $h\ (<l)$ above the surface of water. Prove that, if s be the sp. gr., the vertical position is unstable if

$$h/l < \sqrt{(1-s)};$$

and find the inclined position of equilibrium.

6. A rectangular block of wood floats immersed to a depth of 6 in, with one pair of faces horizontal, these faces being squares of one foot in the side; find the height of the metacentre above the base. [5 in.]

7. Prove that the equilibrium of a solid of uniform density floating with an edge or a corner just emerging is unstable.

8. Prove that if a solid of uniform density float with a flat face just above water, the equilibrium is stable.

9. A prism whose section is a right-angled isosceles triangle floats with its length horizontal and its equal sides inclined at 45° to the horizontal; and the edge where these faces meet is above water. Prove that this position is stable if the sp. gr. is less than $\frac{3}{4}$.

10. A uniform solid octahedron floats in a liquid of twice its own density. Prove that the position in which a diagonal is vertical is stable, whilst that in which two edges are vertical is unstable.

11. Prove that a circular cylinder floating with its axis horizontal will be in stable equilibrium if its length exceed the breadth of the water-line section.

12. A regular tetrahedron floats with one face horizontal and above water. Prove that this position is stable if the sp. gr. of the solid exceeds ·512.

13. A uniform solid circular cylinder of radius a and height h can float in stable equilibrium, with its axis vertical, if $h/2a < {\cdot}707$.

If the ratio $h/2a$ exceed this value, prove that the equilibrium will be stable only if the specific gravity lies outside the limits

$$\frac{1}{2}\left\{1 \mp \sqrt{\left(1 - 2\frac{a^2}{h^2}\right)}\right\}.$$

14. An elliptic cylinder floats with its length horizontal; prove that there are four positions of equilibrium, or two, according to the position of the centre of gravity, which is supposed not to be on the axis.

Examine the stability of the different positions.

15. A thin hollow cylinder of radius a and height h is open at both ends. Prove that it cannot float upright if its sp. gr. lies between the values

$$\frac{1}{2} \mp \sqrt{\left(\frac{1}{4} - \frac{a^2}{h^2}\right)},$$

it being assumed that $h > 2a$. Explain the case of $h < 2a$.

16. Prove that the cylinder of the preceding Ex. can float with its axis horizontal provided

$$\frac{h}{2a} > \sqrt{3} \sin s\pi,$$

where s is the sp. gr. of the material.

17. Prove that any segment of a uniform sphere, of substance lighter than water, can float in stable equilibrium with its plane surface horizontal and immersed.

18. A thin vessel in the form of a solid of revolution can turn freely about a horizontal axis cutting the axis of the vessel at right angles in the point O. If the equilibrium of the vessel when empty be neutral, prove that if a little water be poured in, the position in which the axis is vertical will be stable or unstable according as the centre of curvature at the vertex is below or above O.

19. A cylindrical can of negligible thickness floats in water. If the upright position is unstable when the can is empty, it can be made stable by pouring in water to a depth equal to the distance between the centre of gravity of the can and the original centre of buoyancy.

20. A vessel carries a tank of oil, of specific gravity s, amidships. Prove that the effect of the fluidity of the oil on the rolling of the vessel is equivalent to a diminution of the metacentric height of amount $A\kappa^2 s/V$, where V is the displacement of the ship, and $A\kappa^2$ the moment of inertia of the surface-area of the tank.

In what ratio will the effect be diminished by inserting a longitudinal partition in the tank?

21. A cylindrical solid of any form of section, having a longitudinal plane of symmetry, is immersed, with this plane vertical, at various inclinations. Prove that the curve of buoyancy is a parabola.

22. Prove that if the metacentre of a solid of revolution, whose axis is vertical, is a fixed point in the body, the immersed surface must be spherical.

23. Find the form of a surface of revolution in order that the height of the metacentre above the centre of buoyancy may be constant for all depths of immersion (with the axis vertical).

24. Apply the formula $HM = A\kappa^2/V$ to shew that in the case of a solid of revolution just dipping into water, with the axis vertical, the metacentre coincides with the centre of curvature at the vertex.

25. A thin cylindrical vessel of sectional area A floats upright, being immersed to a depth h, and contains water to a depth k. Find the work required to pump out the water. [$\rho Ak(h-k)$.]

26. A sphere of radius a is just immersed in water which is contained in a cylindrical vessel whose axis is vertical and radius R. Prove that if the sphere is raised just clear of the water, the loss of potential energy of the water is

$$Wa\left(1-\frac{2}{3}\frac{a^2}{R^2}\right),$$

where W is the weight of water originally displaced by the sphere.

27. A sphere of radius a, weight W, and density s, rests on the bottom of a cylindrical vessel of radius R, which contains water to a depth $h\ (>2a)$. Prove that the work required to lift the sphere out of the vessel is less than if the water had been absent, by the amount

$$\left(h-a-\frac{2}{3}\frac{a^3}{R^2}\right)\frac{W}{s}.$$

28. A solid sphere of weight W and radius a is held just above the surface of a large sheet of water. If it be depressed through a space $h\ (<2a)$ prove that the nett loss of potential energy is

$$Wh-\tfrac{1}{3}\pi\rho h^3\left(a-\tfrac{1}{4}h\right);$$

and deduce the condition of equilibrium when the sphere floats freely.

If h now refer to this position, shew that the gain of potential energy when the sphere is further lowered or raised through a *small* space z is

$$\pi\rho h\left(a-\tfrac{1}{2}h\right)z^2,$$

approximately.

29. A solid floats in equilibrium in a liquid, wholly immersed. Prove that if it be slightly compressible its position is unstable.

Prove that if the liquid be compressible, but the solid incompressible, the equilibrium is stable.

30. If the surface of a liquid be disturbed so as to be no longer plane, prove that the potential energy is increased by

$$\tfrac{1}{2}\rho\iint\zeta^2\,dx\,dy,$$

where ζ is the elevation above the undisturbed level, the axes of x, y being horizontal, and the integration extending over the disturbed area.

If the disturbed surface has the form of a train of waves

$$\zeta=c\sin kx,$$

prove that the mean potential energy per unit area is the same as if a plane stratum of thickness $\tfrac{1}{2}c$ had been raised through a height $\tfrac{1}{2}c$.

CHAPTER XII

GENERAL CONDITIONS OF EQUILIBRIUM OF A FLUID

111. General Formula for Pressure-Gradient.

We now contemplate the case of a fluid placed in a field of force of a more general character than that due to ordinary gravity.

Let δs be a linear element PQ drawn in any direction, and let F be the component of the force exerted by the field, per unit mass, in the direction PQ. Imagine a cylinder to be constructed having its length along PQ, and a cross-section ω whose dimensions are small compared with δs. The difference of the pressures on the two ends will be

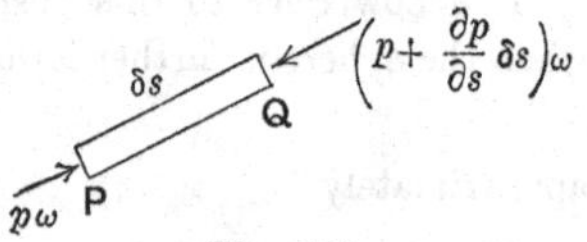

Fig. 142.

$$p\omega - \left(p + \frac{\partial p}{\partial s}\delta s\right)\omega = -\frac{\partial p}{\partial s}\omega\,\delta s,$$

where $\partial p/\partial s$ is of the nature of a partial differential coefficient expressing the rate of variation of the pressure-intensity in the direction PQ. Again, if ρ be the density, the component force of the field on the fluid bounded by the cylinder is $F\rho\omega\delta s$. Since the total force in the direction PQ must vanish, we have

$$\frac{\partial p}{\partial s} = \rho F. \qquad (1)$$

Since ρ is the mass per unit volume, and F the force per unit mass, this equation expresses that the gradient of p in any assigned direction is equal to the component force *per unit volume* in that direction. The space-variation of p is therefore most rapid in the direction of the resultant force of the field.

The formula (1) is independent of the unit of force adopted, provided this be the same for p as for F. In some questions it is convenient to adopt dynamical units*.

112. Surfaces of Equal Pressure. Equipotential Surfaces.

We may imagine a series of surfaces to be drawn through the fluid over each of which p is constant. If the element δs be taken along one of these, we have $\partial p/\partial s = 0$, and therefore $F = 0$. That is, the surfaces of equal pressure must be everywhere at right angles to the direction of the resultant force. Hence a fluid cannot be in equilibrium under an arbitrarily assigned distribution of force, unless this distribution possesses the geometrical property that the lines of force are orthogonal to a system of surfaces.

If the given field of force be 'conservative' (Art. 49), and if V denote the potential energy of unit mass, considered as a function of position in the field, we have

$$F.PQ = V_P - V_Q, \qquad \text{.......................(1)}$$

or

$$F = -\frac{\partial V}{\partial s}, \qquad \text{.......................(2)}$$

and therefore

$$\frac{\partial p}{\partial s} = -\rho\frac{\partial V}{\partial s}. \qquad \text{.......................(3)}$$

Hence $\partial V/\partial s$ vanishes with $\partial p/\partial s$; i.e. V is constant over a surface of equal pressure. In other words, in a fluid in equilibrium the surfaces of equal pressure must coincide with the surfaces of equal potential.

Again, if δp and δV be the increments of p and V when we pass from one surface of equal pressure to a consecutive one, we have

$$\delta p = -\rho\delta V. \qquad \text{...........................(4)}$$

Since δp and δV are constants for this pair of surfaces, ρ must also be constant, i.e. the density must be uniform over any surface of equal potential.

Thus in the case of ordinary gravity the equipotential surfaces

* The unit force being that which generates unit momentum in unit time.

are horizontal planes. Hence in a fluid in equilibrium the pressure and the density must be uniform over any such plane.

In the case of a force tending to a fixed point the equipotential surfaces are concentric spheres. The fluid must therefore be arranged in spherical strata of equal density.

When ρ is a constant, we have by integration of (3)

$$p = -\rho V + C. \qquad (5)$$

Ex. To find the pressure in the interior of a liquid globe due to the mutual attraction of the parts of the fluid.

We assume from the theory of Attractions that the intensity of gravity in the interior of a homogeneous globe varies as the distance (r) from the centre, and denote it accordingly by gr/a, where g is the force per unit mass at the surface ($r=a$). Hence p will be a function of r only, and

$$\frac{dp}{dr} = -g\rho\frac{r}{a}. \qquad (6)$$

Hence

$$p = \frac{1}{2}\frac{g\rho}{a}(a^2 - r^2), \qquad (7)$$

where the constant of integration has been determined so as to make p vanish for $r=a$. The pressure-intensity at the centre is therefore $\frac{1}{2}g\rho a$.

In the case of a homogeneous liquid globe of the same size and mass as the Earth, we should have $g=1$, in ordinary gravitational measure. Hence the pressure-intensity at the centre would be that due to a column of about $5\frac{1}{2}$ times the density of water, and about 2000 miles high, or say (roughly) $1\frac{3}{4}$ million atmospheres.

113. Moving Liquid in Relative Equilibrium.

Problems in which a fluid, although in motion, is in *relative* equilibrium, can be reduced to statical ones in virtue of the following principle.

If m denote the mass of a fluid element, and f its acceleration (supposed given in magnitude and direction), the resultant of all the forces acting on the element, including the pressures of the surrounding fluid, must be mf in dynamical measure. Hence the actual forces, together with a fictitious force $-mf$, would be in equilibrium.

Ex. A tank containing water has a constant horizontal acceleration f; to find the form of the free surface when there is relative equilibrium.

The distribution of pressure will be the same as if the fluid were at rest under gravity and a horizontal force f per unit mass in the direction opposite to the acceleration. The effect is the same as if the direction of gravity were turned through an angle θ backwards from the vertical, such that

$$\tan\theta = f/g, \quad \text{.........(1)}$$

and its intensity increased in the ratio $\sec\theta$. The free surface is therefore a plane tilted down through an angle θ.

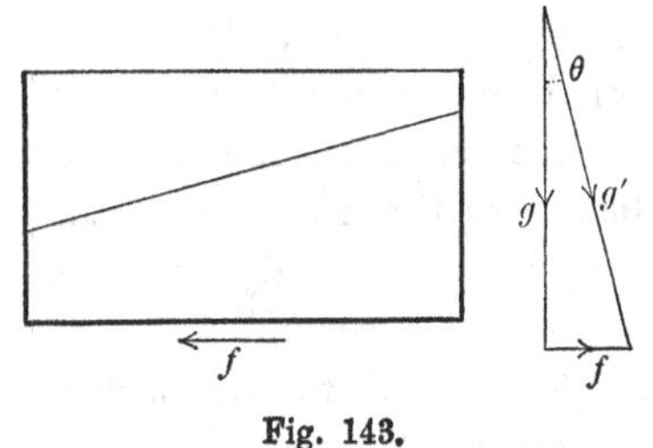

Fig. 143.

114. Rotating Liquid.

When a liquid is rotating with a constant and uniform angular velocity ω about a vertical axis, a particle m at a distance r from this axis is describing a horizontal circle with an acceleration $\omega^2 r$ towards the centre. Hence the fictitious force on each particle, which we must imagine to be introduced in order to transform the problem into a statical one, is a force $m\omega^2 r$, in dynamical units, acting outwards from the axis. There is no harm in speaking of this as a 'centrifugal' force, provided its artificial character be remembered.

If the axis of z be drawn vertically upwards, we have in the statical problem

$$\frac{\partial p}{\partial z} = -g\rho, \quad \frac{\partial p}{\partial r} = \omega^2 \rho r, \quad \text{.....................(1)}$$

by Art. 111 (1). Hence*

$$p = \rho\left(\tfrac{1}{2}\omega^2 r^2 - gz\right) + \text{const.}, \quad \text{..................(2)}$$

and the surfaces of equal pressure, including the free surface if any, are given by

$$z = \frac{\omega^2}{2g} r^2 + C. \quad \text{.......................(3)}$$

* For if we put $$p = \rho\left(\tfrac{1}{2}\omega^2 r^2 - gz\right) + \chi,$$
where χ is a function of z and r as yet undetermined, we find on substitution in (1)

$$\frac{\partial \chi}{\partial z} = 0, \quad \frac{\partial \chi}{\partial r} = 0;$$

i.e. χ is independent both of z and of r and is therefore a constant.

They are therefore equal paraboloids of revolution, the latus-rectum being $2g/\omega^2$.

The parabolic form of the free surface, or of any surface of equal pressure, may also be inferred from the principle that the surface must be perpendicular to the direction of the resultant of gravity and of the fictitious centrifugal force. If the normal to the surface at any point P meet the axis of rotation in G, and PN be the ordinate to this axis, we have

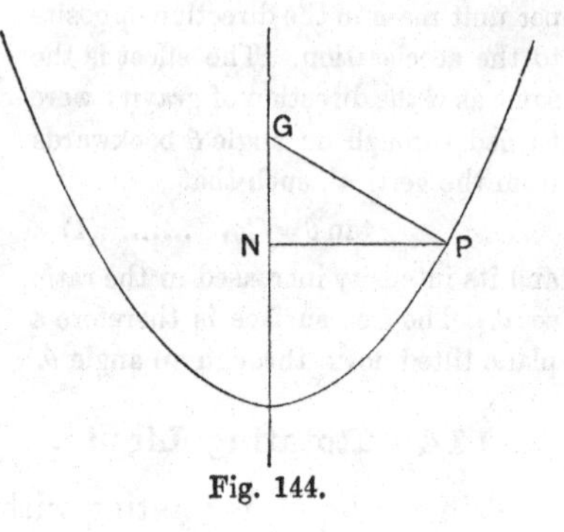

Fig. 144.

$$GN : NP = g : \omega^2 . NP, \quad\text{.....................(4)}$$

whence

$$GN = g/\omega^2. \quad\text{..........................(5)}$$

The subnormal of the meridian curve is therefore constant, and the curve a parabola.

As in Art. 98 it is plain that the effect of the fluid pressures on any solid which is wholly or partially immersed, and rotates with the fluid, may be found from a consideration of the forces which would maintain in relative equilibrium the mass of fluid displaced by the solid. In the case of a solid partially immersed this mass must be supposed limited by the parabolic surface, continued through the space occupied by the solid.

Ex. 1. A circular cylinder floats upright, its axis being coincident with the axis of rotation.

The upward pressure of the fluid must be equal to the weight of the fluid displaced. If a be the radius of the cylinder, and h the height of the water-line above the base, the volume displaced is less than the cylindrical volume $\pi a^2 h$ by that of a paraboloidal segment of height $\frac{1}{2}\omega^2 a^2/g$ and base πa^2. The upward pressure is therefore

$$g\rho\,\pi a^2 (h - \tfrac{1}{4}\omega^2 a^2/g).$$

This must be equal to the weight of the solid. Hence if h' be the depth immersed when there is no rotation, we have

$$h = h' + \tfrac{1}{4}\omega^2 a^2/g. \quad\text{.................................(6)}$$

Ex. 2. If m be the mass of a small submerged body at a distance r from the axis, and m' that of the fluid displaced by it, the fluid pressures on it are equivalent to a force $m'\omega^2 r$ towards the axis, and $m'g$ upwards. Hence in order that the body may revolve in relative equilibrium, a force $(m-m')\,\omega^2 r$ inwards, and a force $(m-m')\,g$ upwards, must be applied to it, e.g. by the tension of a string. If the body is lighter than water, $m-m'$ is negative, and the requisite forces act outwards and downwards, respectively.

Hence particles heavier than water, e.g. grains of sand, tend to collect near the outer edge of the base of the containing vessel; whilst particles lighter than water, e.g. fragments of cork, tend to the surface and the axis.

Ex. 3. Suppose we have an ocean of water covering a spherical attracting nucleus, the whole rotating in relative equilibrium about an axis through the centre, with angular velocity ω.

It is easily seen that the effect of centrifugal force is to add to the potential energy per unit mass a term $-\frac{1}{2}\omega^2 r^2$, if r denote distance from the axis. If R denote distance from the centre, the equation (5) of Art. 112 takes the form

$$\frac{p}{\rho}=\frac{\gamma E}{R}+\tfrac{1}{2}\omega^2 r^2+C, \quad\text{..............................(7)}$$

where E is the mass of the nucleus, and γ the constant of gravitation. (Cf. Art. 49, Ex. 1.)

Hence if a, b be the equatoreal and polar radii, respectively, of the free surface, we have

$$\frac{\gamma E}{a}+\tfrac{1}{2}\omega^2 a^2=\frac{\gamma E}{b}, \quad\text{..............................(8)}$$

whence

$$\frac{a-b}{a}=\frac{1}{2}\frac{\omega^2 a^2 b}{\gamma E}. \quad\text{..............................(9)}$$

If we take data appropriate to the case of the Earth, the expression on the right-hand is a small fraction. In it we may therefore ignore the difference between a and b. Hence, writing $g=\gamma E/a^2$, we have

$$\frac{a-b}{a}=\frac{1}{2}\frac{\omega^2 a}{g}. \quad\text{..............................(10)}$$

This gives the 'ellipticity' of the free surface. In the case of the Earth we have $\omega^2 a/g=\frac{1}{289}$, and the ellipticity would therefore be $\frac{1}{578}$. The actual ellipticity of the mean surface of the ocean is much greater, being about $\frac{1}{300}$. The discrepancy is due mainly to the fact that the nucleus is not exactly spherical, having been deformed (presumably) by centrifugal force, so that it does not attract like a point-centre of force.

EXAMPLES. XX.

1. A barometer is in a railway carriage. How will the reading be affected when the train has a given acceleration, (1) if the barometer hang freely from the roof, (2) if it be fixed relatively to the carriage, in a vertical position?

2. A locomotive runs down an incline (α) with acceleration f; prove that the average level of the water in the boiler makes with the inclined plane an angle θ given by

$$\tan\theta = \frac{f - g\sin\alpha}{g\cos\alpha}.$$

3. A tank slides down a rough incline (α), the coefficient of friction being $\tan\lambda$. Find the inclination of the free surface of the water to the incline, in relative equilibrium.

4. Assuming that the density of a liquid globe of radius a at a distance r from the centre is

$$\rho = \rho_0\left(1 - \beta\frac{r^2}{a^2}\right),$$

find the pressure-intensity at the centre, having given that the intensity of gravity at the surface is g.

If the mean density be twice the surface density, prove that the result is $1\frac{5}{8}$ as great as if the globe had been homogeneous, of the same mean density.

5. A cylindrical vessel, of diameter 1 ft., containing water, is in steady rotation about its axis, which is vertical. If the level of the water in the middle be 1 in. below that at the edge, find the speed, in revolutions per minute. [44·1.]

6. A leaden plummet is immersed in water which is rotating in relative equilibrium about a vertical axis with angular velocity ω, being suspended from a point on this axis by a string of length l. Prove that the vertical position of the plumb-line is stable or unstable according as $l \lesseqgtr g/\omega^2$. Also that when the vertical position is unstable there is an inclined position in which the string is normal to the surface of equal pressure passing through the plummet.

7. A vessel of any form, with a plane horizontal lid, is just filled with liquid of density ρ, and the whole rotates about a vertical axis. Prove that the upward thrust of the fluid on the lid is $\frac{1}{2}\omega^2\rho A\kappa^2$, where $A\kappa^2$ is the quadratic moment of the area with respect to the axis of rotation.

8. A liquid is rotating in relative equilibrium in a spherical vessel which it fills, about a vertical diameter. Prove that the pressure-intensity on the wall of the vessel is greatest at a depth g/ω^2 below the centre.

Also prove that the thrusts on the lower and upper hemispheres are

$$\tfrac{5}{4}Mg + \tfrac{3}{16}M\omega^2 a, \quad \tfrac{1}{4}Mg + \tfrac{3}{16}M\omega^2 a,$$

where M is the mass of the liquid, and a the radius of the vessel.

9. A closed cylindrical vessel of radius a and height h, filled with liquid, rotates about its axis, which is vertical, with uniform angular velocity ω. Prove that the total pressure across a plane through the axis, due to the rotation, is $\frac{1}{3}\rho\omega^2 a^3 h$, and that the resultant outward pressure, due to the rotation, on either half of the curved surface is $\rho\omega^2 a^3 h$.

10. A cubical vessel filled with water is rotating about a vertical axis through the centres of two opposite faces. Prove that in consequence of the rotation the thrust on a side is increased by the amount

$$\tfrac{1}{8}\rho\omega^2 a^4,$$

where a is the length of the edge of the cube, and ω is the angular velocity of rotation.

Also find the centre of pressure.

11. A liquid is rotating about a vertical axis with an angular velocity ω, which is a function of the distance (r) from the axis; prove that the form of the free surface is given by

$$gz = \int \omega^2 r\,dr + C.$$

Sketch the form in the case of $\omega = \omega_0 a^2/r^2$. (Rankine's 'free vortex.')

12. If the angular velocity be equal to ω_0 for $r < a$, and equal to $\omega_0 a^2/r^2$ for $r > a$, find the form of the free surface. (Rankine's 'combined vortex.')

13. Water is rotating in circles about a vertical axis, the angular velocity at a distance r from this axis being

$$\frac{\omega_0 a^2}{a^2 + r^2};$$

find the form of the free surface.

14. Prove that the equation of the meridian curve, in the problem of Art. 114, Ex. 3, is of the form

$$r = a(1 + \epsilon \sin^2 \theta),$$

approximately, where θ is the colatitude, and ϵ the ellipticity.

15. A closed vessel filled with water is rotating with constant angular velocity ω about a horizontal axis; prove that in the state of relative equilibrium the surfaces of equal pressure are circular cylinders whose common axis is at a height g/ω^2 above the axis of rotation.

CHAPTER XIII

EQUILIBRIUM OF GASEOUS FLUIDS

115. Laws of Gases.

The properties of a 'perfect' gas, i.e. one which is far removed from the physical conditions under which it can be liquefied, are summed up in the statement

$$pv = R\theta, \quad \text{.........................(1)}$$

where p is the pressure-intensity, v is the volume of unit mass, θ is the 'absolute' temperature as measured by an air-thermometer, and R is a constant depending on the nature of the gas. The quantity v is of course the reciprocal of the density (ρ); it is sometimes called the 'specific volume,' or the 'bulkiness,' of the substance.

The above statement rests on two experimental laws, which are found to hold with considerable accuracy. The first of these, called after its discoverer* 'Boyle's Law,' is to the effect that the pressure-intensity of any portion of a gas kept at a constant temperature varies inversely as the volume. Hence the product pv is constant for the same gas at the same temperature, so that we may write

$$pv = f(\theta), \quad \text{.........................(2)}$$

where $f(\theta)$ is a function which may (so far) be different for different gases, and will in any case depend on the manner in which the temperature θ is defined. If we take air as our standard thermometric substance, i.e. if we agree to register temperatures in terms of the volume occupied by a mass of air at some standard pressure, $f(\theta)$ will (for air) be proportional to θ, and we have the formula (1). The constant R will however still depend on the

* Robert Boyle (1627–1691); the date of publication is 1662. The same law was formulated by Mariotte in 1676, and is sometimes called by his name.

length of the degree on the temperature scale. Usually this is chosen so that the interval from the freezing point to the boiling point* of water shall be 100 degrees.

The second gaseous law† asserts that different gases expand under constant pressure, between any two given temperatures, in the same proportion, i.e. the increment of volume bears the same ratio to the original volume in each case. Hence a formula of the type (1) will hold for each, but the constant R will vary with the particular gas. The actual ratio of the volumes occupied by a gas at the boiling and freezing points of water is found to be 1·366, so that if θ_0 be the temperature of freezing on the present scale, we have, if we adopt the Centigrade degree,

$$\frac{\theta_0 + 100}{\theta_0} = 1{\cdot}366, \quad \theta_0 = 273. \quad \ldots\ldots\ldots\ldots\ldots(3)$$

It may be well to repeat that the formula (1) has only a limited validity. For great pressures combined with low temperatures the 'gaseous laws' cease to apply to real substances.

The density of atmospheric air at the freezing point, under a pressure of 76 cm. of mercury, is found to be ·00129. Hence if we take the centimetre as unit of length, and the weight of a gramme as the unit of force, we have, for air,

$$R = \frac{76 \times 13{\cdot}6}{{\cdot}00129 \times 273} = 2930,$$

about.

To reduce to absolute (C.G.S.) units we multiply by g (=981), and obtain

$$R = 2{\cdot}88 \times 10^6.$$

To obtain the value of R for any other gas we divide by the ratio of its density to that of air under like conditions of pressure and temperature.

116. Mixture of Gases.

Boyle's law may be stated as follows‡: If, at a given temperature, successive portions of the same gas be introduced into a closed vessel, each produces its own pressure; i.e. the final

* That is, the temperature at which water boils under a certain specified atmospheric pressure.

† Attributed, on the authority of Gay-Lussac (1802), to J. A. C. Charles (1746–1822), the date assigned being about 1787. The law was first published by Dalton in 1801.

‡ This form of statement is due to Rankine.

pressure-intensity is the sum of the pressure-intensities due to each portion separately. If we extend this statement to mixtures of different gases we have the law formulated by Dalton*, viz. that the pressure-intensity in the mixture is the sum of the pressure-intensities due to the several gases alone, provided these do not act chemically on one another.

Thus, if quantities of various gases, whose volumes are $V_1, V_2, \ldots,$ and pressure-intensities $p_1, p_2, \ldots,$ respectively, be taken and introduced into a vessel of volume V, the temperature being the same in each case, the pressure-intensities due to the several gases would be

$$p_1 V_1/V, \quad p_2 V_2/V, \ldots,$$

by Boyle's law. By Dalton's law the final pressure-intensity p will be the sum of these, so that

$$pV = p_1 V_1 + p_2 V_2 + \ldots = \Sigma (p_r V_r). \quad \ldots\ldots\ldots\ldots(1)$$

If the gases were originally at temperatures $\theta_1, \theta_2, \ldots,$ respectively, and if the final temperature be θ, the pressure-intensities of the several gases when brought to the volume V and temperature θ would be

$$\frac{p_1 V_1}{V} \cdot \frac{\theta}{\theta_1}, \quad \frac{p_2 V_2}{V} \cdot \frac{\theta}{\theta_2}, \ldots,$$

whence
$$\frac{pV}{\theta} = \frac{p_1 V_1}{\theta_1} + \frac{p_2 V_2}{\theta_2} + \ldots = \Sigma \left(\frac{p_r V_r}{\theta_r} \right). \quad \ldots\ldots\ldots(2)$$

It may be observed that Dalton's law is implied in such statements as this, that atmospheric air consists of so many parts 'by volume' of oxygen, and so many of nitrogen. If the oxygen contained in a volume V of air, when isolated and brought to atmospheric pressure (p_0), would occupy a volume V_1, the partial pressure due to it when diffused in the space V is $p_0 V_1/V$, by Boyle's law; and on a similar understanding the partial pressure due to the nitrogen is denoted by $p_0 V_2/V$. By Dalton's law, the actual pressure is the sum of these, whence

$$V_1 + V_2 = V. \quad \ldots\ldots\ldots\ldots\ldots\ldots\ldots\ldots(3)$$

117. Indicator Diagram. Isothermal and Adiabatic Lines.

The properties of any fluid may be mapped out, without any special hypothesis, by taking any two of the quantities p, v, θ as

* John Dalton (1766–1844), the discoverer of the atomic law in Chemistry.

independent variables, the third being then necessarily dependent upon them. Usually it is convenient to take v as abscissa, and p as ordinate, as in Watt's 'indicator diagram.' Any particular state is then represented by a point on the diagram, and any succession of states by a continuous line, as in Fig. 145.

For instance, if the succession of states be defined by the condition that the temperature is constant, the corresponding line is called an 'isothermal' line. In the case of a perfect gas the isothermal lines are the rectangular hyperbolas

$$pv = \text{const.} \qquad \text{.........................(1)}$$

In another important type of succession of states the condition imposed is that there shall be no absorption or emission of heat during the process. This would be realized in the case of a gas contained in a cylinder whose volume could be varied by means of a piston, provided the walls and the piston could be made absolutely impervious to heat. For this reason such a succession is said to be 'adiabatic*,' and the representative line on the indicator diagram is called an adiabatic line. The equation of the adiabatic lines of a perfect gas will be obtained presently.

The 'specific heat' of a substance in a given state is defined as the amount of heat† required to raise unit mass one degree in temperature, under prescribed conditions. In the case of a gas the precise specification of the condition under which the change of temperature takes place is important; and in particular we have to distinguish between the specific heat 'at constant volume,' and the specific heat at 'constant pressure.'

If the volume do not vary, as when the gas is enclosed in a rigid vessel, we have, by Art. 115 (1), $\delta\theta/\theta = \delta p/p$. Hence if c denote the specific heat at constant volume, the amount of heat absorbed by unit mass when the pressure increases by δp is

$$c\,\delta\theta = c\theta\,\delta p/p. \qquad \text{.......................(2)}$$

Again, if the pressure be kept constant, as in the case of a gas free to expand by pushing out a piston against a constant

* This name was introduced by Rankine.

† The unit of heat being (usually) the amount of heat required to raise the temperature of unit mass of water at about 4° C. by one degree.

external pressure, we have $\delta\theta/\theta = \delta v/v$. Hence if c' be the specific heat at constant pressure, the amount of heat absorbed by unit mass when the volume increases by δv is

$$c'\delta\theta = c'\theta\,\delta v/v. \quad \text{.......................(3)}$$

Hence if we take v and p as our independent variables, the heat absorbed in a small change denoted by δv, δp will be

$$\theta\left(c\frac{\delta p}{p} + c'\frac{\delta v}{v}\right) \quad \text{.......................(4)}$$

per unit mass.

For instance, in an *isothermal* expansion of a gas we have pv = const., and therefore $\delta p/p = -\,\delta v/v$. The expression (4) then takes the form

$$(c' - c)\,\theta\frac{\delta v}{v}. \quad \text{.......................(5)}$$

It appears from thermodynamical theory that the difference $c' - c$ is constant for a perfect gas. Hence, by integration, we find that the total heat absorbed during an expansion from volume v_0 to volume v at constant temperature θ is

$$(c' - c)\,\theta\log\frac{v}{v_0}. \quad \text{.......................(6)}$$

If a gas be subject to the *adiabatic* condition, we have, from (4),

$$\frac{\delta p}{p} + \gamma\frac{\delta v}{v} = 0, \quad \text{.......................(7)}$$

where $\gamma = c'/c$. This ratio γ of the two specific heats is also approximately constant. Hence, integrating, we find

$$\log p + \gamma\log v = \text{const.},$$

or

$$pv^\gamma = \text{const.} \quad \text{.......................(8)}$$

This is therefore the equation of the adiabatic lines on the indicator diagram. See Fig. 146, p. 259, where the dotted curve represents an adiabatic line, and the continuous curve an isothermal.

The ratio γ is always greater than unity. Its value for air and several other gases (oxygen, hydrogen, nitrogen) is about 1·41.

The change of temperature involved in an adiabatic process is found from (7) combined with the relation

$$\frac{\delta\theta}{\theta} = \frac{\delta p}{p} + \frac{\delta v}{v}, \quad \text{.......................(9)}$$

which follows from Art. 115 (1). We have

$$\frac{\delta\theta}{\theta} = -(\gamma-1)\frac{\delta v}{v}, \quad\text{.....................(10)}$$

whence $$\log\theta = -(\gamma-1)\log v + \text{const.},$$

or $$\frac{\theta}{\theta_0} = \left(\frac{v_0}{v}\right)^{\gamma-1}. \quad\text{........................(11)}$$

Ex. If air at 0° C. be compressed adiabatically to one-tenth of its original volume, we have

$$\log_{10}(\theta/\theta_0) = \gamma - 1 = \cdot 410\,; \quad \theta/\theta_0 = 2{\cdot}57,$$

whence $$\theta - \theta_0 = 1{\cdot}57\theta_0 = 429°.$$

118. Work done in Expansion. Elasticity.

Let us imagine unit mass of any fluid to be enclosed in a deformable envelope, and that an infinitesimal change of volume is produced by a displacement of the boundary. If ν be the normal component, reckoned positive when outwards, of the displacement of a surface-element δS, the work done by the contained gas is $\Sigma(p\delta S.\nu)$, or $p\delta v$, since $\Sigma(\nu\delta S) = \delta v$. Hence the work done in a succession of changes represented by a given line on the indicator

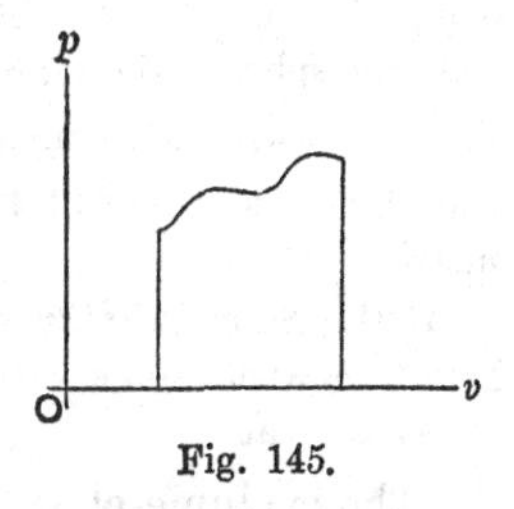

Fig. 145.

diagram will be $\int p\,dv$, i.e. it is represented by the area included between the curve, the axis of v, and the first and last ordinates. Care must of course be taken to attribute the proper sign to this area.

Thus for an isothermal expansion of a perfect gas from v_0 to v, we have, since $pv = p_0v_0$,

$$\int_{v_0}^{v} p\,dv = p_0v_0\int_{v_0}^{v}\frac{dv}{v} = p_0v_0\log\frac{v}{v_0}. \quad\text{...........(1)}$$

This does not of course give the loss of intrinsic energy by the gas, since heat is absorbed during the process, to the amount given by Art. 117 (6)*. The quantity whose diminution is given by (1) is sometimes called the 'free energy' by writers on Thermodynamics.

* It is a consequence of the laws of Thermodynamics that in a perfect gas there is *no* change of intrinsic energy in an isothermal expansion, the work done being exactly compensated by the heat absorbed.

In an adiabatic expansion we have $pv^\gamma = p_0 v_0{}^\gamma$, whence

$$\int_{v_0}^{v} p\,dv = p_0 v_0{}^\gamma \int_{v_0}^{v} \frac{dv}{v^\gamma} = \frac{p_0 v_0 - pv}{\gamma - 1} = \frac{R}{\gamma - 1}(\theta_0 - \theta). \quad \ldots\ldots(2)$$

The change of temperature is given by Art. 117 (11). The expression (2) measures the actual loss of intrinsic energy in the process, since there is now no transfer of heat.

The unit of work (per unit mass) implied in (1) or (2) will depend of course on the unit of force employed in the specification of the pressure-intensity p. For instance, if this be expressed in grammes per square centimetre, the unit of (1) or (2) will be the gramme-centimetre.

If we wish to find the work done in the expansion of any given quantity of a gas, we have only to replace the symbol v in the above calculations by V, the actual volume of the mass under consideration. Thus, to find the work required to fill a vessel whose capacity is 1 litre with air at a pressure of 100 atmospheres, the temperature being assumed constant, we put, in (1),

$$p_0 = 1034 \text{ gm./cm.}^2, \quad V_0 = 10^5 \text{ cm.}^3, \quad \log(V_0/V) = 4{\cdot}650,$$

and obtain as the result $4{\cdot}81 \times 10^8$ gramme-centimetres, or 4810 kilogramme-metres.

If the same *quantity* of air could be forced in adiabatically, we should have $p/p_0 = 660$, and substituting in (2) we should obtain a result about three times as great.

The 'volume-elasticity' of a substance in a given state may be defined as the ratio of an increment δp of pressure-intensity to the accompanying 'compression,' as measured by the ratio $(-\delta v/v)$ of the diminution of volume to the original volume. Hence, denoting it by κ, we have

$$\kappa = -v\frac{dp}{dv}. \quad \ldots\ldots\ldots\ldots\ldots\ldots(3)$$

To make the definition precise it is necessary, however, to specify the condition under which the variation of volume is supposed to take place. In the case of a solid or a liquid this makes as a rule little difference, but in a gas the matter is important.

If, through the point P on the indicator diagram which represents the initial state, we draw a tangent to the curve which shews the prescribed mode of variation, to meet the axis of p in U, and if NU be the projection of PU on this axis, we have

$$-v\frac{dp}{dv} = NP.\tan NPU = NU. \quad \ldots\ldots\ldots\ldots(4)$$

Hence this projection represents the volume-elasticity under the given condition.

If the changes be isothermal, we have $pv =$ const., and

$$\kappa = p, \qquad (5)$$

i.e. the elasticity is numerically equal to the pressure-intensity. The equality of ON and NU in this case is a familiar property of the rectangular hyperbola.

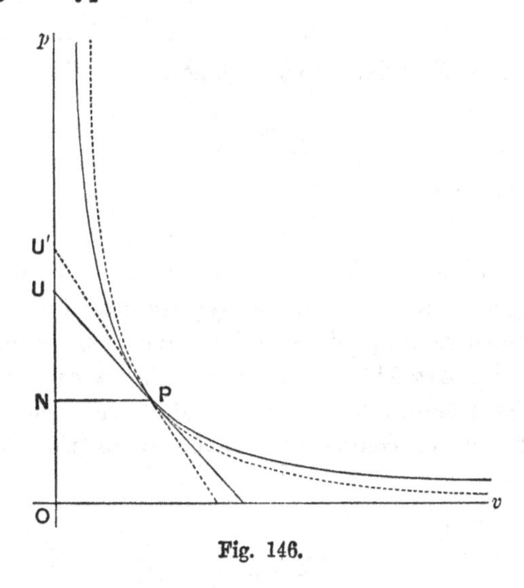

Fig. 146.

The volume-elasticity of a gas subject to the adiabatic condition is

$$\kappa = -v\frac{dp}{dv} = \gamma p, \qquad (6)$$

by Art. 117 (7), and is therefore greater than the isothermal elasticity in the ratio γ. It is represented by NU' in Fig. 146.

The distinction between the two elasticities is important in the theory of Sound, where the expansions and contractions, and consequent variations of temperature, alternate so rapidly that there is no time for equalization of temperature by conduction. Consequently, it is the adiabatic rather than the isothermal elasticity which is effective.

119. Law of Pressure in Isothermal Atmosphere.

To find the distribution of pressure in an atmosphere of uniform temperature, let z denote height above some standard level. We have then the hydrostatic equation

$$\frac{dp}{dz} = -\rho, \quad \text{.......................(1)}$$

in gravitational units, together with Boyle's law

$$p/p_0 = \rho/\rho_0, \quad \text{.......................(2)}$$

where the zero suffix relates to the plane $z = 0$. Hence

$$\frac{1}{p}\frac{dp}{dz} = -\frac{\rho_0}{p_0}. \quad \text{.......................(3)}$$

It is convenient to write

$$p_0 = \rho_0 H, \quad \text{.......................(4)}$$

so that H denotes the height of a 'homogeneous atmosphere,' i.e. the height of a column of a hypothetical fluid of *uniform* density ρ_0 which would produce by its gravitation the pressure p_0. Comparing with Art. 115 (1) we see that, for a particular gas, H varies as the absolute temperature but is independent of the density. It will of course vary inversely as the intensity of gravity.

We have, then,

$$\frac{1}{p}\frac{dp}{dz} = -\frac{1}{H}, \quad \text{.......................(5)}$$

whence

$$\log p = -\frac{z}{H} + \text{const.}, \quad \text{.......................(6)}$$

or

$$p = p_0 e^{-z/H}, \quad \text{.......................(7)}$$

if the arbitrary constant be adjusted so as to make $p = p_0$ for $z = 0$. Hence the pressure diminishes in geometric progression as the altitude increases in arithmetic progression.

The formula (7) may be used to determine differences of altitude by means of the barometer. We have

$$z_2 - z_1 = H \log (p_1/p_2) = 2\cdot3026\, H \log_{10} (p_1/p_2), \quad \text{......(8)}$$

where the suffixes refer to any two stations, and the numerical factor is that required to reduce logarithms to base e to ordinary logarithms. If H_0 denote the value of H at freezing point we may put

$$H = H_0(1 + \cdot 00366\tau), \quad\quad (9)$$

where τ is the temperature Centigrade. The formula (8) is based on the hypothesis of uniform temperature; but the variation of temperature with altitude, if not too great, may be allowed for by assuming a mean temperature for the stratum included between the two levels at which the readings are taken. If the difference of temperature is considerable, observations may be taken at a number of intermediate altitudes, and the thickness of each stratum calculated separately, on the basis of its estimated mean temperature.

The calculation assumes of course that the air may be treated as a simple gas, the variation of its composition with altitude being neglected. This is legitimate, for a considerable range of altitude, owing to the mixing which takes place by currents.

If the density of air at 0° C. be ·00129 under an atmospheric pressure of 76 cm. of mercury, we have

$$H_0 = \frac{p_0}{\rho_0} = \frac{76 \times 13\cdot 6}{\cdot 00129} = 8 \times 10^5 \text{ cm.} = 8 \text{ km.}$$

120. Atmosphere in Convective Equilibrium.

In the actual atmosphere the temperature as a rule diminishes upwards. This is due in a great measure to the action of convection currents. When a mass of air ascends, its temperature falls in consequence of the expansion due to diminished pressure, and similarly if it descends its temperature rises. The final condition which would be brought about by the free play of convection currents, without conduction or radiation of heat, would be such that if equal masses at different levels were interchanged the equilibrium would not be disturbed, i.e. each portion would by adiabatic expansion or contraction assume the temperature and therefore also the pressure proper to its new locality. This state is described by Lord Kelvin as one of 'convective equilibrium.'

In such a state we have

$$\frac{p}{\rho^\gamma} = \frac{p_0}{{\rho_0}^\gamma}, \quad\quad (1)$$

where the zero suffix refers to one particular level, say $z = 0$.

Hence $$\frac{\delta p}{p} = \gamma \frac{\delta \rho}{\rho}. \quad \text{.......................(2)}$$

When combined with the relation

$$\frac{\delta p}{p} = \frac{\delta \rho}{\rho} + \frac{\delta \theta}{\theta}, \quad \text{.......................(3)}$$

this gives $$\frac{\delta \theta}{\theta} = \frac{\gamma - 1}{\gamma} \cdot \frac{\delta p}{p}, \quad \text{.......................(4)}$$

the relations being in fact as in Art. 117 (7), (9), with ρ written for $1/v$.

The hydrostatic equation is

$$\frac{dp}{dz} = -\rho, \quad \text{.......................(5)}$$

as usual, whence

$$\frac{d\theta}{dz} = -\frac{\gamma - 1}{\gamma} \frac{\rho\theta}{p} = -\frac{\gamma - 1}{\gamma} \cdot \frac{\rho_0 \theta_0}{p_0}, \quad \text{...........(6)}$$

so that the temperature diminishes upwards with a uniform gradient. If we put

$$p_0 = \rho_0 H_0, \quad \text{.......................(7)}$$

so that H_0 is the height of the homogeneous atmosphere corresponding to the temperature θ_0 at the level $z = 0$, we have

$$\frac{d\theta}{dz} = -\frac{\gamma - 1}{\gamma} \cdot \frac{\theta_0}{H_0}, \quad \text{.....................(8)}$$

or $$\theta = \theta_0 \left(1 - \frac{\gamma - 1}{\gamma} \frac{z}{H_0}\right). \quad \text{.................(9)}$$

Again, $$\frac{p}{\rho} = \frac{p_0}{\rho_0} \frac{\theta}{\theta_0} = \frac{p_0}{\rho_0} \left(1 - \frac{\gamma - 1}{\gamma} \frac{z}{H_0}\right),$$

or $$z = \frac{\gamma}{\gamma - 1} \left(\frac{p_0}{\rho_0} - \frac{p}{\rho}\right), \quad \text{...............(10)}$$

which gives, with (1), the relation between p and z.

If $\gamma = 1{\cdot}410$, $H_0 = 8 \times 10^5$ cm., $\theta_0 = 273$, we find

$$\frac{d\theta}{dz} = -{\cdot}000099,$$

in degrees Centigrade per centimetre. The gradient is therefore very nearly

1 degree per 100 metres*. It is to be noticed that the formula (9) would assign a definite limit to the height of the atmosphere, viz. we should have $\theta = 0$ for

$$z = \frac{\gamma}{\gamma - 1} H_0. \quad \text{.................................(11)}$$

With the above numerical values this comes out at $2{\cdot}75 \times 10^6$ cm., or 27·5 km. The gaseous laws on which the calculation is based cannot, however, be assumed to hold up to this limit.

121. Compressibility of Liquids.

The effect of compressibility on the vertical distribution of density in a *liquid* can be calculated by a similar process. For most purposes it is sufficiently accurate, however, to assume that the mean density in a vertical column is the arithmetic mean of the densities at the top and bottom.

We assume, in accordance with the usual law of elasticity (Art. 137), that the small variations of density are connected with the variations of pressure by a formula of the type

$$p - p_0 = \kappa s, \quad \text{..........................(1)}$$

where s denotes what is called the 'condensation,' i.e. the ratio of the increment of density to the standard density ρ_0, which we may take to be the density at the atmospheric pressure (p_0). In symbols

$$s = \frac{\rho - \rho_0}{\rho_0}. \quad \text{..........................(2)}$$

The coefficient κ may be taken to be under ordinary conditions approximately constant; it is called the 'elasticity of volume.' It is easy to shew that when *small* variations from atmospheric pressure are considered this definition of κ is equivalent to that of Art. 118 (3).

The pressure at a depth h will be $p_0 + \rho_0 h$, approximately, and the density accordingly $\rho_0(1 + \rho_0 h/\kappa)$, so that the mean density of a column of depth h extending downwards from the free surface is $\rho_0(1 + \frac{1}{2}\rho_0 h/\kappa)$. If h' be the height which the same column would occupy if of uniform density ρ_0, we have

$$(1 + \tfrac{1}{2}\rho_0 h/\kappa)\, h = h', \quad \text{....................(3)}$$

or

$$\frac{h' - h}{h} = \frac{1}{2}\frac{\rho_0 h}{\kappa}. \quad \text{........................(4)}$$

* The observed gradient is considerably less. The difference is attributed to the influence of aqueous vapour.

This gives the depression $h'-h$ of the surface of the water in a lake of depth h, due to compressibility.

The more formal calculation is as follows. Substituting from (1) and (2) in the hydrostatic equation

$$\frac{dp}{dz}=-\rho, \quad\text{.....................(5)}$$

we have

$$\frac{\dfrac{dp}{dz}}{1+(p-p_0)/\kappa}=-\rho_0. \quad\text{.....................(6)}$$

Hence if the free surface be at a height h above the origin of z,

$$\kappa\log\left(1+\frac{p-p_0}{\kappa}\right)=\rho_0(h-z). \quad\text{.....................(7)}$$

Putting $z=0$, $p-p_0=\rho_0 h'$, we have

$$\log\left(1+\frac{\rho_0 h'}{\kappa}\right)=\frac{\rho_0 h}{\kappa}, \quad\text{.....................(8)}$$

or

$$1+\frac{\rho_0 h'}{\kappa}=e^{\rho_0 h/\kappa}=1+\frac{\rho_0 h}{\kappa}+\frac{1}{2}\left(\frac{\rho_0 h}{\kappa}\right)^2+\ldots \quad\text{.....................(9)}$$

In all cases of interest $\rho_0 h/\kappa$ is a small fraction. If we stop at the term last written, we find

$$h'=h\left(1+\frac{1}{2}\frac{\rho_0 h}{\kappa}\right), \quad\text{.....................(10)}$$

in agreement with (4).

Thus for a sheet of water a kilometre in depth, putting $\rho_0=1$, $h=10^5$ cm., $\kappa=2\cdot26\times10^7$, we find $h'-h=2\cdot21$ metres. It appears from (4) that the depression varies as the *square* of the depth.

EXAMPLES. XXI.

(Boyle's Law, &c.)

1. Find the ratio of the whole mass of the Earth's atmosphere to that of the Earth itself, assuming that the mean density of the Earth is 5·5, the mean height of the barometer 76 cm., and the Earth's radius $6\cdot38\times10^8$ cm.

[$8\cdot8\times10^{-7}$.]

2. The densities of oxygen and nitrogen as compared with atmospheric air at the same pressure and temperature are 1·1056 and ·9714, respectively. Find the percentages of oxygen and nitrogen in the atmosphere, (1) by weight, and (2) by volume. [(1) 23·6, 76·4 ; (2) 21·4, 78·6.]

3. A barometer tube originally containing air at atmospheric pressure is weighted and lowered into the sea, with the closed end uppermost. When it is raised it is found that the water had ascended through ·832 of the length of the tube. Find the depth reached in fathoms. (Height of barometer = 30 in.; density of mercury = 13·6; density of sea-water = 1·026.) [27·3.]

4. A kilogramme of air is contained in a vessel whose capacity is 10 cubic decimetres; find its pressure in gms. per sq. cm., assuming that the atmospheric density is ·00129, and that the height of the barometer is 76 cm. [$8·01 \times 10^4$.]

5. The pressure of saturated aqueous vapour at 50° C. is about one-eighth of an atmosphere. A bell jar is inverted over water at 50° C., and then depressed until half its volume is occupied by water; find the pressure in the space above the water. [$1\frac{7}{8}$ atm.]

6. How much must a column of air 12 in. long be suddenly compressed in order that its temperature may rise from 20° C. to 400° C.? [To 1·6 in.]

7. Two equal cylindrical diving-bells, closed at the top and open below, are just immersed in water, and the water inside stands respectively 4 ft. and 6 ft. below the level outside. If communication be established between the interiors by a pipe, find the new level at which the water will stand. (Height of water-barometer = 33 ft.) [5 ft.]

8. A cylindrical diving-bell 8 ft. high, originally full of air, is lowered until the water rises 3 ft. in the interior; what is the depth of the top of the bell below the surface?

Also, how many cubic feet of air at atmospheric pressure must be pumped in, in order that the water may be expelled from the interior, the sectional area of the bell being 12 sq. ft.? [14·8 ft.; 66·2 c. ft.]

9. A closed cylindrical canister of height h, whose walls are of negligible thickness, contains air at atmospheric pressure. It floats partially immersed in water to a depth k, the axis being vertical. If water leaks in through a small hole, prove that the canister will be in equilibrium when the depth of water inside is

$$\frac{hk}{H+k}$$

where H is the height of the water-barometer.

Is this condition stable?

10. A vertical tube having a uniform section of ·4 sq. in. opens at the top into a bulb whose capacity is 36 c. in. The lower end is open and dips into a wide cistern of mercury; the mercury in the tube stands at 20 in. above the level in the cistern; and the portion of the tube above the mercury is 10 in. long. If the tube be depressed vertically through 10 in., find the level at which the mercury will stand. (Height of barometer = 30 in.) [19 in.]

11. A vertical barometer tube is constructed, of which the upper portion is closed at the top and has a sectional area a^2, the middle portion is a bulb of volume b^3, and the lower portion has a sectional area c^2, and is open at the bottom; the mercury fills the bulb and part of the upper and lower portions of the tube, and is prevented from running out below by means of a float against which the air presses; and the upper part of the tube is a vacuum. Find the change of position of the upper and lower ends of the mercurial column due to a given alteration of the pressure of the atmosphere.

Shew that if the whole volume of mercury be c^2h, where h is the height of the barometer, the upper surface will be unaffected by changes of temperature.

12. Prove that if B be the whole volume of the barrel of an ordinary air-pump, A that of the receiver, and if there be an untraversed space C at the end of the barrel, then

$$(A+B)\rho_n = A\rho_{n-1} + C\rho_0,$$

where ρ_n denotes the density of the air in the receiver after n strokes.

Hence shew that

$$\frac{\rho_n}{\rho_0} = \frac{C}{B} + \left(\frac{A}{A+B}\right)^n \left(1 - \frac{C}{B}\right).$$

13. If in an ordinary condensing pump the volume of the barrel be B, that of the receiver A, and that of the untraversed space in the barrel C, prove that

$$(A+C)\rho_n = A\rho_{n-1} + B\rho_0.$$

Hence shew that $$\frac{\rho_n}{\rho_0} = \frac{B}{C} - \left(\frac{B}{C} - 1\right)\left(\frac{A}{A+C}\right)^n.$$

14. A hollow vessel full of water of density ρ is immersed to a depth z. Prove that the work required to expel the water by pumping in air is

$$\rho(z+h)V\log\left(1+\frac{z}{h}\right) + \rho z V,$$

where V is the volume of the vessel and h the height of the water-barometer.

15. A vertical pipe of height h and sectional area ω dips into a wide cistern of water. Its upper end is connected with a small pump which removes a volume V of air per unit time. Prove that the time the water takes to rise to a height x in the pipe is

$$\left\{2x + (H-h)\log\left(1-\frac{x}{H}\right)\right\}\frac{\omega}{V},$$

where H is the height of the water-barometer.

EXAMPLES. XXII.

(Atmospheric Problems.)

1. If the barometer reading at the Earth's surface be 76 cm., calculate what would be the reading at a height of one kilometre if the temperature were uniform, having given that the density of the air at the surface is ·00129 and that of mercury 13·6. [67·1.]

2. Prove that the fraction of the whole mass of an isothermal atmosphere which is included between the ground and a horizontal plane at a height z is

$$1-e^{-z/H}.$$

Evaluate this for $z=H$, $2H$, $3H$, respectively. [·632, ·865, ·950.]

3. Calculate the percentages (by weight) of oxygen and nitrogen in the air at a height of 5 times that of a homogeneous atmosphere, on the assumption that the density of each gas varies as if it alone were present, and that the temperature is uniform. (See Ex. XXI. 2.) [13·7, 86·3.]

4. Prove that if the temperature in an atmosphere in equilibrium diminish upwards with a certain uniform gradient, the density will be uniform; and find the gradient in question in degrees C. per 100 metres. Would this condition be stable? [3·4.]

5. If the (absolute) temperature diminish upwards in the atmosphere according to the law

$$\frac{\theta}{\theta_0}=1-\frac{z}{c},$$

prove that

$$\frac{p}{p_0}=\left(1-\frac{z}{c}\right)^{c/H}$$

6. If the (absolute) temperature θ diminish upwards in the atmosphere according to the law

$$\theta=\frac{\theta_0}{1+\beta z},$$

where β is a constant, the pressure at a height z is given by

$$\frac{p}{p_0}=e^{-z/H-\frac{1}{2}\beta z^2/H}.$$

7. Prove that if the (absolute) temperature θ be any given function of the altitude z, the vertical distribution of pressure in the atmosphere is given by the formula

$$\log\frac{p}{p_0}=\frac{\theta_0}{H_0}\int_z^{z_0}\frac{dz}{\theta}.$$

8. Prove that in an atmosphere arranged in horizontal strata, the work (per unit mass) required to interchange two thin strata of equal mass without disturbance of the remaining strata would be

$$\frac{1}{\gamma-1}(\rho_2{}^{\gamma-1}-\rho_1{}^{\gamma-1})\left(\frac{p_1}{\rho_1{}^{\gamma}}-\frac{p_2}{\rho_2{}^{\gamma}}\right),$$

where the suffixes refer to the initial states of the two strata.

Hence shew that for stability the ratio p/ρ^{γ} must increase upwards.

9. Prove that if the Earth were surrounded by an atmosphere of uniform temperature, the pressure at a distance r from the centre would be given by

$$p/p_0=e^{\frac{a^2}{H}\left(\frac{1}{r}-\frac{1}{a}\right)},$$

where a is the Earth's radius.

What would be the pressure at infinity if $a=21\times10^6$ ft., $H=25000$ ft.?

[$1{\cdot}56p_0\times10^{-365}$.]

10. Prove that if the whole of space were occupied by air at uniform temperature θ, the densities at the surfaces of the various planets would be proportional to the corresponding values of the expression

$$e^{ga/R\theta},$$

where a is the radius of a planet, and g the intensity of gravity at its surface.

11. A closed tube AB containing air is made to rotate uniformly in a horizontal plane about the end A. Prove that when the air is in relative equilibrium the density at B exceeds that at A in the ratio

$$e^{v^2/2gH},$$

where v is the velocity of the end B, and H is the height of the homogeneous atmosphere.

Work out the result for the case of $v=25$ metres per sec., and $H=8$ km.

[1·004.]

12. Prove that the lowering of the level of water in a lake of variable depth, due to compressibility, is $\frac{1}{2}\rho_0/\kappa$ multiplied by the mean square of the depth, approximately.

CHAPTER XIV

CAPILLARITY

122. Hypothesis of Surface Tension.

There is a certain class of statical phenomena presented by liquids which are not accounted for by the preceding theory. The globular form of a dewdrop, for example, is an obvious exception to the statement that the free surface of a liquid in equilibrium is a horizontal plane. Again the rise of water, or the depression of mercury, in a capillary tube is in opposition to the theorem as to the uniformity of level in communicating vessels; and even in the case of a liquid contained in a wide open vessel, where the surface is for the most part indistinguishable from a plane by any test which we can apply, a sharp curvature is observed near the edge, upwards or downwards as the case may be.

It is found that such cases admit of explanation, and of mathematical calculation, on the hypothesis that a film at the surface, of exceedingly minute thickness, is in a peculiar state of stress, similar to that of a uniformly stretched membrane. The stress across any line drawn on the surface is assumed to be everywhere perpendicular to this line, and in the tangent plane. Hence if ABC be a small triangular portion of the film, the equilibrium of the forces in the tangent plane requires that the tensions across the sides should be respectively proportional to those sides, by Art. 23. The state

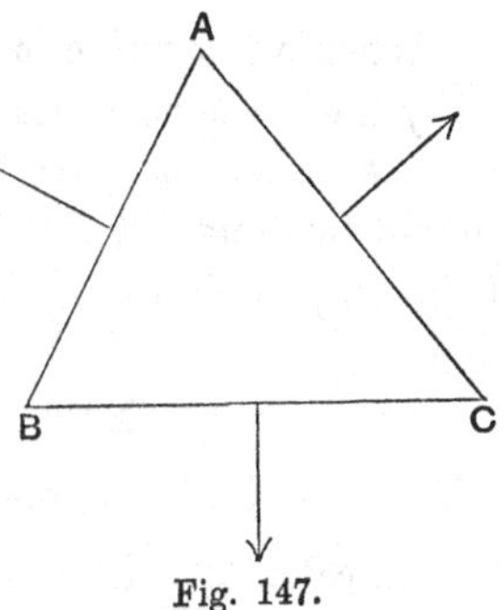

Fig. 147.

of stress at any point of the film is therefore completely specified by a single symbol T, viz. the tension per unit length of any line across which it acts. It is further assumed that this 'surface-tension' is a physical constant depending only on the nature of the liquid, and on the temperature*. It is usually expressed, in gravitation measure, in grammes per (linear) centimetre.

Similar statements are applicable to the surface of separation of two liquids which do not mix. The surface-tension, in this case, depends on the nature of the two liquids, and on the temperature.

The phenomena of surface-tension are most conspicuous when the surface is large in proportion to the total mass, as in the case of a globule of water or mercury, or a soap-bubble. The spherical shape assumed in such cases is due to the tendency of the surface to contract as much as possible, the sphere being the geometrical form of least surface for a given volume.

The following are a few numerical values of the surface-tension for different liquids at 0° C., in grammes per linear centimetre.

Water	·0773
Mercury	·450
Alcohol	·0258
Ether	·0197
Olive oil	·034

The exact determination is a matter of some difficulty, and in the cases of water and mercury the observed tension is greatly reduced by a very slight contamination.

123. Superficial Energy.

When any portion of the surface is extended, work is done on it by the tensions at its boundary. Considering in the first place the case of a *small* extension, let ν denote the displacement parallel to the tangent plane and at right angles to an element δs of the bounding curve, the positive direction being outwards. The work done is

$$\Sigma (T\delta s.\nu) = T.\Sigma (\nu\delta s) = T.\delta S, \quad \text{..............(1)}$$

where S is the total area of the portion.

To calculate the work done in a *finite* extension it would in general be necessary to take account of the fact that a film when

* It diminishes as the temperature rises.

extended has its temperature lowered, unless heat be supplied to it, and that consequently the surface-tension will be increased. But if we suppose the changes to be effected so slowly that a flow of heat can take place so as to maintain the temperature sensibly constant, the total work will be equal to the surface-tension T multiplied by the increment of area.

Conversely, a film of area S has a capacity for doing work in contracting at constant temperature, against resistance, which is measured by the product TS. It is therefore customary to regard the film as the seat of a special form of energy; and from this point of view T is called the 'superficial energy' (per unit area) at the given temperature. The spherical form assumed by a mass of liquid free from other forces is thus explained as being the configuration of least potential energy.

It should be observed, however, that T is not to be identified with the 'intrinsic energy' of the film, which is increased by the amount of heat absorbed during the extension. It is rather of the nature of what is called 'free energy' by writers on Thermodynamics. When a material system is brought from a state A to another state B, the work done on it, together with the heat absorbed, is independent of the manner in which the change is made, being equal to the increment of the energy. If the change be made at constant temperature, and if the processes are *reversible*, it may be shewn that the amount of heat absorbed is itself independent of the manner of the transition. In that case, the work required is also determined solely by the nature of the two states, and is reckoned as an excess of 'free' energy in the state B over that in the state A*.

124. Discontinuity of Pressure.

The hypothesis of surface-tension involves in general a difference in the values of the fluid pressure-intensity on the two sides of the surface. This is easily calculated in the case of a spherical or a cylindrical surface.

Consider, first, a spherical film of radius r. If p_0 denote the excess of pressure-intensity on the inside, the resultant of the

* The question has already been met with in another form in Art. 118.

fluid pressures on either half of the surface cut off by a diametral plane will be $p_0 . \pi r^2$, by Art. 96. If we neglect the weight of the

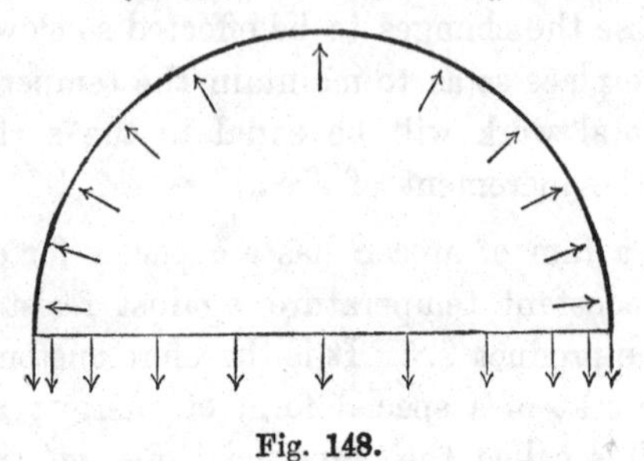

Fig. 148.

surface-film itself, as practically infinitesimal, this resultant must be balanced by the tension $T . 2\pi r$ across the circular edge of the hemisphere; see Fig. 148. Hence, equating,

$$p_0 = 2T/r. \quad \text{.........................(1)}$$

Again, consider the portion of a cylindrical film, of radius r, included between two planes drawn perpendicular to the axis,

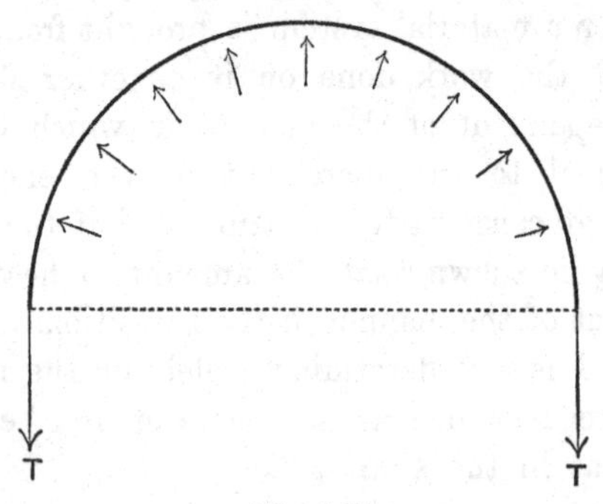

Fig. 149.

at unit distance apart, and a plane through the axis. The fluid pressures have a resultant $p_0 . 2r$, whilst the tensions opposing this, across the straight edges of the portion considered, give $2T$ (Fig. 149). Hence

$$p_0 = T/r. \quad \text{.........................(2)}$$

The general formula, of which (1) and (2) are particular cases, may be noticed, although it is hardly needed for the purposes of this book. If we compare the curvatures, at any point of a surface, of the various normal sections through that point, it is shewn in books on Solid Geometry that there are two positions of the plane of section, mutually perpendicular, for which the curvature is stationary. If R_1, R_2 be the corresponding radii of curvature, the formula is

$$p_0 = T\left(\frac{1}{R_1} + \frac{1}{R_2}\right), \qquad \text{.............................(3)}$$

provided R_1, R_2 be reckoned positive or negative according as the respective centres of curvature are on the side on which the excess of pressure p_0 lies, or the opposite. If we put $R_1 = R_2 = r$, we get the formula (1); whilst in the case of (2) we have $R_1 = r$, $R_2 = \infty$.

125. Angle of Contact.

Where the free surface of a given liquid abuts against a given solid, as in the case of water in a glass tube, there is a definite 'angle of contact.' To account for this we must assume the existence of a certain amount of energy, per unit area, in the boundary, or interface, between the liquid and the solid, and in that between the solid and the air, as well as in the common boundary of the air and the liquid.

In the annexed figure, which represents a section perpendicular

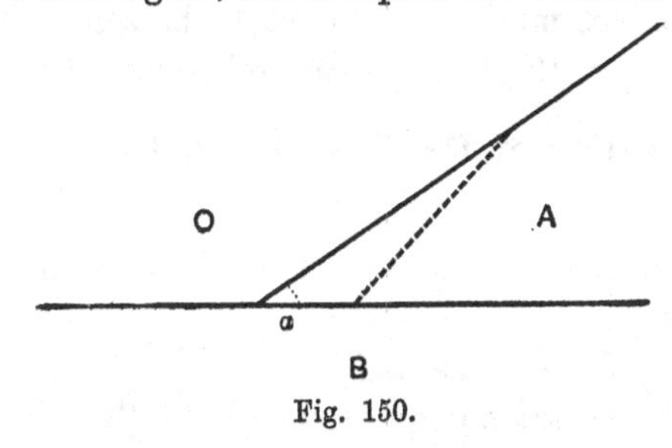

Fig. 150.

to the edge in which the three boundaries meet, A denotes the region occupied by the liquid, O that occupied by the air, and B that occupied by the solid. The values of the superficial energy corresponding to the three surfaces may be denoted by T_{OA}, T_{OB}, T_{AB}. If the boundaries be slightly modified in the neighbourhood of the edge, as shewn by the dotted line, the area of contact of the air with the solid is increased, and that of the liquid with the solid is diminished, by an amount which we will denote by δs, per unit breadth perpendicular to the plane of the figure. The area of contact of the liquid with the air is diminished by $\delta s \cos \alpha$, where α is the angle of contact on the side of the liquid. The total increment of superficial energy is therefore

$$T_{OB}\,\delta s - T_{AB}\,\delta s - T_{OA}\,\delta s \cos\alpha. \quad\text{..............(1)}$$

Hence since, for equilibrium, the energy must be stationary, we must have

$$\cos\alpha = \frac{T_{OB} - T_{AB}}{T_{OA}}. \quad\text{....................(2)}$$

The same formula would follow from a consideration of surface-tensions, if this idea were really appropriate to the case where one of the media concerned is a solid. We have merely to resolve parallel to the surface of the solid, in the direction perpendicular to the edge.

Similar considerations obviously apply to the case where the common surface of two liquids abuts against a solid wall.

Determinations of the angle of contact of water and glass range from about 25° to 29°. A good deal depends on the purity of the water surface and the cleanness of the glass; and it is not impossible that if these qualities could be secured in perfection the angle of contact might even prove to be zero. The angle of contact of mercury with glass appears to be about 127°.

126. Elevation or Depression of a Liquid in a Capillary Tube.

The value of the surface-tension, together with that of the angle of contact, leads to an expression for the elevation of water in a capillary tube. If z be the elevation in question, and a the radius of the cross-section, we have a column of weight $\pi a^2 z \,.\, \rho$ sustained by the tension T acting at the edge of a circle of radius a, in a direction making an angle α with the vertical. The atmospheric pressure $p_0 \,.\, \pi a^2$ is here omitted, because it is operative both on the top and the bottom of the column. Hence

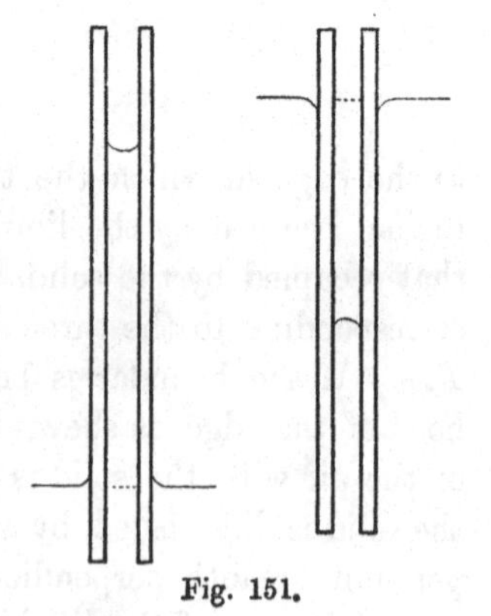

Fig. 151.

$$\pi a^2 z \,.\, \rho = T \cos\alpha \,.\, 2\pi a, \quad\text{...(1)}$$

or
$$z = \frac{2T \cos \alpha}{\rho a}, \quad \text{......................(2)}$$

varying inversely as the radius of the section.

The same result follows from a consideration of the discontinuity of the pressure at the curved surface. The pressure-intensity in the column, just beneath the upper surface, is $p_0 - \rho z$, whilst just above it is p_0. Hence, by Art. 124 (1), we have

$$\rho z = 2T/r, \quad \text{........................(3)}$$

where r is the radius of curvature of the surface, which is assumed to be approximately spherical. Since $a = r \cos \alpha$, this agrees with (2).

In the case of mercury and glass the angle α is obtuse; $\cos \alpha$ is negative, and we have a depression.

If we have two parallel vertical plates at a distance b apart, the weight of the supported column, per unit of horizontal breadth, is $\rho b z$, and the resolved part of the tension is $2T \cos \alpha$. Thus

$$z = \frac{2T \cos \alpha}{\rho b}, \quad \text{........................(4)}$$

the same as in a tube of *radius* b. This formula may be deduced also from Art. 124 (2).

The pressure-intensity between the plates is less than the atmospheric pressure by an amount whose mean value is $\frac{1}{2}\rho z$. Hence the plates experience an apparent attraction $\frac{1}{2}\rho z^2$ per unit breadth.

127. The Capillary Curve.

In the preceding elementary investigation it was assumed that the radius of curvature of the surface is small compared with the elevation z, and sensibly uniform. If we abandon this restriction the question becomes more difficult, and we shall accordingly notice only the case where the surface is cylindrical (in the general sense) with generating lines horizontal, so that the problem is virtually a two-dimensional one.

Let the axes of x and y be taken horizontal and vertical, respectively, in a plane perpendicular to the generating lines, and let the axis of x be at the level at which the pressure-intensity in the interior of the liquid has the atmospheric value. The surface is now sufficiently represented by its section with the plane xy. Let s denote the arc of this curve, and ψ the angle which the tangent, drawn in the direction of s increasing, makes with the axis of x.

Consider the forces acting on a portion of the surface-film bounded by two planes parallel to xy at unit distance apart, and two straight edges perpendicular to these planes. If we resolve parallel to x, the tensions on the two straight edges contribute a component

$$T \cos \psi_2 - T \cos \psi_1,$$

where the suffixes refer to the two edges.

We will suppose that the liquid occupies the region lying to the right of the profile as this is traversed in the direction of s increasing. Since ρy measures the defect of pressure-intensity on the side of the fluid, the difference of pressures on the two faces of the portion of the film considered gives a component

$$\int \rho y \sin \psi \, ds = \rho \int y \, dy = \tfrac{1}{2} \rho \left(y_2^2 - y_1^2\right)$$

parallel to x. Hence along the profile we have

$$T \cos \psi + \tfrac{1}{2} \rho y^2 = \text{const.}, \quad \text{.................}(1)$$

or

$$y^2 = C - 2b^2 \cos \psi, \quad \text{.......................}(2)$$

if

$$b^2 = T/\rho. \quad \text{.........................}(3)$$

If we differentiate (1) with respect to the arc s, we have, since $dy/ds = \sin \psi$,

$$Ry = b^2, \quad \text{...........................}(4)$$

where R denotes the radius of curvature $(ds/d\psi)$*. This result might have been obtained at once from the consideration that the tensions on the ends of an element δs are equivalent to a normal

* The curves defined by (4) present themselves again in the theory of the finite flexure of a thin rod (see Art. 150), and are accordingly known also as the 'elastic curves.'

force $T\delta\psi$, as in the case of a string (Art. 80), and that this is balanced by the excess of pressure, $\rho y\,\delta s$, on the atmospheric side. The formula is in fact a particular case of Art. 124 (3), obtained by making $p_0 = \rho y$, $R_1 = R$, $R_2 = \infty$.

The curve assumes a variety of forms according to the value of the constant C in (2). In the particular case where the free surface is mainly plane, the axis of x is an asymptote; whence, putting $y = 0$, $\psi = 0$, we find $C = 2b^2$, and

$$y = \pm\, 2b \sin \tfrac{1}{2}\psi. \quad \text{........................(5)}$$

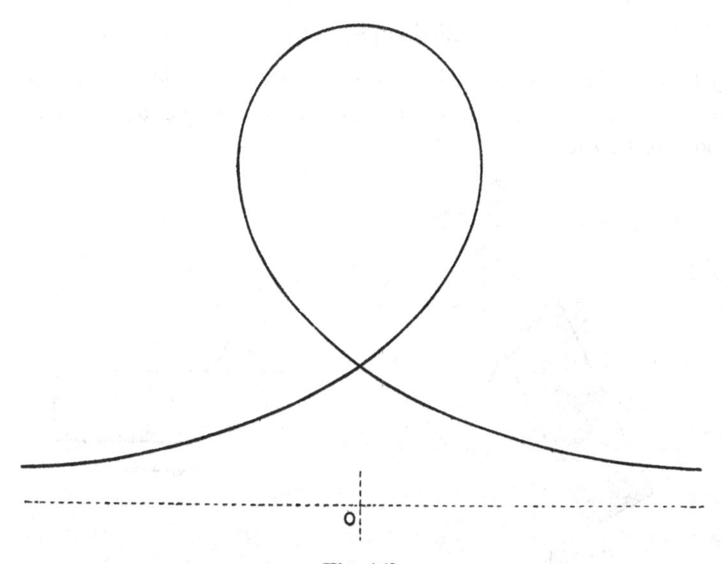

Fig. 152.

To find the relation between x and ψ we have, taking the upper sign,

$$\frac{dx}{d\psi} = \frac{dx}{ds}\frac{ds}{d\psi} = \frac{b^2 \cos\psi}{y} = \tfrac{1}{2} b \operatorname{cosec} \tfrac{1}{2}\psi - b \sin \tfrac{1}{2}\psi, \quad \text{...(6)}$$

and therefore

$$x = b \log \tan \tfrac{1}{4}\psi + 2b \cos \tfrac{1}{2}\psi, \quad \text{..............(7)}$$

provided the origin of x correspond to $\psi = \pi$.

If we put $$\tan \tfrac{1}{4}\psi = e^{-u}, \qquad (8)$$

we find $$x/2b = \tanh u - \tfrac{1}{2}u, \quad y/2b = \operatorname{sech} u. \qquad (9)$$

The curve (Fig. 152) is easily traced with the help of a table of the hyperbolic functions.

A line cutting the curve at an angle equal to the angle (α) of contact may be taken to represent a solid boundary; and the portion of the curve on one side of the point of contact will then give the profile of the liquid surface as modified by the presence of the solid. Thus where the liquid is in contact with a vertical wall we have $\psi = \tfrac{1}{2}\pi - \alpha$, and the elevation above the general level is

$$2b \sin(\tfrac{1}{4}\pi - \tfrac{1}{2}\alpha), \qquad (10)$$

by (5)*. The case of a plate dipping in an inclined position into water is illustrated by Fig. 153, where the angle of contact is taken to be zero.

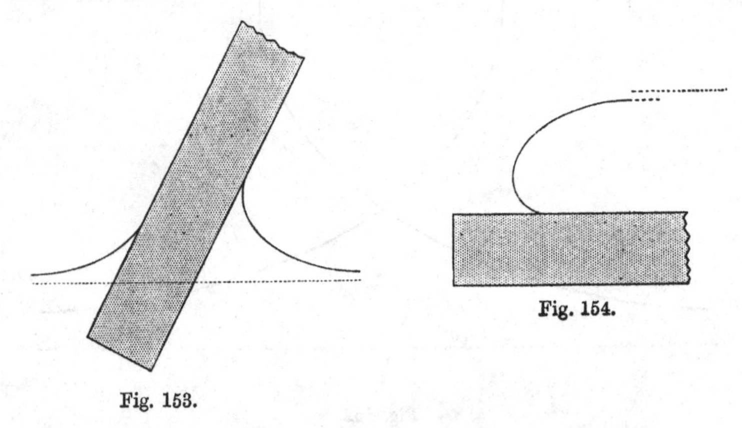

Fig. 153.

Fig. 154.

In the application to the case of a large drop of liquid resting on a horizontal plate we take the lower sign in (5). The investigation will in other respects hold, so far as the problem can be regarded as a two-dimensional one; and the figure has merely to be inverted. See Fig. 154.

Thus in the case of a large flat drop of mercury resting on a glass plate, if we ignore the effect of the curvature in a horizontal

* In the case of $\rho = 1$, $T = \cdot 077$, we have $2b = \cdot 55$ cm., $\sqrt{2} \, . \, b = \cdot 40$ cm.

sense, the height h of the drop, where its surface is sensibly plane, above the plate, is given by (5), viz. we have

$$h = 2b \cos \tfrac{1}{2}\alpha', \quad \text{.......................(11)}$$

where α' is the supplement of the obtuse angle of contact*.

When the constant C in (2) has any value other than $2b^2$ the expression for x in terms of ψ involves elliptic integrals.

If $C > 2b^2$, we may write

$$y^2 = \frac{4b^2}{k^2}(1 - k^2 \sin^2 \phi), \quad \text{..........................(12)}$$

where

$$\phi = \tfrac{1}{2}\pi + \tfrac{1}{2}\psi, \quad \text{..............................(13)}$$

and k is less than unity. Then

$$\frac{dx}{d\phi} = \frac{dx}{dy}\frac{dy}{d\phi} = -\cot 2\phi \, . \, \frac{2kb \sin\phi \cos\phi}{\Delta(k, \phi)} = \frac{2b}{k}\left\{\frac{1-\frac{1}{2}k^2}{\Delta(k, \phi)} - \Delta(k, \phi)\right\}. \quad \text{...(14)}$$

Hence†

$$\left.\begin{aligned} \frac{x}{b} &= \left(\frac{2}{k} - k\right) F(k, \phi) - \frac{2}{k} E(k, \phi), \\ \frac{y}{b} &= \frac{2}{k}\Delta(k, \phi). \end{aligned}\right\} \quad \text{..................(15)}$$

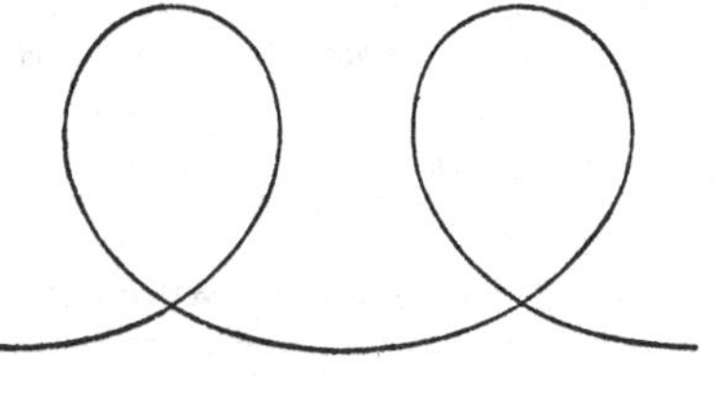

Fig. 155.

* By proper treatment of the glass surface the angle α' can be made equal to 0, and the formula then gives a method of determining b, and thence the surface-tension T. Another method consists in determining the depth h_1 of the points where the tangent plane is vertical; this is connected with b by the relation $h_1 = \sqrt{2} . b$. This observation, combined with (11), gives both b and α'.

† The notation is $\Delta(k, \phi) = \sqrt{(1 - k^2 \sin^2 \phi)}$,

$$E(k, \phi) = \int_0^\phi \Delta(k, \phi)\, d\phi, \quad F(k, \phi) = \int_0^\phi \frac{d\phi}{\Delta(k, \phi)}.$$

The integrals F, E are known as the elliptic integrals of the first and second kinds, respectively, and the parameter k which they involve is called the 'modulus.' The upper limit ϕ is called the 'amplitude.' Extensive tables of the integrals were calculated by Legendre. Useful abbreviations are given by Hoüel, *Recueil de Formules et de Tables Numériques*, and by J. B. Dale, *Five-Figure Tables...*, London, 1903.

The curve in Fig. 155 is drawn for the case of $k=\sin 75°$. It serves to illustrate (for example) the form assumed by a liquid surface between two parallel vertical plates when these are too far apart for the curvature to be uniform, but not sufficiently so for their effects to be independent. As the figure stands, it applies to the case where the angles of contact are acute; if they are obtuse we have only to invert it. In Fig. 156 the angle of contact is assumed to be acute.

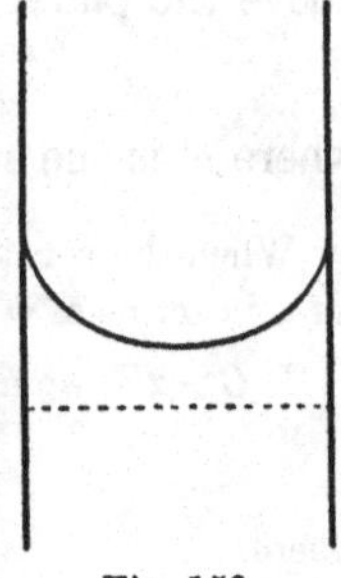

Fig. 156.

The relation between the modulus k of the formulæ and the distance (d) between the parallel plates is found as follows. For simplicity we assume the angle of contact to be zero.

As ψ increases from $-\frac{1}{2}\pi$ to 0, ϕ will increase from $\frac{1}{4}\pi$ to $\frac{1}{2}\pi$, whilst x increases by $\frac{1}{2}d$. Hence

$$\frac{d}{2b}=\left(\frac{2}{k}-k\right)\{F(k, \tfrac{1}{2}\pi)-F(k, \tfrac{1}{4}\pi)\}-\frac{2}{k}\{E(k, \tfrac{1}{2}\pi)-E(k, \tfrac{1}{4}\pi)\}. \quad \text{...(16)}$$

The simplest way of dealing with this equation would be to tabulate the values of the right-hand member for a series of values of k. Since b is known from (3), this gives the corresponding values of d. The corresponding elevations of the liquid, midway between the plates, are then given by the expression

$$\frac{2b}{k}\Delta(k, \tfrac{1}{2}\pi). \quad \text{.................................(17)}$$

When k is sufficiently small successive loops overlap, and in the limit they become indistinguishable from circles. In this way we get a transition to the state of things postulated in Art. 126. We have, in fact, for small values of k,

$$E(k, \phi)=\phi-\tfrac{1}{4}k^2(\phi-\sin\phi\cos\phi),$$
$$F(k, \phi)=\phi+\tfrac{1}{4}k^2(\phi-\sin\phi\cos\phi),$$

approximately. Substituting in (16) we find

$$k=d/b, \quad \text{.......................................(18)}$$

and the expression (17) then becomes

$$=\frac{2b^2}{d}=\frac{2T}{\rho d}, \quad \text{.................................(19)}$$

in agreement with Art. 126 (4).

If we were to carry the approximation a step further we should find

$$\frac{d}{b}=k\left\{1+(\tfrac{1}{4}+\tfrac{1}{16}\pi)\frac{d^2}{b^2}\right\}, \quad \text{.........................(20)}$$

and, in place of (19),

$$\frac{2T}{\rho d}-(\tfrac{1}{2}-\tfrac{1}{8}\pi)\,d. \quad \text{.............................(21)}$$

If, in (2), $C < 2b^2$, y will vanish for certain values of ψ, and we write accordingly

$$y^2 = 2b^2(\cos\beta - \cos\psi)$$
$$= 4b^2(\cos^2\tfrac{1}{2}\beta - \cos^2\tfrac{1}{2}\psi). \quad \text{...(22)}$$

Hence, putting

$$\cos\tfrac{1}{2}\psi = \cos\tfrac{1}{2}\beta \sin\phi, \quad \text{...(23)}$$

we have $\quad y = 2b\cos\tfrac{1}{2}\beta\cos\phi, \quad$ ......(24)

and

$$\sin\tfrac{1}{2}\psi = \Delta(k, \phi), \quad [k = \cos\tfrac{1}{2}\beta]. \quad \text{...(25)}$$

Also

$$\frac{dx}{d\phi} = \frac{dx}{dy}\frac{dy}{d\phi} = -\cot\psi \,.\, 2b\cos\tfrac{1}{2}\beta\sin\phi$$
$$= -\frac{b\cos\psi}{\sin\tfrac{1}{2}\psi}$$
$$= b\left\{2\Delta(k, \phi) - \frac{1}{\Delta(k, \phi)}\right\}. \quad \text{......(26)}$$

Hence

$$\left.\begin{aligned} x &= 2bE(k, \phi) - bF(k, \phi), \\ y &= 2kb\cos\phi, \end{aligned}\right\} \quad \text{...(27)}$$

provided the origin of x correspond to $\phi = 0$.

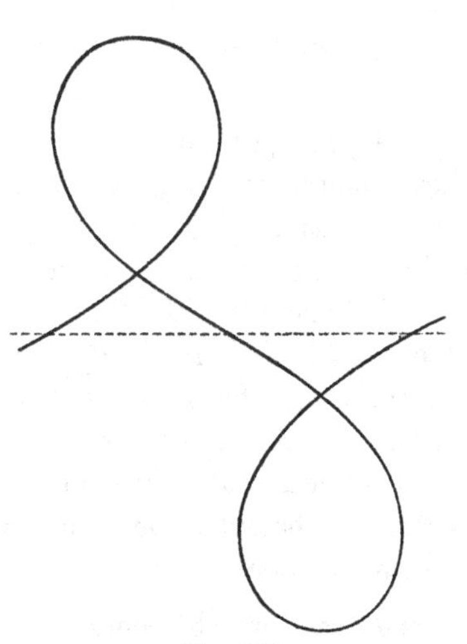

Fig. 157.

The curve assumes a variety of forms according to the value of β. The case of $\beta = 30°$ is shewn in Fig. 157. This serves to illustrate, for example, the form assumed by a liquid between two parallel vertical plates when the angle of contact is in one case acute, and in the other obtuse.

When β exceeds a certain value the curve has a wave-like form, and as β approaches 180° it approximates to a curve of sines (cf. Art. 150).

128. Soap-films. Minimal Surfaces.

The most striking phenomena of surface-tension are those exhibited by soap-films under various conditions as to the shape of the boundary, and the difference of pressure on the two sides. The forms produced are extremely beautiful in themselves, and have furnished problems of great interest to mathematicians*.

We shall denote by T the *total* tension of the film; this is equal to the sum of the tensions of *both* faces when (as is usually the case) these are independent.

* They were studied in great detail by J. A. F. Plateau (1801–83), professor of physics at Ghent, 1835–71. Similar phenomena were obtained in the case of masses of olive oil suspended in a mixture of water and alcohol of the same specific gravity, the influence of gravity being thus eliminated.

The form assumed is determined by the equation (3) of Art. 124, which connects the sum of the curvatures with the excess (p_0) of pressure on one side, and by the shape of the boundary*. If the pressure on both sides is the same, we have

$$R_1 = -R_2, \quad \text{..........................(1)}$$

i.e. the principal curvatures are equal and opposite. It may be shewn mathematically that this is the condition that the area of a surface having a given boundary should be stationary for small deformations. This is in agreement with the principle that the potential energy of the film must be stationary in a configuration of equilibrium. Practically, the only forms which can be realized are the stable forms for which the energy is a *minimum*. The surfaces possessing the property (1) are accordingly known to mathematicians as 'minimal' surfaces. The simplest instance is where the boundary is a plane curve, and the film itself consequently plane.

The case which comes next in interest is where the surface is of revolution. This can be treated without reference to the general surface-condition, as follows. If the coordinates x, y refer to the generating curve, the axis of x being along the axis of symmetry, we have, since the resultant of the tensions across any circular section must be the same,

$$T \cos \psi . 2\pi y = \text{const.} \quad \text{......................(2)}$$

or

$$y = c \sec \psi. \quad \text{..........................(3)}$$

We have already met with this relation in the case of the uniform catenary, and it is easy to see that it is characteristic of that curve. Thus, differentiating with respect to s we find

$$\frac{ds}{d\psi} = c \sec^2 \psi, \quad \text{........................(4)}$$

whence

$$s = c \tan \psi, \quad \text{........................(5)}$$

if the origin of s be at the point for which $\psi = 0$. It was from this formula (5) that the properties of the catenary were developed in Art. 83. The surface in question is therefore that generated by the revolution of a catenary about its directrix, and

* Since the arithmetic mean of the principal curvatures R_1^{-1}, R_2^{-1} is constant the surfaces in question are called surfaces 'of constant mean curvature.'

is accordingly known as the 'catenoid.' Of the two principal radii of curvature, one is the radius of curvature of the generating curve, and the other is the normal intercepted by the directrix. It is known (Art. 84) that these are numerically equal, but on opposite sides of the surface, in accordance with the formula (1).

129. Soap-films symmetrical about an Axis.

The forms which the surfaces of revolution may assume when there is a constant difference of pressure on the two sides are also of interest. The particular cases of the sphere and the cylinder are covered by the formulæ (1) and (2) of Art. 124. In the more general case, the portion of the film included between two circular sections of radii y_1 and y_2 is in equilibrium under the tensions across these circles and a uniform pressure p_0, which may be positive or negative, in the interior. The latter has, on the principles of Art. 96, a resultant $-p_0(\pi y_1^2 - \pi y_2^2)$ parallel to the axis, whence

$$T\cos\psi_1 . 2\pi y_1 - T\cos\psi_2 . 2\pi y_2 = p_0(\pi y_1^2 - \pi y_2^2). \quad \text{...(1)}$$

The equation of the meridian curve is therefore

$$\frac{y^2}{a} = 2y\cos\psi - C, \quad \text{.....................(2)}$$

if

$$a = T/p_0. \quad \text{...........................(3)}$$

The equation (2), which is virtually a differential equation of the first order, has a simple interpretation*. If we put

$$y = p, \quad \cos\psi = p/r, \quad \text{.....................(4)}$$

we obtain the tangential-polar equation of a curve which, rolling

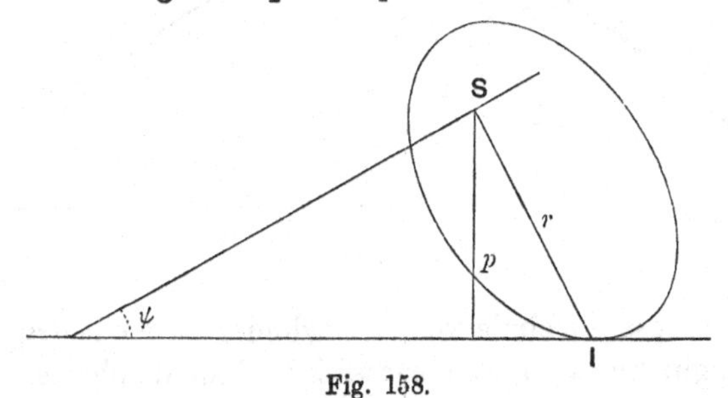

Fig. 158.

on the axis of x, would generate the required profile, the origin of the radius vector r being the tracing point.

* Due to C. E. Delaunay (1841).

The equation thus obtained from (2), viz.

$$\frac{C}{p^2} = \frac{2}{r} - \frac{1}{a}, \qquad \text{(5)}$$

where the sign of a depends on that of p_0, is seen to be identical with the tangential-polar equation of a conic referred to a focus. Thus in the case of the ellipse we have

$$\frac{l}{p^2} = \frac{2}{r} - \frac{1}{a}, \qquad \text{(6)}$$

where l is the half latus-rectum, and a the major semi-axis; for the branch of a hyperbola surrounding the focus in question we have

$$\frac{l}{p^2} = \frac{2}{r} + \frac{1}{a}; \qquad \text{(7)}$$

whilst for the opposite branch

$$\frac{l}{p^2} = -\frac{2}{r} + \frac{1}{a}. \qquad \text{(8)}$$

The transition between the cases (6) and (7) is furnished by the parabola

$$\frac{l}{p^2} = \frac{2}{r}. \qquad \text{(9)}$$

If the conic is an ellipse, the focus generates a wavy curve, and the corresponding surface of revolution is called an 'unduloid.'

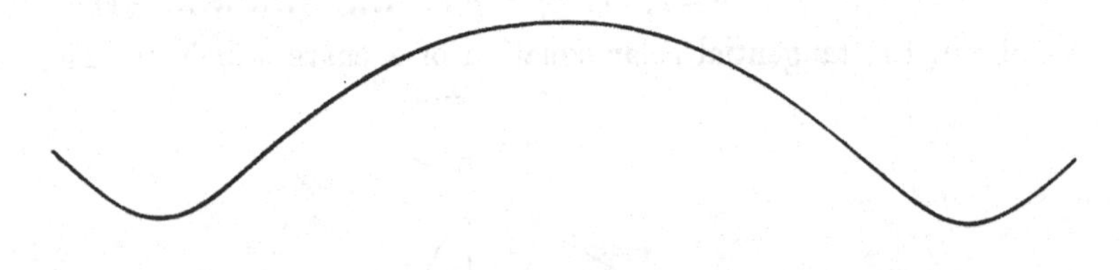

Fig. 159.

When the ellipse is a circle we get a cylinder; when it degenerates into a straight line we have a succession of equal spheres.

When the conic is a parabola, the curve traced out is a catenary. This corresponds to the case of $a = \infty$, or $p_0 = 0$, in accordance with our previous result (Art. 128).

When the conic is a hyperbola, we must imagine the two branches to roll alternately on the straight line, the point of contact changing from one branch to the other at infinity, when an asymptote coincides with the fixed straight line. The complete curve has a succession of nodes, and the surface generated is called a 'nodoid.' It will be found on examination that the excess of pressure is everywhere on the concave side of the curve.

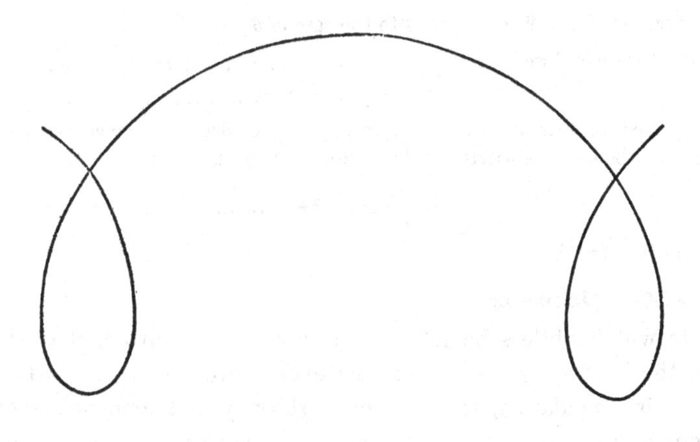

Fig. 160.

To calculate the forms of the various curves we may proceed as follows.

In the case of the *unduloid*, if α, β be the maximum and minimum values of y, the equation (2) may be written

$$y^2=(\alpha+\beta)\,y\cos\psi-\alpha\beta, \qquad (10)$$

whence

$$\cot\psi=\pm\frac{y^2+\alpha\beta}{\sqrt{\{(\alpha^2-y^2)(y^2-\beta^2)\}}}. \qquad (11)$$

Since y^2 ranges between α^2 and β^2 we may put

$$y^2=\alpha^2\cos^2\phi+\beta^2\sin^2\phi=\alpha^2(1-k^2\sin^2\phi), \qquad (12)$$

where ϕ is an auxiliary variable, and

$$k^2=(\alpha^2-\beta^2)/\alpha^2. \qquad (13)$$

Hence

$$y\,dy=-k^2\alpha^2\sin\phi\cos\phi\,d\phi, \qquad (14)$$

and therefore

$$\frac{dx}{d\phi}=\frac{dx}{dy}\frac{dy}{d\phi}=\mp\left(y+\frac{\alpha\beta}{y}\right). \qquad (15)$$

Hence, if we restrict ourselves to that part of the curve for which $dx/d\phi$ is positive,

$$\left.\begin{aligned} x &= aE(k, \phi) + \beta F(k, \phi), \\ y &= a\Delta(k, \phi), \end{aligned}\right\} \quad \text{.......................(16)}$$

the origin of x being at a point where $\phi = 0$, and therefore $y = a$.

If e be the eccentricity of Delaunay's rolling ellipse, we have

$$e = \frac{a - \beta}{a + \beta} = \tan^2 \tfrac{1}{2}\theta, \quad \text{if } k^2 = (a^2 - \beta^2)/a^2 = \sin^2\theta. \quad \text{...........(17)}$$

The curve in Fig. 159 corresponds to the case of $\theta = 75°$.

For the *nodoid* we have, assuming $y = a$ for $\psi = 0$, and $y = \beta$ for $\psi = \pi$,

$$y^2 = (a - \beta)\, y \cos\psi + a\beta. \quad \text{..........................(18)}$$

The preceding formulæ will therefore apply, provided we change the sign of β. If e be the eccentricity of the rolling hyperbola, we find

$$e = \frac{a + \beta}{a - \beta} = \cot^2 \tfrac{1}{2}\theta. \quad \text{............................(19)}$$

In Fig. 160, $\theta = 75°$.

130. Cohesion.

It will doubtless be felt, and it must be admitted, that the hypothesis of a surface-tension, however convenient as a starting point for calculation, is somewhat arbitrary and artificial when regarded as a physical assumption. A more fundamental view ascribes the phenomena of Capillarity to the existence of a 'cohesive force' in liquids (as well as in solids). This force of mutual attraction is supposed to act with very great intensity at very short distances, but to fall off very rapidly as the distance between the attracting particles increases, so that the actual range (ϵ, say) within which it is sensible is still very minute. This force is distinct from the mutual gravitation of the parts of a liquid, whose effects, in masses of ordinary size, are quite insensible. On the other hand the theory of cohesion, in its main outlines at least, does not necessarily involve any recourse to molecular hypotheses; for the range ϵ, although actually extremely minute, may be supposed to be very great compared with intra-molecular distances, so that the liquid may still, from the present standpoint, be treated as continuous.

The full development of this view would require the methods of the theory of Attractions, but a few consequences may be noticed.

In the first place, a given mass of liquid will possess a certain quantity of potential energy in virtue of the cohesive forces alone. Moreover, it is plain that all particles whose shortest distance from the surface exceeds the range ϵ will be under the same conditions, since a sphere of radius ϵ described about any such particle as a centre will include all the matter which exerts any force upon it. In the residual surface film, of thickness ϵ, the conditions will be different, since a sphere of radius ϵ described about any point within this film as centre will not be completely occupied by attracting matter. Moreover, if the radii of curvature of the surface film be very great compared with ϵ, equal small areas (whose linear dimensions are large compared with ϵ) will be under similar conditions, so that the term in the expression of the potential energy which is due to the film will be simply proportional to its area. We have here the explanation of the 'superficial energy,' which leads mathematically, as we have seen, to the same consequences as the more artificial hypothesis of surface-tension.

A reason is also apparent why capillary phenomena should be so remarkably sensitive to the influence of 'contamination.' The introduction of a film of foreign matter, of thickness comparable with ϵ, greatly alters the attractive forces which are operative.

131. Intrinsic Pressure.

Again, it appears that in the interior of a liquid there is a certain 'intrinsic pressure,' due to the cohesive forces, which is superposed on the pressure calculated by the rules of the preceding chapters. Suppose, for example, we have a mass of liquid with a plane horizontal free surface at which the pressure is p_0; and consider the equilibrium of a vertical column, of unit sectional area, extending downwards from the surface to a depth z, which we will suppose greater than ϵ. The pressure p on the base of this column has to balance the atmospheric pressure p_0 on the top, the gravity ρz of the column, and the attraction exerted on a film of thickness ϵ at the base by the whole mass of fluid below the plane of the base. This latter attraction is a constant K, depending only on the law of attraction, whence we have

$$p = p_0 + K + \rho z, \quad \text{.......................(1)}$$

in place of Art. 92 (3). The argument does not apply if $z < \epsilon$; within a depth ϵ the pressure must vary rapidly from p_0 to $p_0 + K$.

The existence of this intrinsic pressure K does not affect any of the conclusions based on the ordinary principles of Hydrostatics. Whether we take account of the cohesive forces and the intrinsic pressure, or whether we ignore both, we are led to the same results. Suppose, for instance, we are calculating the forces exerted by a liquid on a vertical wall. To put the argument in as simple a form as possible, imagine a plane drawn parallel to the wall at a distance ϵ from it, on the side of the fluid, and let us reckon the film thus separated as forming part of the wall. Then on every part of the 'wall,' as thus understood, we have a pressure $p_0 + K + \rho z$, and an attraction K, per unit area in each case, and the result is equivalent to a pressure $p_0 + \rho z$, as in Art. 92. The calculations of resultant pressure, and of centres of pressure, are therefore unaffected.

It follows that the magnitude of the intrinsic pressure K cannot be determined by hydrostatic experiments. The estimates which have been made by various physicists are based on a consideration of the energy which is required to overcome the cohesion of a liquid in the process of conversion into vapour, or on other thermodynamical data. The estimates agree in assigning very high values to the constant; thus in the case of water Prof. van der Waals infers that K is about 11,000 atmospheres.

132. Influence of Curvature of the Surface.

So far, the upper surface has been assumed to be plane and horizontal. A solid wall modifies the form of the free surface in its immediate neighbourhood, by its own attraction. The way in which the angle of contact is determined, by considerations of superficial energy, has already been indicated in Art. 125. It remains to account for the variation of pressure in the neighbourhood of a *curved* surface of liquid, as dependent on the curvature. The detailed calculation would be out of place here, but an indication can be given. We have seen (Art. 111) that the pressure-gradient in a fluid, in any direction, is equal to the force per unit volume in that direction. Now as we pass inwards in the direction of the normal, from (say) a convex surface, the force is *greater* than in the case of a plane surface, owing to the absence of the attraction of the matter which would fill the space between the surface and its tangent plane. The pressure therefore increases more rapidly, so that at a depth ϵ it is now greater than $p_0 + K$ by an amount which is evidently greater, the greater the curvature. The result is found to be

$$p_0 + K + T\left(\frac{1}{R_1} + \frac{1}{R_2}\right), \quad \text{..................(1)}$$

where T is the superficial energy, and R_1, R_2 are the principal radii of curvature. The case of a concave surface, or of an anticlastic curvature, is included if we attribute the proper signs to R_1, R_2. The pressure at a depth z is found, as usual, by adding a term ρz.

EXAMPLES. XXIII.

1. If n equal spherules of water coalesce so as to form a single drop, prove that the superficial energy is diminished in the ratio $n^{-\frac{1}{3}}$.

2. Find the elevation of water ($T = {\cdot}077$) in a vertical glass tube 1 mm. in diameter, assuming the angle of contact to be 27°. [2·75 cm.]

How will the result be affected if the tube is inclined?

3. How is the principle of Archimedes affected by capillarity? A circular cylinder of radius a, height h, and density ρ floats upright in water; find the depth of the base below the general level of the water surface.

$$\left[\rho h + \frac{2T}{a}\cos\alpha.\right]$$

4. Prove that if a film of liquid be raised by capillary attraction between two vertical plates which make a very acute angle with one another, the edge of the film has the form of a rectangular hyperbola.

5. A film of water is included between two parallel plates of glass at a small distance d apart. Prove that the apparent attraction between the plates is

$$\frac{2AT\cos\alpha}{d} + lT\sin\alpha,$$

where A is the area of the film, l its perimeter, and α the angle of contact.

Examine the case of a film of mercury.

6. Prove that the intrinsic equation of the capillary curve of Art. 127 (9) is

$$s = b\log\tan\tfrac{1}{4}\psi.$$

7. Prove that if a fine thread form part of the boundary of a plane soap-film, the thread will form an arc of a circle.

8. Prove that if the superficial energy E of a very thin soap-film varies with the thickness t, the surface-tension is connected with E by the formula

$$T = E - t\frac{dE}{dt}.$$

9. Prove that a soap-film cannot exist between two equal coaxial circular rings if the ratio of the distance between the planes of the rings to their diameter exceeds ·6625.

[This is the value of cosech u, where u is the positive root of the equation $\tanh u = 1/u$.]

10. Prove that if the surface of a sheet of water be slightly corrugated the superficial energy is increased by

$$\tfrac{1}{2}T\int\left(\frac{d\zeta}{dx}\right)^2 dx,$$

per unit of breadth of the corrugations, the axis of x being horizontal and perpendicular to the corrugations, and ζ denoting the elevation above the mean level.

If the disturbed surface has the form of a train of ripples

$$\zeta = c \sin kx,$$

prove that the average increment of surface energy per unit area is $\tfrac{1}{4}Tk^2c^2$.

Comparing with Ex. XIX. 30 (p. 243), shew that gravitational or capillary energy is the more important according as the wave-length of the corrugations is $\gtrless 2\pi\sqrt{(T/\rho)}$.

11. A mass of liquid revolves, under surface-tension alone, about a fixed axis, with small angular velocity ω, so as to assume a slightly ellipsoidal shape. Prove that the ellipticity is

$$\frac{1}{8}\frac{\rho\omega^2 a^3}{T},$$

if T be expressed in dynamical measure.

CHAPTER XV

STRAINS AND STRESSES

133. Introduction.

In the theory of the equilibrium of solid bodies, as developed in the earlier portions of this book, only slight and occasional reference has been made to the internal forces, and accompanying small deformations, which are called into play. The fundamental principles of the subject may in fact be understood as applying, without any sort of qualification, to bodies as they are in their actual equilibrium condition. Whether, or how much, this differs from the normal condition is really irrelevant.

The study of internal stress is, however, important for several reasons. In practice it is essential that the stresses in a given structure should not be greater than the material can safely bear; on the other hand it is desirable, for economical reasons, that the parts should not be unduly massive, and out of all proportion to the effort which may be demanded of them. Again, as already indicated (Art. 25), it is only by taking the elasticity of actual bodies into account that we can hope to obtain a real solution of problems which, on the ordinary principles of Statics, are 'indeterminate.'

The theoretical treatment of such questions, as might be expected, is difficult; and in the following pages a brief sketch of the more elementary portions of the subject is all that is attempted. This will be sufficient, however, for the discussion of a number of problems, all of them interesting and of practical importance. It is hoped, moreover, that even this outline may be acceptable to the theoretical student, as tending to dispel the suspicion of unreality which sometimes attaches to the subject of Statics, as apparently dealing only with fictitious 'rigid' bodies enduring and transmitting forces with absolute insensibility.

134. . Homogeneous Strain.

The investigation of the various kinds of deformation, or 'strain,' which a body can undergo, apart from any consideration of its physical constitution, or of the forces which are in action, is a matter of pure Geometry. It is usual to begin with the case of 'homogeneous' strain, which is characterized by the property that any two lines in the substance which were originally straight and parallel remain straight and parallel after the deformation. It follows that a parallelogram remains a parallelogram, although its angles and the directions of its sides are generally altered. Consequently the lengths of all finite parallel straight lines are altered in the same ratio; but this ratio will in general be different for different directions in the substance. If PQ, $P'Q'$ denote any straight line in the substance, before and after the strain, the ratio of the *increment* of length to the original length, viz.

$$\epsilon = \frac{P'Q' - PQ}{PQ}, \quad \text{.......................(1)}$$

is called the 'extension.' It will, on the present hypothesis, depend only on the direction of PQ in the substance, and not on its actual position.

In a homogeneous strain there is one, and in general only one, set of three mutually perpendicular directions in the substance which remain mutually perpendicular after the deformation. For consider the various directions through any point O. Since the extensions of lines lying originally along these are confined between finite limits, there must be one such direction, at least, for which the extension is a maximum. Let OA be a line drawn in this direction, and OP a line of equal length making an

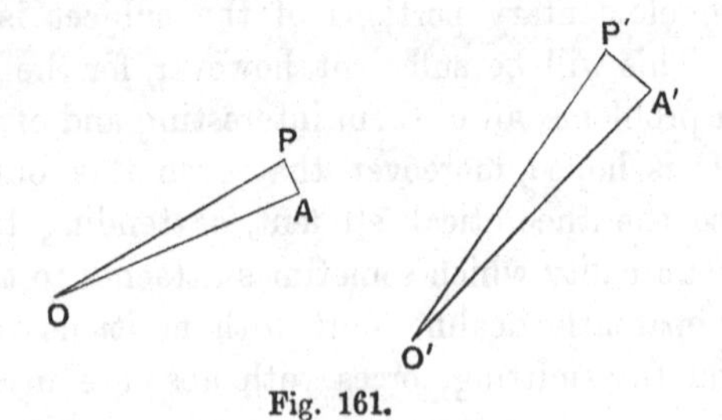

Fig. 161.

infinitely small angle with it. The lengths $O'A'$, $O'P'$ of the same lines after the strain will then be equal, to the first order, so that $A'P'$ is ultimately perpendicular to $O'A'$. We infer that all lines and planes which were originally perpendicular to OA remain perpendicular to $O'A'$. Next consider the various directions OQ lying in a plane perpendicular to OA. For one of these, say OB, the extension must be a maximum, and it follows in the same way that a line OC which was originally perpendicular to both OA and OB will after the strain be perpendicular to both $O'A'$ and $O'B'$. Hence there is in general one definite set of three mutually perpendicular lines OA, OB, OC, such that the corresponding lines $O'A'$, $O'B'$, $O'C'$ after the strain are mutually perpendicular. These are called the 'principal axes' of the strain. The corresponding extensions are called the 'principal extensions'; we shall denote them by ϵ_1, ϵ_2, ϵ_3.

If we denote the coordinates of a point P relative to OA, OB, OC as coordinate axes by x, y, z, and those of the corresponding point P' in the strained state, relative to $O'A'$, $O'B'$, $O'C'$, by x', y', z', we have

$$x'=(1+\epsilon_1)\,x,\quad y'=(1+\epsilon_2)\,y,\quad z'=(1+\epsilon_3)\,z, \quad\ldots\ldots\ldots\ldots(2)$$

by (1). Hence the points which originally lay upon a sphere

$$x^2+y^2+z^2=r^2 \quad\ldots\ldots\ldots\ldots(3)$$

will after the strain lie on the ellipsoid

$$\frac{x'^2}{a'^2}+\frac{y'^2}{b'^2}+\frac{z'^2}{c'^2}=1, \quad\ldots\ldots\ldots\ldots(4)$$

where $$a'=(1+\epsilon_1)\,r,\quad b'=(1+\epsilon_2)\,r,\quad c'=(1+\epsilon_3)\,r\,; \quad\ldots\ldots\ldots\ldots(5)$$

cf. Art. 76.

The ratio of the increase of volume to the original volume is called the 'dilatation'; we denote it by Δ*. Considering the change of volume of a unit cube whose edges are parallel to the principal axes, we have

$$1+\Delta=(1+\epsilon_1)(1+\epsilon_2)(1+\epsilon_3)\ldots\ldots\ldots\ldots(6)$$

135. Simple types of Strain. Uniform Extension. Shear.

So far, the strains contemplated may be of any magnitude; but in the application to elastic solids the extensions are always so

* If ρ be the density, and ρ_0 the density in the unstrained state, we have $\rho_0=\rho\,(1+\Delta)$. Hence if s be the 'condensation' (Art. 121), we have $(1+\Delta)\,(1+s)=1$. In the case of small strains this reduces to $s=-\Delta$.

very minute that the squares and products of ϵ_1, ϵ_2, ϵ_3 may be neglected in comparison with these quantities themselves. We therefore write, for example,

$$\Delta = \epsilon_1 + \epsilon_2 + \epsilon_3. \quad \text{.......................(1)}$$

There are certain types of homogeneous strain which are of special importance.

First, suppose that the principal extensions are all equal, say

$$\epsilon_1 = \epsilon_2 = \epsilon_3 = \epsilon. \quad \text{.......................(2)}$$

Any portion of the substance which was originally spherical then remains spherical, and the extension is accordingly uniform in all directions. Also, if the strains are small,

$$\Delta = 3\epsilon, \text{..............................(3)}$$

i.e. the dilatation of volume is three times the linear extension.

The next type to be considered may be described as a differential sliding of a system of parallel planes. These planes undergo no deformation in themselves, and no alteration in their mutual distances; but each is shifted relatively to a fixed plane of the system, in a fixed direction, by an amount proportional to its distance from that plane. This kind of strain is called a 'shear*.' The amount (η) of the shear is specified by the relative displacement of two planes of the system which are at unit distance apart.

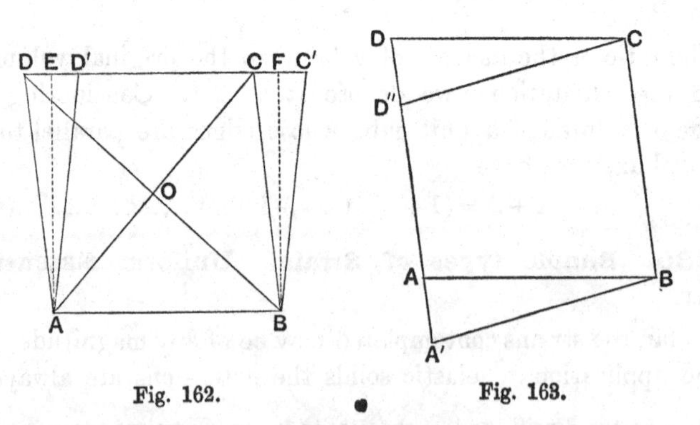

Fig. 162. Fig. 163.

* The action of an ordinary pair of shears is in fact to set up a strain somewhat of this type, of amount so great that the material finally gives way.

The plane of the annexed Fig. 162 is supposed to be perpendicular to the aforesaid system of parallel planes, and to contain the direction of relative displacement. AB is a line of unit length in the fixed plane of reference, and $AD = AD' = BC = BC' = AB$. Hence, considering a shear which converts the rhombus $ABCD$ into the equal rhombus $ABC'D'$, we have $\eta = DD'/AE$, where E is the middle point of DD'. Also, if O be the intersection of the diagonals AC, BD, it is plain that OA, OB, being lines of the substance which are at right angles before and after the strain, are the directions of two of the principal axes, whilst the third principal axis is normal to the plane of the figure. Hence, with our former notation,

$$\frac{OB}{OA} = \frac{BD}{AC} = \frac{AC'}{AC} = 1 + \epsilon_1, \quad \text{.................(4)}$$

and similarly

$$\frac{OA}{OB} = \frac{AC}{BD} = \frac{BD'}{BD} = 1 + \epsilon_2, \quad \text{.................(5)}$$

so that

$$(1 + \epsilon_1)(1 + \epsilon_2) = 1; \quad \text{.......................(6)}$$

whilst, of course, $\epsilon_3 = 0$. Again,

$$\tan ABC = \tan 2ABO = \frac{2(1 + \epsilon_2)}{1 - (1 + \epsilon_2)^2} = \frac{2}{\epsilon_1 - \epsilon_2}, \quad \text{......(7)}$$

by (5) and (6), whence

$$\eta = 2 \cot ABC = \epsilon_1 - \epsilon_2. \quad \text{.................(8)}$$

These results are exact, but in the case of *small* strains we have, by (6),

$$\epsilon_1 = -\epsilon_2, = \epsilon, \quad \text{..........................(9)}$$

say, and therefore

$$\eta = 2\epsilon. \quad \text{.........................(10)}$$

Hence an infinitely small shear η may be resolved into a uniform extension $\frac{1}{2}\eta$ in one direction together with an equal contraction in a perpendicular direction.

If we imagine the rhombus $ABC'D'$ to be moved in its own plane until BC' coincides with BC, it will assume a position such as $A'BCD''$ in Fig. 163. Since this does not affect the *relative* displacements, we see that the strain may also be described as a differential sliding of planes parallel to BC. There is therefore a second system of parallel planes in the substance which undergo no deformation in themselves. In the case of an infinitesimal shear, the two systems of parallel planes are at right angles.

136. Homogeneous Stress. Simple types.

The term 'stress' was used in the earlier portions of this book to denote, in a general sense, the mutual actions between the various parts of a body or a mechanical system. We have now to consider the mutual action between two portions of a body separated by an ideal surface, or interface, S. This question has already been discussed in Art. 90, by way of introduction to the particular case of hydrostatic stress; and it was seen that the force exerted across any small area of S may be taken to be ultimately proportional to this area, and that the intensity of the stress across S at any point is accordingly to be specified by the force per unit area.

In a solid body this stress may be of the nature of a pressure or a tension, and it may be normal or oblique, or even tangential to the area. It will moreover in general be different for surfaces S drawn in different directions through the same point. The complete specification of stress in a solid body is therefore a much more complicated matter than in the case of a fluid in equilibrium.

A stress is said to be 'homogeneous' throughout a body when it is uniform, and the same in every respect, over any two parallel planes. It may be shewn that there are then three mutually perpendicular systems of parallel planes such that the stress across each is in the direction of the normal, although it is usually of different intensity for the several systems. The planes in question are called 'principal planes' of the stress, and the three mutually perpendicular directions determined by their intersections are called the 'principal axes.' The corresponding stress-intensities are called the 'principal stresses.' An elementary proof of the theorem might be given, but is hardly necessary for our purpose. In the case of an 'isotropic' solid (Art. 137) its truth can be inferred indirectly. The principal stresses are usually reckoned positive when of the nature of *tensions*; we denote them by p_1, p_2, p_3.

There are certain special types of stress to be noticed, analogous to the special types of strain considered in Art. 135.

First, let $p_1 = p_2 = p_3$. The stress across *any* plane is then in the direction of the normal, and of uniform intensity, as in Hydrostatics (Art. 91).

We take next the case of a pure 'shearing stress.' This is characterized by the property that there are two systems of parallel planes, mutually perpendicular, such that the stress over each is wholly tangential, and at right angles to the common intersections. In Fig. 164 $ABCD$ represents a section of a unit cube, having two pairs of its faces parallel to the aforesaid planes. By taking moments about an axis perpendicular to the plane of

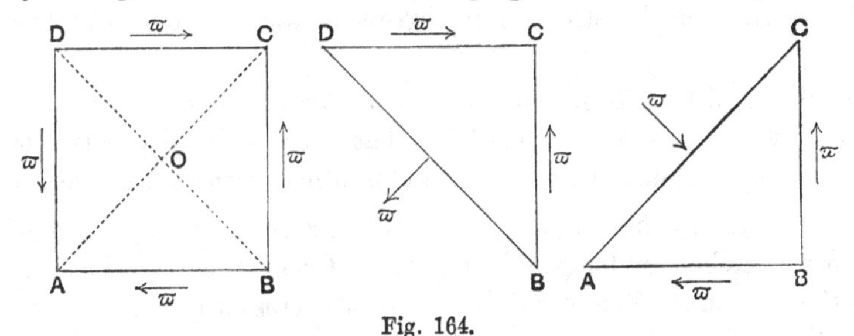

Fig. 164.

the figure we see that the stress-intensity (ϖ) must be the same for both systems. Again, the portion of the cube included between the faces AB, BC and the diagonal plane AC is in equilibrium under three forces. Two of these forces are perpendicular and proportional to AB and BC, viz. the tangential tractions on BC, AB, respectively. Hence the third force, viz. the total stress on the plane AC, must be perpendicular and proportional to CA. Hence the stress across AC is wholly normal, and equal to ϖ per unit area. A similar argument applies to the diagonal plane BD. Hence OA, OB, and a perpendicular through O to the plane of the paper are principal axes of the stress; and we may write

$$p_1 = -p_2 = \varpi, \quad p_3 = 0. \qquad \text{.....................(1)}$$

Hence a shearing stress ϖ is equivalent to a uniform tension ϖ in one direction combined with an equal pressure ϖ in a perpendicular direction.

137. Stress-Strain Relations. Young's Modulus. Poisson's Ratio.

When a solid body previously free from stress is deformed in any manner internal stresses are called into play. As to the relation of these stresses to the strains some physical hypothesis is necessary. We assume, in the first place, that the strains are *definite* functions of the stresses, so that the body when released from the action of force returns exactly to its original state. In other words the substance is assumed to be 'perfectly elastic.' A further natural assumption is that the strains, when sufficiently small, are *linear* functions of the stresses. This is known as 'Hooke's law'*; like the former assumption, though true in the limit when the stresses are infinitely small, it ceases to hold when the strains (or the stresses) transgress certain values called the 'elastic limits.'

We shall further assume, for the most part, that the bodies of which we treat are 'isotropic,' i.e. that there is no distinction of properties between one direction and another in the substance.

In practice, the simplest type of elastic deformation is that of a bar stretched by longitudinal force. If ϵ be the extension, and p the stretching force per unit area of the cross-section, we have, on the basis of Hooke's law,

$$p = E\epsilon, \qquad (1)$$

where E is a constant depending on the nature of the material, known as 'Young's modulus'†. Since ϵ is a mere ratio, the value of E is specified in terms of force per unit area. Some numerical results, expressed in gravitation measure (grammes per square centimetre) are given in the table on p. 301.

The bar at the same time undergoes a lateral change of dimensions, usually a contraction. The ratio of this lateral contraction to the longitudinal extension ϵ is called 'Poisson's ratio'‡; we denote it by σ. It will appear in a moment that the elastic properties of an isotropic substance are completely defined by these two constants E and σ.

138. General Stress-Strain Relations in an Isotropic Substance.

We proceed to the general expressions for the principal strains in terms of the stresses. It is evident from considerations of

* Robert Hooke (1635–1703), professor of geometry at Gresham College.

† Thomas Young (1773–1829), 'mathematician, physician, physicist, and Egyptologist.'

‡ S. D. Poisson (1781–1840), one of the founders of the mathematical theory of elasticity.

symmetry that in an isotropic body the principal axes of the strain will also be the principal axes of the accompanying stress; and further that the principal strain ϵ_1 will involve the principal stresses p_2, p_3 symmetrically, and so on. The general formulæ will therefore be of the type

$$\left.\begin{aligned} \epsilon_1 &= Ap_1 + B(p_2+p_3), \\ \epsilon_2 &= Ap_2 + B(p_3+p_1), \\ \epsilon_3 &= Ap_3 + B(p_1+p_2), \end{aligned}\right\} \quad \text{.................(1)}$$

involving two constants A and B. These are easily expressed in terms of E and σ. For if we put $p_2 = p_3 = 0$, as in the case of the stretched bar, we have

$$\epsilon_1 = Ap_1, \quad \epsilon_2 = \epsilon_3 = Bp_1, \text{ ....(2)}$$

whence $$E = p_1/\epsilon_1 = 1/A, \quad \sigma = -\epsilon_2/\epsilon_1 = -B/A. \text{ ....(3)}$$

The formulæ thus take the shape

$$\left.\begin{aligned} E\epsilon_1 &= p_1 - \sigma(p_2+p_3), \\ E\epsilon_2 &= p_2 - \sigma(p_3+p_1), \\ E\epsilon_3 &= p_3 - \sigma(p_1+p_2). \end{aligned}\right\} \quad \text{.................(4)}$$

There are two other important constants which express the behaviour of the substance under the two special types of stress considered in Art. 136.

In the case of a stress uniform in all directions we have

$$p_1 = p_2 = p_3, = p, \text{ ....(5)}$$

say. It follows from (4) that

$$\epsilon_1 = \epsilon_2 = \epsilon_3, = \tfrac{1}{3}\Delta, \text{ ....(6)}$$

and that $$p = \kappa\Delta, \text{ ....(7)}$$

where $$\kappa = \frac{E}{3(1-2\sigma)}. \text{ ....(8)}$$

The coefficient κ, which is the ratio of the uniform stress to the uniform dilatation which it involves, is called the 'volume-elasticity,' or 'cubical elasticity,' of the substance.

Next, suppose that

$$p_1 = -p_2 = \varpi, \quad p_3 = 0, \text{ ....(9)}$$

so that the stress is a pure shearing stress of intensity ϖ. It follows from (4) that

$$\epsilon_1 = -\epsilon_2, \quad \epsilon_3 = 0, \text{ ....(10)}$$

so that the consequent strain is a pure shear of amount $\eta = 2\epsilon_1$. Further, we have

$$\varpi = \mu\eta, \qquad \text{.........................(11)}$$

where

$$\mu = \frac{E}{2(1+\sigma)}. \qquad \text{.........................(12)}$$

This constant μ, which measures the ratio of a shearing stress to the shearing strain which it produces, is called the 'rigidity' of the substance.

Another case of some interest is where

$$p_1 = p_2, = p, \quad p_3 = 0, \qquad \text{..................(13)}$$

as in the case of a plate uniformly stressed laterally, but free from stress on the two faces. We find

$$p = \frac{E}{1-\sigma}\epsilon, \qquad \text{..........................(14)}$$

where $\epsilon = \epsilon_1 = \epsilon_2$, and

$$\frac{\epsilon_3}{\epsilon} = -\frac{2\sigma}{1-\sigma}. \qquad \text{......................(15)}$$

From (4) we can derive expressions for p_1, p_2, p_3 in terms of ϵ_1, ϵ_2, ϵ_3. It is evident from symmetry that these will have the forms

$$\left.\begin{aligned} p_1 &= \lambda(\epsilon_1 + \epsilon_2 + \epsilon_3) + 2\mu\epsilon_1, \\ p_2 &= \lambda(\epsilon_1 + \epsilon_2 + \epsilon_3) + 2\mu\epsilon_2, \\ p_3 &= \lambda(\epsilon_1 + \epsilon_2 + \epsilon_3) + 2\mu\epsilon_3, \end{aligned}\right\} \qquad \text{..............(16)}$$

and it is easily verified that here as before the symbol μ denotes the rigidity. In terms of the constants λ, μ we find by simple considerations

$$E = \frac{(3\lambda + 2\mu)\mu}{\lambda + \mu}, \quad \kappa = \lambda + \frac{2}{3}\mu, \quad \sigma = \frac{\lambda}{2(\lambda+\mu)}. \quad \text{...(17)}$$

As to the values which the various constants can assume, it is evident that κ, μ, and E are necessarily positive, since the substance would otherwise be in an unstable condition. Thus if κ were negative, a uniform pressure would produce an *expansion*, and the greater the pressure the greater the expansion. It follows from (8) and (12) that σ must lie between -1 and $\frac{1}{2}$. There does not appear to be any further necessary restriction on the value of σ, although, so far as observations go, it appears in isotropic substances to be positive.

The value of E is most readily found by observations on the stretching of a wire*, or the flexure of a bar (Art. 145); and that of μ by torsion experiments (Art. 154). The values of κ and σ are then deduced by the formulæ

$$\kappa = \frac{\mu E}{9\mu - 3E}, \quad \sigma = \frac{E}{2\mu} - 1, \quad \text{..............(18)}$$

which follow from (8) and (12).

On a particular hypothesis as to the ultimate structure of an elastic solid Poisson was led to the conclusion that the two elastic constants of an isotropic substance are not really independent, but are connected by an invariable relation, equivalent to $\sigma = \frac{1}{4}$. This would make

$$\kappa = \tfrac{5}{3}\mu, \quad E = \tfrac{5}{2}\mu. \quad \text{..............................(19)}$$

On experimental grounds Wertheim proposed in 1848 the value $\sigma = \frac{1}{3}$, which gives

$$\kappa = \tfrac{8}{3}\mu, \quad E = \kappa. \quad \text{...............................(20)}$$

All recent experiments confirm the view maintained by Stokes and other leading physicists that there is no necessary relation between the constants.

The annexed table gives the results of a few determinations by Everett (1867), the values of E, μ, κ being expressed in grammes per square centimetre.

	E	μ	κ	σ
Steel	$2{\cdot}18 \times 10^9$	$8{\cdot}35 \times 10^8$	$1{\cdot}88 \times 10^9$	·310
Wrought iron	2·00 ,,	7·84 ,,	1·48 ,,	·275
Copper	1·26 ,,	4·56 ,,	1·72 ,,	·378
Glass (1)	·615 ,,	2·45 ,,	·423 ,,	·258
Glass (2)	·585 ,,	2·40 ,,	·354 ,,	·229

139. Potential Energy of a Strained Elastic Solid.

The potential energy (W), per unit volume, of an isotropic substance homogeneously strained is found by calculating the work done by the surrounding matter on the faces of a unit cube whose edges are parallel to the principal axes, as the strains increase from zero to their final values.

In the case of a longitudinal extension, with free lateral contraction, the only stress is p_1. As the extension increases from

* It is to be observed however that a substance drawn into the form of a wire is not always isotropic.

zero to its final value ϵ_1, the stress increases proportionally from zero to its final value p_1, its mean value being $\frac{1}{2}p_1$. Hence

$$W = \tfrac{1}{2}p_1\epsilon_1 = \tfrac{1}{2}E\epsilon_1^2. \quad \text{......................(1)}$$

In the general case, no work is done by the stress p_1 during the extensions ϵ_2, ϵ_3, and so on. The result is therefore

$$\begin{aligned} W &= \tfrac{1}{2}(p_1\epsilon_1 + p_2\epsilon_2 + p_3\epsilon_3) \\ &= \tfrac{1}{6}\{(p_1+p_2+p_3)(\epsilon_1+\epsilon_2+\epsilon_3) + (p_2-p_3)(\epsilon_2-\epsilon_3) \\ &\qquad + (p_3-p_1)(\epsilon_3-\epsilon_1) + (p_1-p_2)(\epsilon_1-\epsilon_2)\}. \quad \text{...(2)} \end{aligned}$$

Now by Art. 138 (4), (8), (12) we have

$$p_1+p_2+p_3 = \frac{E}{1-2\sigma}(\epsilon_1+\epsilon_2+\epsilon_3) = 3\kappa\Delta, \quad \text{........(3)}$$

$$p_2-p_3 = 2\mu(\epsilon_2-\epsilon_3),\ p_3-p_1 = 2\mu(\epsilon_3-\epsilon_1),\ p_1-p_2 = 2\mu(\epsilon_1-\epsilon_2). \ (4)$$

We thus obtain

$$W = \tfrac{1}{2}\kappa\Delta^2 + \tfrac{1}{3}\mu\{(\epsilon_2-\epsilon_3)^2 + (\epsilon_3-\epsilon_1)^2 + (\epsilon_1-\epsilon_2)^2\} \quad \text{.....(5)}$$

as the expression for the potential energy in terms of the strains. It is here in evidence again that for stability the constants κ and μ must be positive, since W must be a minimum in the unstrained condition.

The potential energy of a pure shear may be found by putting $\epsilon_1 = -\epsilon_2 = \frac{1}{2}\eta$, $\epsilon_3 = 0$; or more simply by considering a unit cube having two pairs of its faces parallel to the planes of purely tangential stress (ϖ). If the strain be supposed to take place in the manner indicated by Fig. 163, we find

$$W = \tfrac{1}{2}\varpi\eta = \tfrac{1}{2}\mu\eta^2. \quad \text{........................(6)}$$

EXAMPLES. XXIV.

1. Shew that a simple longitudinal stress is equivalent to a tension uniform in all directions, together with two shearing stresses having a common principal axis in the line of the given stress, and their other two principal axes any two lines at right angles to one another and to it.

Hence shew that $\quad E = \dfrac{9\kappa\mu}{3\kappa+\mu}, \quad \sigma = \dfrac{3\kappa-2\mu}{2(3\kappa+\mu)}.$

2. If Ox, Oy, Oz be the principal axes of stress, prove that the normal and tangential components of the stress (per unit area) on a plane whose normal makes an angle θ with Ox in the plane xy are

$$p_1\cos^2\theta + p_2\sin^2\theta, \quad \text{and} \quad (p_1-p_2)\cos\theta\sin\theta.$$

3. The principal axes of a two-dimensional strain make angles θ with the axes of x and y, respectively, and the extensions along them are ϵ_1 and ϵ_2. Prove that the strain is equivalent to extensions

$$\epsilon_1 \cos^2\theta + \epsilon_2 \sin^2\theta, \quad \epsilon_1 \sin^2\theta + \epsilon_2 \cos^2\theta$$

parallel to Ox, Oy, together with a shear

$$(\epsilon_1 - \epsilon_2)\sin 2\theta.$$

4. Prove that if the conic

$$\frac{x^2}{p_1} + \frac{y^2}{p_2} = \text{const.}$$

be constructed, the stress across a plane parallel to Oz, through any diameter, is in the direction of the conjugate diameter, and proportional to ϖr, where ϖ is the perpendicular from the centre on a parallel tangent, and r is the length of the conjugate semi-diameter.

Examine the case of a pure shear ($p_1 = -p_2$, $p_3 = 0$).

5. Prove that in a homogeneous stress the potential energy per unit volume is

$$W = \frac{1}{18\kappa}(p_1 + p_2 + p_3)^2 + \frac{1}{12\mu}\{(p_2 - p_3)^2 + (p_3 - p_1)^2 + (p_1 - p_2)^2\}.$$

CHAPTER XVI

EXTENSION OF BARS

140. Extension of Bars.

Consider a bar of isotropic material, of uniform sectional area ω. If it undergoes longitudinal extension, being free to contract laterally, the stress per unit area of any cross-section is $E\epsilon$*, where ϵ is the local extension, and the total tension is

$$T = E\omega\epsilon. \quad \text{.........................(1)}$$

If the bar is free from external force (except at the ends), T and therefore also ϵ has the same value throughout, and the total increase of length is

$$\epsilon l = \frac{T}{E\omega} \cdot l, \quad \text{.........................(2)}$$

where l is the unstrained length.

To examine the case of a variable extension, let O be a fixed point of reference in the bar; and let PQ denote an element δx of the length in the unstrained state, $P'Q'$ the same element in

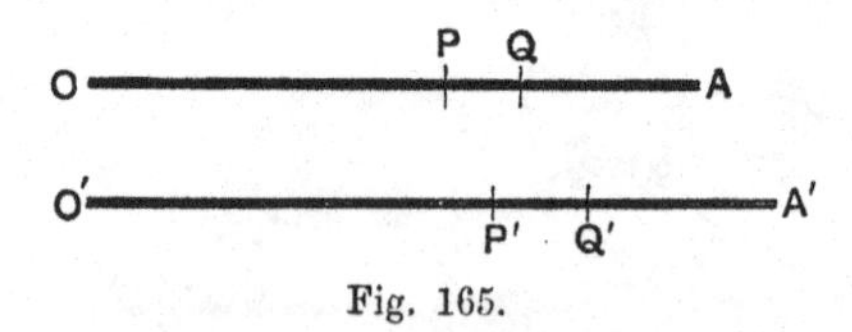

Fig. 165.

the altered condition. Let $OP = x$, $O'P' = x + \xi$, so that ξ denotes the *displacement* of P, relative to O. Then

$$O'Q' = x + \xi + \delta(x + \xi) = x + \xi + \left(1 + \frac{d\xi}{dx}\right)\delta x,$$

* It is not essential, in the present question, that the material of the bar should be isotropic, provided the constant E be determined by experiments on the longitudinal extension of the bar itself.

whence
$$P'Q' = \left(1 + \frac{d\xi}{dx}\right) PQ, \quad \text{.....................(3)}$$

or
$$\epsilon = \frac{d\xi}{dx}, \quad \text{...........................(4)}$$

by the definition of ϵ (Art. 134). The tension is therefore
$$T = E\omega \frac{d\xi}{dx}. \quad \text{.........................(5)}$$

Ex. We may apply this to the case of a bar hanging vertically, and stretched by its own weight.

If we measure x downwards from the upper end O, the tension at any point is
$$T = \rho\omega (l - x), \quad \text{.................................(6)}$$
where ρ is the density, and l the total length. Hence
$$E\frac{d\xi}{dx} = \rho (l - x), \quad \text{.................................(7)}$$
$$E\xi = \rho (lx - \tfrac{1}{2}x^2), \quad \text{..........................(8)}$$
no additive constant being necessary, since ξ vanishes for $x = 0$. Putting $x = l$, we find for the total increase of length
$$\xi_1 = \tfrac{1}{2}\rho l^2/E. \quad \text{....................................(9)}$$
This is one-half that which would be produced by a load $\rho\omega l$ attached at the lower end if the bar were itself without weight.

If a load W is attached at the lower end, the equation (6) is replaced by
$$T = W + \rho\omega (l - x), \quad \text{.............................(10)}$$
whence we find
$$\xi_1 = \frac{W}{E\omega} l + \frac{1}{2}\frac{\rho l^2}{E}. \quad \text{.............................(11)}$$

141. Deformation of Frames.

The stresses in the various bars of a just rigid frame which is subject to a given system of external forces acting at the joints are determined (if we exclude critical forms) uniquely by these forces. If S denote the stress in any bar, reckoned positive when a tension, the small increment of length of this bar will be
$$e = \lambda S, \quad \text{............................(1)}$$
where λ is a constant coefficient which may be called the 'extensibility' of the particular bar. We have seen (Art. 140) that in the case of a uniform bar $\lambda = l/E\omega$, where l is the length and ω the cross-section.

The work done in extending the bar, and the consequent elastic energy, is
$$\tfrac{1}{2}Se = \tfrac{1}{2}\lambda S^2 = \tfrac{1}{2}\frac{e^2}{\lambda}. \quad \text{.....................(2)}$$

It is a matter of some interest and importance to be able to find the resulting *displacements* of the various points of the structure; for instance, to find the vertical deflection at any point of a lattice girder loaded in a given manner. The direct calculation would in all but the simplest cases be very tedious, and the accuracy precarious. An elegant and systematic procedure has however been devised by Maxwell which avoids these difficulties.

The method depends on the principle of virtual velocities. In the case of a just rigid frame, the stresses $S_1, S_2, \ldots$ in the various bars, due to given loads, may be supposed to be known from a diagram of forces, or otherwise. We determine also the stresses $s_1, s_2, \ldots$ which would be produced in the same bars by a *unit* load at the point P whose deflection (z) is required; this involves in general the construction of a second diagram of forces, or some equivalent procedure. Since the forces $s_1, s_2, \ldots$, considered as mutual actions between the joints of the frame, together with the corresponding reactions of the supports (assumed to be unyielding) are in equilibrium with the unit load, the total work done by them and by the unit load in any arbitrary system of small displacements of the joints will be zero. We choose as our system of displacements those which actually take place in the given problem. Since these imply increments $e_1, e_2, \ldots$ in the mutual distances of the joints, we have

$$-s_1e_1 - s_2e_2 - \ldots + 1.z = 0,$$

or

$$z = \Sigma(es) = \Sigma(\lambda Ss). \quad \ldots\ldots\ldots\ldots\ldots\ldots(3)$$

In particular, to find the deflection of P due to a single load W at P itself, we have $S_1 = Ws_1$, $S_2 = Ws_2, \ldots$, and therefore

$$z = W\Sigma(\lambda s^2). \quad \ldots\ldots\ldots\ldots\ldots\ldots(4)$$

Ex. 1. A weight W hangs from a bracket consisting of a horizontal bar AB and a wire stay AC, B and C being fixed points in the same vertical.

If α be the inclination of the wire to the horizontal, we have obviously

$$s_1 = \text{cosec}\,\alpha, \quad s_2 = -\cot\alpha. \quad \ldots(5)$$

Hence the deflection of the end of the bracket is

$$z = (\lambda_1 \text{cosec}^2\alpha + \lambda_2 \cot^2\alpha)\,W. \quad \ldots(6)$$

In this case the result is easily verified by a direct calculation.

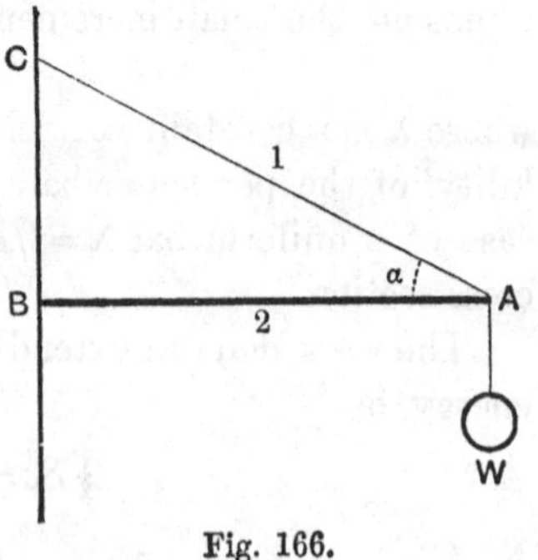

Fig. 166.

Ex. 2. A frame composed of two equilateral triangles, as in Fig. 167, is supported at the two points A, D, at the same level, and carries a load W at C.

The upper diagram of forces in the figure refers to the case of a unit load at C. We find

$$s_1 = s_2 = -\tfrac{1}{2}, \quad s_3 = s_4 = \tfrac{1}{2}, \quad s_5 = \tfrac{1}{2}. \quad \text{.....................(7)}$$

Hence if the extensibility λ be the same for each bar, the deflection of the point C is

$$W\Sigma(\lambda s^2) = \tfrac{5}{4}\lambda W. \quad \text{..................................(8)}$$

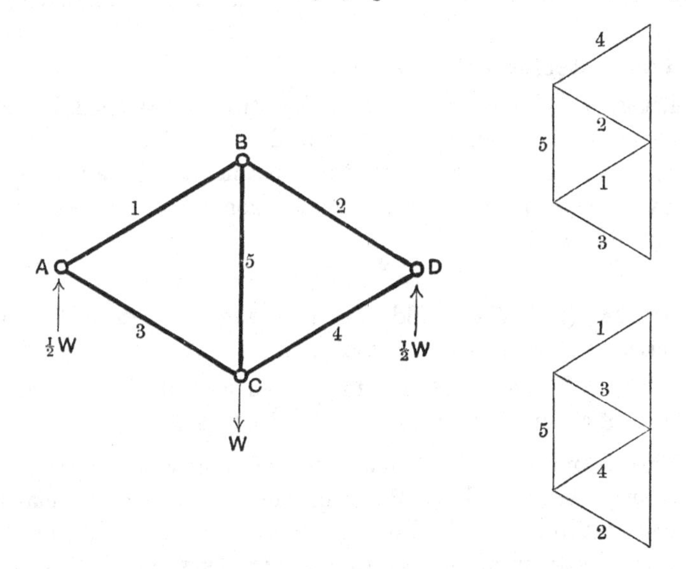

Fig. 167.

To find the deflection at B we construct a diagram of forces (viz. the lower diagram of the figure) for a unit load at B. We find

$$s_1 = s_2 = -\tfrac{1}{2}, \quad s_3 = s_4 = \tfrac{1}{2}, \quad s_5 = -\tfrac{1}{2}, \quad \text{.....................(9)}$$

whilst $$S_1 = S_2 = -\tfrac{1}{2}W, \quad S_3 = S_4 = \tfrac{1}{2}W, \quad S_5 = \tfrac{1}{2}W. \quad \text{...............(10)}$$

The deflection of B is therefore

$$\lambda\Sigma(Ss) = \tfrac{3}{4}\lambda W. \quad \text{..................................(11)}$$

The difference between (8) and (11) is due to the lengthening of the bar BC, which is seen independently to be $\tfrac{1}{2}\lambda W$.

Ex. 3. If the same frame be subject to two equal and opposite forces P acting outwards at A and D, we find

$$S_1 = S_2 = S_3 = S_4 = \frac{1}{\sqrt{3}}.P, \quad S_5 = -\frac{1}{\sqrt{3}}.P, \quad \text{..............(12)}$$

whilst the values of s_1, s_2, s_3, s_4, s_5 corresponding to a pair of unit forces at these points differ only by omission of the factor P. Hence the increase in the distance AD is, by (4),

$$\tfrac{5}{3}\lambda P. \qquad\qquad (13)$$

On the other hand, if two unit forces act inwards at B and C we have

$$s_1 = s_2 = s_3 = s_4 = 0, \quad s_5 = -1. \qquad\qquad (14)$$

The contraction in the diagonal bar, due to the two outward forces at A and D, is therefore

$$\frac{1}{\sqrt{3}}\lambda P. \qquad\qquad (15)$$

142. Reciprocal Theorem.

Let $s_1, s_2, \ldots$ denote as before the stresses produced in the various bars of a frame by a unit load at any joint A, and let $s_1', s_2', \ldots$ be the stresses produced in the same bars by a unit load at any other joint A'. Then the deflection at A due to a unit load at A' will be

$$\Sigma(\lambda s s'), \qquad\qquad (1)$$

by Art. 141 (2). We should obtain the same expression for the deflection at A' due to a unit load at A. Hence:

The deflection at A due to a given load at A' is equal to the deflection at A' due to an equal load at A.

This is merely a particular case of a general theorem of reciprocity, formulated by Maxwell, which holds for any elastic system whatever. Let P, Q denote external forces applied, in any assigned directions, at two points A, B of the system. We assume that the remaining external forces, which are called into play to balance these (for instance the reactions of rigid supports) do no work in the small deformations considered. Let p, q denote the component displacements of the points A, B in the specified directions. These quantities will, by Hooke's Law, be linear functions of P, Q, say

$$p = \lambda P + \mu Q, \quad q = \mu' P + \nu Q. \qquad\qquad (2)$$

We will first suppose a force to be applied at A only, gradually increasing from zero to the final value P. Since the component deflection at A is λP, the work done in this process is $\tfrac{1}{2}\lambda P^2$. Next, let a force be applied at B, gradually increasing from zero to its final value Q, the force at A remaining constant. This will produce additional component deflections μQ and νQ at A and B, respectively. The work done at A will therefore be μPQ, and that at B will be $\frac{1}{2}\nu Q^2$. The total work is therefore

$$\tfrac{1}{2}\lambda P^2 + \mu PQ + \tfrac{1}{2}\nu Q^2.$$

If the order of the operations had been reversed, we should have obtained the expression

$$\tfrac{1}{2}\lambda P^2 + \mu' PQ + \tfrac{1}{2}\nu Q^2.$$

Since these must be equal, we see that $\mu = \mu'$; i.e. the component deflection at A due to a unit force at B is equal to the component deflection at B due to a unit force at A.

The potential energy of the strained system, so far as it depends on the forces P, Q, has the form

$$V = \tfrac{1}{2}(\lambda P^2 + 2\mu PQ + \nu Q^2); \quad \text{..............(3)}$$

and we notice that

$$p = \frac{\partial V}{\partial P}, \qquad q = \frac{\partial V}{\partial Q}. \quad \text{.....................(4)}$$

By means of the relations (2), with $\mu' = \mu$, we can obtain P, Q as linear functions of p, q, so that the energy can also be expressed (as is otherwise obvious) as a quadratic function of p, q, say

$$V = \tfrac{1}{2}(Ap^2 + 2Hpq + Bq^2), \quad \text{..................(5)}$$

and we should find

$$P = \frac{\partial V}{\partial p}, \qquad Q = \frac{\partial V}{\partial q}. \quad \text{..................(6)}$$

These relations also follow at once from the principle of virtual velocities, which gives in our case

$$P\,\delta p + Q\,\delta q = \delta V. \quad \text{....................(7)}$$

143. Stresses in an Over-Rigid Frame.

Maxwell has also shewn how to determine the stresses produced by external forces in the bars of a frame which is over-rigid. On purely statical principles the problem is, as we have seen, indeterminate, but the theory of elasticity now comes to our assistance.

Such a frame may of course be in a state of stress independently of the external forces. This will depend (Art. 39) on the relations between the unstrained lengths, and on the elasticities of the several bars; but we are here concerned with the *additional* stresses due to the external forces.

It is to some extent a matter of choice as to which bars shall be regarded as 'redundant,' and therefore capable of being removed without impairing the rigidity of the frame. When the choice has been made the remaining bars may be referred to as 'essential.'

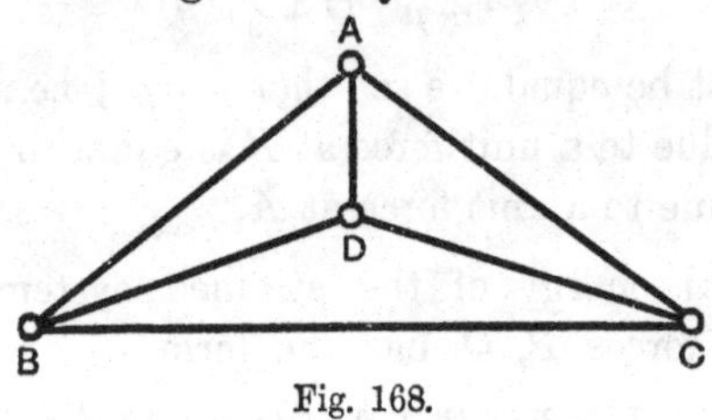

Fig. 168.

We will assume, to begin with, that there is only one redundant bar, as e.g. BC in the figure.

First suppose the redundant bar to be removed, so that the frame becomes just rigid; and let $S_1, S_2, \ldots$ be the stresses which would be produced by the given external forces in this modified frame, and $s_1, s_2, \ldots$ the stresses in the same bars, which would be produced by two opposite unit forces acting inwards at B and C. With our previous notation, the diminution in the distance BC would be

$$\Sigma(\lambda S s). \quad \text{.........................(1)}$$

Now if T be the actual tension in the bar BC, the actual stresses in the remaining bars will be, by superposition,

$$S_1 + Ts_1, \quad S_2 + Ts_2, \ldots,$$

and the actual diminution in the distance BC will be, by Art. 141 (3),

$$\Sigma\{\lambda(S + Ts)s\} = \Sigma(\lambda S s) + T\Sigma(\lambda s^2). \quad \text{...........(2)}$$

This must be equal to $-\mu T$, where μ is the extensibility of the bar BC. Hence, equating,

$$\{\mu + \Sigma(\lambda s^2)\}T + \Sigma(\lambda S s) = 0, \quad \text{..............(3)}$$

which gives the value of T. The stresses in the remaining bars will be given by expressions of the type

$$S + Ts = S - \frac{s\Sigma(\lambda S s)}{\mu + \Sigma(\lambda s^2)}. \quad \text{.................(4)}$$

The deformations in the frame are to be found from the *actual* extensions in the bars of the frame, as modified by the omission of the bar BC, by the method of Art. 141.

Ex. Suppose that in the frame shewn in Fig. 167 (p. 307) an additional bar is inserted connecting the joints A and D.

In the preceding notation we have

$$S_1 = S_2 = -\tfrac{1}{2} W, \quad S_3 = S_4 = S_5 = \tfrac{1}{2} W, \quad \text{.....................(5)}$$

$$s_1 = s_2 = s_3 = s_4 = -\frac{1}{\sqrt{3}}, \quad s_5 = \frac{1}{\sqrt{3}} \quad \text{.....................(6)}$$

Hence if μ be the extensibility of AD, supposed initially unstrained, its tension is

$$T = -\frac{\lambda W}{2\sqrt{3}\,(\mu + \frac{5}{3}\lambda)}. \quad \text{..............................(7)}$$

To find the deflection at C we refer to the formula (3) of Art. 141, viz.

$$z = \Sigma\,(es). \quad \text{..................................(8)}$$

The values of e are obviously

$$\lambda\,(S_1 + Ts_1), \quad \lambda\,(S_2 + Ts_2), \ldots, \quad \text{........................(9)}$$

whilst the values of s in (8) are to be taken from Art. 141 (7). The result is

$$z = \frac{5}{4}\lambda W + \frac{1}{2\sqrt{3}}\lambda T, \quad \text{.............................(10)}$$

where T is given by (7).

If there are two redundant bars, let μ, T, s_1, s_2, ... have the above meanings for one of these, and let the corresponding quantities for the other be distinguished by accents. The actual stresses in the essential bars of the frame will therefore be

$$S_1 + Ts_1 + T's_1', \quad S_2 + Ts_2 + T's_2', \ldots,$$

and the actual diminution in the length of the first redundant bar will be

$$\Sigma\,\{\lambda\,(S + Ts + T's')\,s\} = \Sigma\,(\lambda Ss) + T\Sigma\,(\lambda s^2) + T'\Sigma\,(\lambda ss'). \quad \text{...(11)}$$

Equating this to $-\mu T$, we have

$$\{\mu + \Sigma\,(\lambda s^2)\}\,T + \Sigma\,(\lambda ss')\,T' + \Sigma\,(\lambda Ss) = 0. \quad \text{......(12)}$$

Similarly

$$\Sigma\,(\lambda ss')\,T + \{\mu' + \Sigma\,(\lambda s'^2)\}\,T' + \Sigma\,(\lambda Ss') = 0. \quad \text{......(13)}$$

These two equations determine T and T'.

The procedure when there are three or more redundant bars will now be evident. In each case we obtain equations equal in number to the unknown tensions.

144. Principle of Least Energy.

The elastic energy of a just rigid frame when strained is, by Art. 141 (2),

$$V = \tfrac{1}{2}(S_1 e_1 + S_2 e_2 + \ldots) = \tfrac{1}{2}\Sigma(Se), \quad \ldots\ldots\ldots\ldots(1)$$

where $S_1, S_2, \ldots$ are the stresses in the several bars, and $e_1, e_2, \ldots$ the increments of length. Introducing the relations

$$e_1 = \lambda_1 S_1, \quad e_2 = \lambda_2 S_2, \ldots, \quad \ldots\ldots\ldots\ldots\ldots\ldots(2)$$

we have $$V = \tfrac{1}{2}(\lambda_1 S_1^2 + \lambda_2 S_2^2 + \ldots) = \tfrac{1}{2}\Sigma(\lambda S^2). \quad \ldots\ldots\ldots(3)$$

In the case of a frame with one redundant bar this formula is replaced by

$$V = \tfrac{1}{2}\{\lambda_1(S_1 + Ts_1)^2 + \lambda_2(S_2 + Ts_2)^2 + \ldots\} + \tfrac{1}{2}\mu T^2$$
$$= \tfrac{1}{2}\Sigma(\lambda S^2) + \Sigma(\lambda Ss)T + \tfrac{1}{2}\{\Sigma(\lambda s^2) + \mu\}T^2, \quad \ldots\ldots(4)$$

where the symbols have the same meanings as in Art. 143. It now appears that the equation (3) of that Art. is equivalent to

$$\frac{\partial V}{\partial T} = 0. \quad \ldots\ldots\ldots\ldots\ldots\ldots\ldots\ldots\ldots(5)$$

Hence, the required value of T is that which makes the elastic energy (when expressed in the above manner) a minimum.

A similar result holds when there are two or more redundant bars. For instance, if there are two such bars, the elastic energy is

$$V = \tfrac{1}{2}\Sigma\{\lambda(S + Ts + T's')^2\} + \tfrac{1}{2}\mu T^2 + \tfrac{1}{2}\mu' T'^2$$
$$= \tfrac{1}{2}\Sigma(\lambda S^2) + \Sigma(\lambda Ss)T + \Sigma(\lambda Ss')T'$$
$$+ \tfrac{1}{2}\{\Sigma(\lambda s^2) + \mu\}T^2 + TT'\Sigma(\lambda ss') + \tfrac{1}{2}\{\Sigma(\lambda s'^2) + \mu'\}T'^2. \ \ldots(6)$$

The equations (12) and (13) of the preceding Art. are seen to be equivalent to

$$\frac{\partial V}{\partial T} = 0, \quad \frac{\partial V}{\partial T'} = 0, \quad \ldots\ldots\ldots\ldots\ldots\ldots\ldots(7)$$

which are the conditions for a minimum of V.

This theorem of 'least energy' as it is called, is due to Castigliano (1873). It depends on the assumption, which we have made throughout, that the stresses are linear functions of the strains, so that the combined effect of two or more distributions of force can be found by simple superposition of the corresponding strains.

The following simple proof from first principles is due to Mr R. V. Southwell. Suppose we start with an over-rigid frame which is free from external force, but self-stressed. This means that there are certain stresses, which we will denote by T_0, in the redundant bars, and consequent stresses S_0 in the essential bars. The elastic energy in this state is, in our previous notation,

$$V_0 = \Sigma(\tfrac{1}{2}\mu T_0^2) + \Sigma(\tfrac{1}{2}\lambda S_0^2). \quad \text{..............(8)}$$

The stresses S_0 are entirely due to the T_0, and are in fact linear functions of these, so that V_0 may be regarded as a homogeneous, essentially positive quadratic function of the quantities T_0. If we now imagine the external forces to be applied, increasing gradually from zero to their final values, the *additional* displacements thus produced will be the same as if the frame had been free from stress to begin with, in virtue of the principle of superposition, and the same amount of work (V_1, say) will therefore be done by the external forces. The total elastic energy is therefore

$$V = V_0 + V_1, \quad \text{..........................(9)}$$

where V_1 is independent of V_0. With a given system of external forces this is least when $V_0 = 0$, which can only be the case if the initial stresses T_0 in the redundant bars are all zero. Now the elastic energy may be regarded as a function of the external forces and the final stresses T in the redundant bars. The present argument shews that the values of the stresses T which make this function a minimum are those which correspond to an initially unstressed frame.

It will be evident from the nature of the above proof that the scope of the theorem can be greatly extended. Suppose for instance that we have any elastic system subject to certain constraints, whether rigid or elastic, which are inactive so long as there are no external forces, and that the system is 'statically indeterminate' (Art. 39), so that the constraining forces which are called into play by given external forces cannot be determined by the rules of pure Statics alone. If we assume arbitrary values for such constraining forces, consistent with equilibrium, as are necessary to make the problem determinate, the elastic energy can be calculated in terms of these and of the external forces. The preceding argument shews that the correct values of these constraining forces will be such as make the energy, as thus expressed, a minimum. An example will be found in Art. 148.

EXAMPLES. XXV.

1. A steel wire $\frac{1}{30}$ in. in diameter and 20 ft. long hangs vertically from one end; find the increase of length when a weight of 50 lbs. is attached at the lower end, assuming that the value of Young's modulus for the wire is 15000 tons per sq. in. [·41 in.]

2. A uniform rod of weight W and length l hangs from one end; prove that its elastic energy is $\frac{1}{6} W^2 l/E\omega$.

3. A uniform bar rotates about one end, in a horizontal plane, with constant angular velocity. Find the radial displacement of any point, due to the rotation.

If l be the length, and v the velocity of the free end, prove that the increase of length is to that which would be produced by its own weight, if the bar were hanging vertically, in the ratio

$$\tfrac{2}{3} v^2/gl.$$

4. A uniform ring rotates in its own plane, about its centre, the velocity of the circumference being v. Prove that the tensile stress is $\rho v^2/g$ in gravitational measure.

What is the limit to the value of v in the case of a steel ring ($\rho = 7{\cdot}8$), if the stress is not to exceed 4×10^6 gm. per sq. cm.? [224 metres per sec.]

5. A frame of the type shewn in Ex. IX. 2 (p. 108) rests on rigid supports at the ends, and carries a load W at the middle lower joint. Assuming that the bars are all equal in length, and have the same extensibility λ, find the deflections at the loaded joint and at the two upper joints, respectively. [$\frac{11}{6}\lambda W$, λW.]

6. In the symmetrical frame of Fig. 168, the bars AB, BD make angles α, β, respectively with the horizontal, and their extensibilities are λ, μ, whilst that of AD is ν. If the bar BC be removed, prove that the deflection at A due to a load W at this point is

$$\frac{W}{\sin^2(\alpha-\beta)}\left(\tfrac{1}{2}\lambda\cos^2\beta + \tfrac{1}{2}\mu\cos^2\alpha + \nu\cos^2\alpha\sin^2\beta\right).$$

CHAPTER XVII

FLEXURE AND TORSION OF BARS

145. Uniform Flexure.

We consider first a state of strain in which the cross-sections of a bar remain plane, and there is no shearing of adjacent sections relatively to one another. Suppose that after the deformation the planes of two consecutive sections, which were originally parallel, meet in a line C. The points in either of these planes at which the distance from the consecutive plane retains its original value will evidently lie in a certain line parallel to C. This is called the 'neutral line' of the section. It is not necessary to assume at present that it meets the contour of the section.

Let us take rectangular axes Ox, Oy in the plane of a section, the axis of x being coincident with the neutral line; and let R denote the distance of this line from C. If θ be the small angle between the adjacent sections, the distance between these sections before the strain, being equal to the actual distance at O, may be denoted by $R\theta$, whilst the actual distance at the point (x, y) is $(R+y)\theta$. The extension in the direction normal to the plane of the cross-section is therefore

$$e = y/R. \quad \text{.........(1)}$$

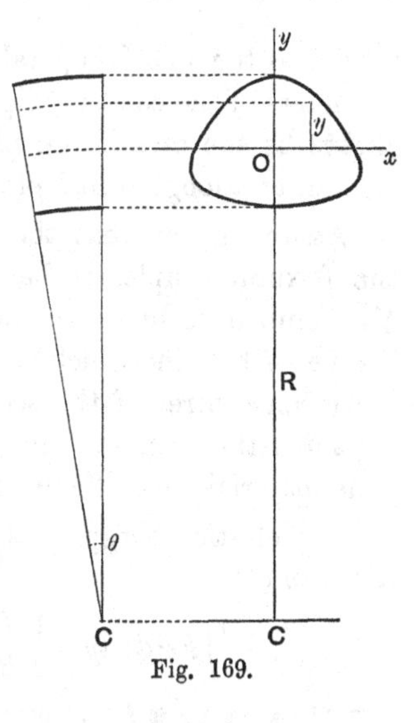

Fig. 169.

The longitudinal stress is therefore Ey/R*, and the total tension across the section is

* There is no restriction here to the case of isotropy, provided the proper value of E be taken.

$$T = \frac{E}{R}\iint y\,dx\,dy = \frac{E\omega\bar{y}}{R}, \quad \text{.................(2)}$$

where ω is the sectional area, and $\bar{y}$ refers to the mean centre of the section.

We will confine ourselves, for the present, to the case of pure flexure, where the tension T vanishes. This requires that $\bar{y} = 0$; i.e. the neutral line must pass through the mean centre of the section. The stresses across the section then reduce to two couples about Ox, Oy, respectively. The first of these is

$$M = \iint \frac{Ey}{R}\,y\,dx\,dy = \frac{E\omega\kappa^2}{R}, \quad \text{.................(3)}$$

where $\omega\kappa^2$ is the quadratic moment of the area with respect to the neutral line*. The second couple is

$$M' = \iint \frac{Ey}{R}\,x\,dx\,dy = \frac{E}{R}\iint xy\,dx\,dy. \quad \text{...........(4)}$$

Hence the couple required to maintain the given flexure will not be parallel to the plane of bending unless

$$\iint xy\,dx\,dy = 0, \quad \text{.......................(5)}$$

i.e. unless the axis C is parallel to one or other of the two principal diameters of the central ellipse of the section (Art. 75). This will always be the case if the plane of flexure be parallel, or perpendicular, to a longitudinal plane of symmetry of the beam.

Assuming now that the condition (5) is fulfilled, we see that the flexural couple, or bending moment (Art. 27), is equal to $E\omega\kappa^2$ multiplied by the curvature ($1/R$) of the 'central line' of the bar, i.e. of the line which in the unstrained state passes through the mean centres of the sections. The factor $E\omega\kappa^2$, or EI (say), measures the 'flexural rigidity' of the beam. For beams of the same material, but different forms of section, it varies as I.

The elastic energy of the bent beam, per unit length, is by Art. 139 (1),

$$\frac{1}{2}\iint E\epsilon^2\,dx\,dy = \frac{1}{2}\frac{E}{R^2}\iint y^2\,dx\,dy = \frac{1}{2}\frac{EI}{R^2} = \frac{1}{2}\frac{M^2}{EI}. \quad \text{......(6)}$$

* The symbol κ is not at present required in its former sense as denoting an elastic constant. The preceding theory is due to C. A. Coulomb (1776).

Exx. For a square section the central ellipse is a circle, and the flexural rigidity is the same in all planes through the axis. If the flexure take place in a plane parallel to a diagonal, the extreme extension and contraction are greater than if it had been in a plane parallel to a pair of sides, in the ratio $\sqrt{2}$.

For a cylindrical tube whose outer and inner radii are a and b, respectively, we have $\kappa^2 = \frac{1}{4}(a^2+b^2)$. The flexural rigidity is therefore greater than in the case of a solid cylindrical rod of the same sectional area, in the ratio

$$(a^2+b^2)/(a^2-b^2).$$

Owing to the varying lateral contraction $(\sigma y/R)$ the cross-section undergoes a slight change of shape. Thus if the section be a rectangle with one pair of sides perpendicular to the plane of flexure, its strained form will be somewhat as shewn in Fig. 170, the inner sides being extended, and the outer sides contracted, to equal amounts. It is not difficult to see that the curvature of these sides is $-\sigma/R$, approximately. The inner and outer faces of the bar are in fact surfaces of 'anticlastic' curvature*.

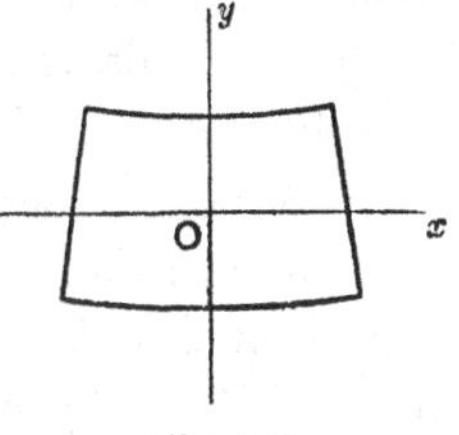

Fig. 170.

The theory of the cylindrical flexure of a *plate* is similar to that of the flexure of a bar, except that it involves an altered elastic constant. It is evident that the contractions and extensions at right angles to the plane of flexure cannot in this case be appreciably developed, except close to the edges, since the consequent anticlastic curvature would involve greatly increased extensions and contractions at right angles to the cross-section. If p_1, p_2, p_3 be the stresses in the directions of the breadth, the thickness, and the length, respectively, we have now practically $\epsilon_1 = 0$, $p_2 = 0$, and therefore

$$E\epsilon_3 = (1-\sigma^2)\,p_3. \qquad \text{.......................(7)}$$

Hence the E of the previous calculation is to be replaced by $E/(1-\sigma^2)$. Since, also, $\kappa^2 = \frac{1}{12}h^2$, where h is the thickness, the bending moment per unit breadth is

$$\frac{Eh^3}{12(1-\sigma^2)R}. \qquad \text{.......................(8)}$$

* If the two curvatures can be measured, the value of σ is obtained directly. This plan was employed by A. Cornu (1869), the curvatures being ascertained by an optical method.

146. Varying Flexure of a Beam.

We consider the case of a horizontal rod or beam slightly bent by vertical forces applied to it. The state of strain is no longer of the simple character appertaining to pure flexure; in particular there will be a relative shearing of adjacent cross-sections, and also a warping of the sections so that these do not remain accurately plane. We shall assume, however, that the additional *strains* thus introduced are on the whole negligible, and consequently that the bending moment is connected with the curvature of the axis by the same formula as before*. A reason in support of this assumption will be indicated presently.

Let F be the shearing stress, and M the bending moment, at any point of a horizontal beam, estimated according to the conventions of Art. 27. Then, taking the axis of x parallel to the length, we have, by Art. 28,

$$\frac{dF}{dx} = -w, \quad \frac{dM}{dx} = -F, \quad \text{.................(1)}$$

where w is the load per unit length, not as yet assumed to be uniform. If y denote the downward deflection of the neutral line, the curvature of the axis will be d^2y/dx^2, approximately†, since by hypothesis dy/dx is small. Hence, by Art. 144 (3),

$$M = EIy'', \quad \text{..........................(2)}$$

where the accents denote differentiations with respect to x. The relations (1) then give

$$F = -\frac{d}{dx}(EIy''), \quad \text{..................(3)}$$

and

$$\frac{d^2}{dx^2}(EIy'') = w. \quad \text{......................(4)}$$

In the case of a *uniform* beam these take the simpler forms

$$F = -EIy''', \text{..........................(5)}$$

$$EIy^{iv} = w. \quad \text{........................(6)}$$

The integration of (4) or (6) introduces four arbitrary constants,

* The assumption that the bending moment varies as the curvature is the basis of the 'Euler-Bernoulli' theory of flexure. This was developed in memoirs by James Bernoulli (1705), D. Bernoulli (1742), L. Euler (1744).

† If ψ denote the downward inclination of the tangent to the horizontal, we have $\psi = \tan\psi = dy/dx$, and $d\psi/ds = d\psi/dx = d^2y/dx^2$, approximately.

which are to be determined from the remaining conditions of the particular question. Thus, suppose we have a uniform beam which is free from concentrated force except at the ends. Then, (i) at a *free* end we have the two conditions $M=0$, $F=0$, or $y''=0$, $y'''=0$; (ii) at a *clamped* end the values of y and y' are prescribed; (iii) at a *supported* end $M=0$ and therefore $y''=0$, whilst the value of y is prescribed. Hence there are in each case four conditions, two for each end.

In particular cases the work can often be shortened by special considerations. For instance, if the value of M can be found as a function of x, as in all statically determinate cases (Chap. III), we may proceed to the integration of (2).

The elastic energy of any portion of the beam is given by the expression

$$\frac{1}{2}\int EIy''^2 dx \quad \text{..........................(7)}$$

taken between the proper limits.

The shearing force F distributed over the section ω implies an average shearing strain $F/\mu\omega$, which is left out of account in the preceding theory. It is, in fact, usually negligible in comparison with the elongation given by Art. 145 (1), the average value of which (without regard to sign) over the section is of the order $M/E\omega\kappa$. The ratio of these two quantities is of the order

$$\kappa \frac{dM}{dx} \bigg/ M,$$

by (1), and therefore comparable with the ratio which the variation of M within a length κ of the beam bears to the value of M itself. Since M is usually continuous, this ratio is generally small except in the neighbourhood of points where $M=0$.

147. Examples of Concentrated Load.

In a uniform beam which is subject to external force at isolated points only we have $w=0$, and therefore

$$y^{\text{iv}}=0, \quad \text{..............................(1)}$$

in the intervening portions of the length. Hence

$$y = Ax^3 + Bx^2 + Cx + D; \quad \text{................(2)}$$

but the constants will change their values at a point of application of force, owing to the discontinuity of F (Art. 27).

Since our equations are linear, the deflections of a beam due to different systems of loads may be superposed. Hence when there are concentrated as well as continuous loads it is convenient to consider the effect of each separately. Some examples of the former kind are appended.

Ex. 1. A beam supported at the ends ($x=0$, $x=l$) carries an isolated load W at the centre.

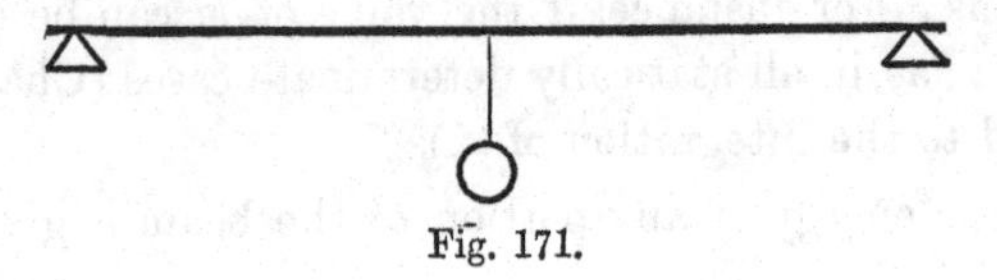

Fig. 171.

Over the left-hand half we have

$$EIy''=M=-\tfrac{1}{2}Wx, \qquad (3)$$

whence
$$EIy'=-\tfrac{1}{4}Wx^2+\tfrac{1}{16}Wl^2, \qquad (4)$$

and
$$EIy=-\tfrac{1}{12}Wx^3+\tfrac{1}{16}Wl^2x$$
$$=\tfrac{1}{48}Wx(3l^2-4x^2), \qquad (5)$$

the constants being determined so as to make $y'=0$ for $x=\frac{1}{2}l$, and $y=0$ for $x=0$. Hence the droop at the centre is

$$\frac{1}{48}\frac{Wl^3}{EI}. \qquad (6)$$

The change of form is the same as if the beam had been firmly clamped in the middle and pressed upwards by a force $\frac{1}{2}W$ at either end. Hence

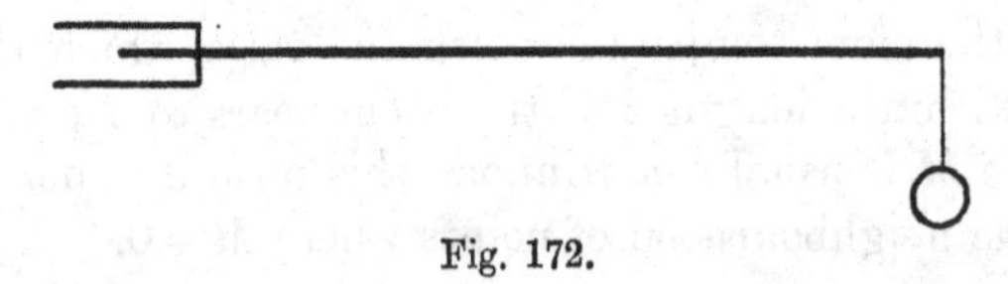

Fig. 172.

writing $2l$ for l and $2W$ for W, we infer that the deflection at the end of a horizontal cantilever due to a weight W suspended from it is

$$\frac{1}{3}\frac{Wl^3}{EI}. \qquad (7)$$

Ex. 2. Next, suppose the load to be placed at any point ($x=x_1$).

The solutions for the two parts of the beam must be conducted separately, since there is a discontinuity in the value of the shearing force F, and therefore of y''', by Art. 145 (5). We arrange the work in parallel columns.

$$[0<x<x_1] \qquad\qquad [x_1<x<l]$$

$$EIy''=-\frac{W}{l}(l-x_1)\,x, \qquad EIy''=-\frac{W}{l}x_1(l-x), \qquad\qquad (8)$$

$$EIy'=-\frac{1}{2}\frac{W}{l}(l-x_1)\,x^2+A, \qquad EIy'=\frac{1}{2}\frac{W}{l}x_1(l-x)^2+B, \qquad\qquad (9)$$

$$EIy=-\frac{1}{6}\frac{W}{l}(l-x_1)\,x^3+Ax, \qquad EIy=-\frac{1}{6}\frac{W}{l}x_1(l-x)^3-B(l-x). \qquad (10)$$

The additive constants which present themselves in the final integrations have been chosen so as to make $y=0$ for $x=0$ and $x=l$, respectively. The remaining constants A, B are to be determined by the consideration that the values of y and y' must be continuous at the point $x=x_1$. Thus

$$-\frac{1}{2}\frac{W}{l}(l-x_1)\,x_1^2+A=\frac{1}{2}\frac{W}{l}x_1(l-x_1)^2+B, \qquad\qquad (11)$$

$$-\frac{1}{6}\frac{W}{l}(l-x_1)\,x_1^3+Ax_1=-\frac{1}{6}\frac{W}{l}x_1(l-x_1)^3-B(l-x_1). \qquad (12)$$

We find

$$A=\frac{Wx_1(l-x_1)}{l}(\tfrac{1}{3}l+\tfrac{1}{6}x_1), \quad B=\frac{Wx_1(l-x_1)}{l}(-\tfrac{1}{6}l-\tfrac{1}{3}x_1). \qquad (13)$$

Hence, for $0<x<x_1$,

$$EIy=\frac{1}{6}\frac{W}{l}(l-x_1)\,x\,\{l^2-(l-x_1)^2-x^2\}, \qquad\qquad (14)$$

and for $x_1<x<l$,

$$EIy=\frac{1}{6}\frac{W}{l}x_1(l-x)\{l^2-x_1^2-(l-x)^2\}. \qquad\qquad (15)$$

The droop at the point $x=x_1$ is therefore

$$\frac{1}{3}\frac{W}{EIl}x_1^2(l-x_1)^2. \qquad\qquad (16)$$

This is, however, not the maximum deflection, unless $x_1=\frac{1}{2}l$.

It may be noticed that if in (14), (15) we write $l-x$ for x_1, and $l-x_1$ for x, the two formulæ become interchanged. This shews that the deflection at a point Q due to a given weight at P is equal to the deflection at P due to an equal weight at W, in accordance with the general reciprocal theorem of Art. 142.

When there are a number of loads, the above plan of solving the differential equation separately for each segment of the beam, and adjusting the constants so as to secure the continuity of y and y' at each point of division, becomes troublesome. A simplified procedure has been devised by Mr W. H. Macaulay. He introduces the notation $\{f(x)\}_{x_1}$ to mean that the value of the enclosed function $f(x)$ is to be replaced by zero so long as $x<x_1$. Supposing that we have loads

$W_1, W_2, \ldots$ situate at the points $x_1, x_2, \ldots$, taken in order from the left, the equation

$$EIy'' = -Px + \{W_1(x-x_1)\}_{x_1} + \{W_2(x-x_2)\}_{x_2} + , \ldots\ldots(17)$$

where P is the pressure on the (supported) end $x = 0$, as determined by the ordinary rules of Statics, will apply to the whole length of the beam. Hence, integrating,

$$EIy' = -\tfrac{1}{2}Px^2 + \{\tfrac{1}{2}W_1(x-x_1)^2\}_{x_1} + \{\tfrac{1}{2}W_2(x-x_2)^2\}_{x_2} + \ldots + A \ldots(18)$$

where the same constant A applies throughout, on account of the continuity of y' at the points $x_1, x_2, \ldots$. A second integration gives

$$EIy = -\tfrac{1}{6}Px^3 + \{\tfrac{1}{6}W_1(x-x_1)^3\}_{x_1} + \{\tfrac{1}{6}W_2(x-x_2)^3\}_{x_2} + \ldots + Ax + B, \quad \ldots\ldots\ldots(19)$$

where B, again, has the same value throughout, on account of the continuity of y. The conditions that $y = 0$ for $x = 0$ and $x = l$ determine the values of A and B.

Thus in the above case of a single load W we have

$$EIy = -\frac{W}{6l}(l-x_1)x^3 + \{\tfrac{1}{6}W(x-x_1)^3\}_{x_1} + Ax + B, \ldots\ldots\ldots\ldots(20)$$

and the terminal conditions make

$$B = 0, \quad A = \frac{W}{6l}(l-x_1)\{l^2-(l-x_1)^2\}. \ldots\ldots\ldots\ldots\ldots\ldots(21)$$

The results (14) and (15) then follow.

If the beam is clamped horizontally at the origin, instead of being merely supported, an unknown quantity, viz. the bending moment M_0 at $x=0$, will occur in (17), but we now have an additional terminal condition, $y'=0$, so that the problem is determinate.

148. Continuous Loads.

We take next some cases of continuous loads.

Ex. 1. A uniformly loaded beam, of length l, supported at the ends.

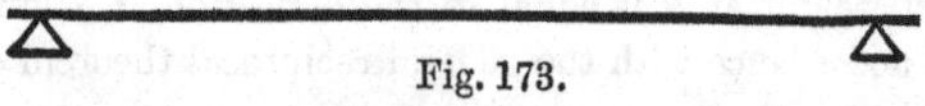

Fig. 173.

Taking the origin at one end, and denoting the length by l, we have

$$EIy'' = M = -\tfrac{1}{2}wx(l-x), \ldots\ldots\ldots\ldots\ldots\ldots\ldots\ldots(1)$$

by Art. 28 (4). Hence

$$EIy' = \tfrac{1}{6}wx^3 - \tfrac{1}{4}wlx^2 + \tfrac{1}{24}wl^3, \ldots\ldots\ldots\ldots\ldots\ldots\ldots(2)$$

where the additive constant has been chosen so as to make $y'=0$ for $x=\tfrac{1}{2}l$, as it must be, by the symmetry of the conditions. The next integration gives

$$EIy = \tfrac{1}{24}wx^4 - \tfrac{1}{12}wlx^3 + \tfrac{1}{24}wl^3x$$
$$= \tfrac{1}{24}wx(l-x)\{l^2+x(l-x)\}, \ldots\ldots\ldots\ldots\ldots\ldots(3)$$

no additive constant being necessary if $y=0$ for $x=0$. Putting $x=\frac{1}{2}l$, we find that the droop in the middle is

$$\frac{5}{384}\frac{wl^4}{EI}. \quad \text{.....(4)}$$

If, to compare with Art. 146, Ex. 1, we put $wl=W$, we see that the effect of distributing the load uniformly over the beam instead of concentrating it at the centre is to reduce the deflection at this point in the ratio 5 : 8.

Ex. 2. A uniformly loaded cantilever; i.e. a beam clamped at one end, free at the other.

Taking the origin at the fixed end, we find

$$EIy''=M=\tfrac{1}{2}w(l-x)^2, \quad \text{.....(5)}$$

whence

$$EIy'=-\tfrac{1}{6}w(l-x)^3+\tfrac{1}{6}wl^3, \quad \text{.....(6)}$$

the constant being adjusted so as to make $y'=0$ for $x=0$. Hence

$$\begin{aligned} EIy &= \tfrac{1}{24}w(l-x)^4+\tfrac{1}{6}wl^3x-\tfrac{1}{24}wl^4 \\ &= \tfrac{1}{24}wx^2(6l^2-4lx+x^2), \quad \text{.....(7)} \end{aligned}$$

a similar adjustment being made. Hence the droop at the end is

$$\frac{1}{8}\frac{wl^4}{EI}, \quad \text{.....(8)}$$

which is three-eighths of the value obtained when the weight wl is concentrated at the free end.

We infer that if a beam of length l be supported at its centre only, the droop at the ends will be

$$\frac{1}{128}\frac{wl^4}{EI}. \quad \text{.....(9)}$$

This is three-fifths of the droop at the centre when the ends are supported.

Ex. 3. A beam, uniformly loaded, rests on three supports at the same level, viz. at the ends and the centre.

Fig. 174.

The pressures on the supports are in this case not determinable by the principles of pure statics alone (see Art. 25), so that we cannot begin by forming the expression for the bending moment. We therefore have recourse to the general equation

$$EIy^{\text{iv}}=w. \quad \text{.....(10)}$$

The two parts of the beam would need to be treated separately, since there is a discontinuity in the value of F, and therefore of y'''; but in the present case it is sufficient to consider either half alone, on account of the symmetry.

We take the origin at one end. From (10) we have

$$EIy'''=wx+A, \quad \text{.....(11)}$$

$$EIy''=\tfrac{1}{2}wx^2+Ax, \quad \text{.....(12)}$$

no additive constant being necessary at this stage since the bending moment vanishes for $x=0$. Hence

$$EIy'=\tfrac{1}{8}w(x^3-a^3)+\tfrac{1}{2}A(x^2-a^2), \quad\text{.....................(13)}$$

the constant being chosen so as to make $y'=0$ at the middle point $(x=a)$. Finally

$$EIy=\tfrac{1}{24}w(x^4-4a^3x)+\tfrac{1}{6}A(x^3-3a^2x), \quad\text{..............(14)}$$

since y is assumed to vanish for $x=0$. By hypothesis, y vanishes also for $x=a$. This determines the value of A, viz.

$$A=-\tfrac{3}{8}wa. \quad\text{..................................(15)}$$

Hence

$$F=w(\tfrac{3}{8}a-x),\quad M=\tfrac{1}{2}wx(x-\tfrac{3}{4}a), \quad\text{.................(16)}$$

$$EIy=\tfrac{1}{24}wx(a-x)^2(x+\tfrac{1}{2}a). \quad\text{........................(17)}$$

It is easily seen from the value of F that the pressures on the three supports are respectively $\tfrac{3}{16}$, $\tfrac{10}{16}$, $\tfrac{3}{16}$ of the whole weight $2wa$. The bending moment changes sign for $x=\tfrac{3}{4}a$, which is a point of inflexion. It has a stationary value $(\tfrac{9}{128}wa^2)$ when $x=\tfrac{3}{8}a$. Its value at the centre is $\tfrac{1}{8}wa^2$.

The pressures on the supports may also be found expeditiously by the extended form of the principle of least energy given in Art. 144. When the ends and the middle point are fixed at the same level we have an instance of an elastic system subject to constraints which are inactive so long as gravity is not operative. If, when gravity comes into play, we assume an arbitrary value P for the pressure of the middle support, the pressures at the ends will be $wa-\tfrac{1}{2}P$, and therefore, along the first half of the beam

$$M=\tfrac{1}{2}wx^2-(wa-\tfrac{1}{2}P)x. \quad\text{..........................(18)}$$

The elastic energy of the whole beam is therefore

$$V=\frac{1}{EI}\int_0^a M^2dx=\frac{a^3}{120EI}(16wa^2-25Pwa+10P^2), \quad\text{.........(19)}$$

which is a minimum for

$$P=\tfrac{5}{4}wa. \quad\text{....................................(20)}$$

The results in this problem might also have been derived by superposition from those of Ex. 1 above and of Art. 146, Ex. 1. Thus if a uniform bar supported at the ends be subject to an upward pressure P at the middle, the droop (η) at this point will be given by

$$\eta=\frac{5}{384}\frac{wl^4}{EI}-\frac{1}{48}\frac{Pl^3}{EI}. \quad\text{..........................(21)}$$

Hence if $\eta=0$, we have

$$P=\tfrac{5}{8}wl, \quad\text{....................................(22)}$$

as above found. If the middle support bears the whole of the weight, so that $P=wl$, we have

$$-\eta=\frac{1}{128}\frac{wl^4}{EI}. \quad\text{..................................(23)}$$

This gives the droop at the ends when a bar is supported at the middle; cf. Ex. 2 above.

It may be noticed that in all these cases the elastic energy of the deflected beam is one-half the potential energy lost by the loads in sinking from the level of the supports. Take, first, the case of a single load W, and suppose that it produces a deflection y of its point of application. If we imagine the beam to be bent by a vertical force which increases gradually from 0 to W, the mean value of the force will be $\frac{1}{2}W$, and the work done by it on the beam will be $\frac{1}{2}Wy$. If we have a system of loads $W_1, W_2, \ldots$, and the deflections of the respective points of application be $y_1, y_2, \ldots$, we may imagine these loads to gradually increase from zero to their final values, preserving always the same ratios to one another. In this way the work done on the beam is seen to be

$$\tfrac{1}{2}(W_1 y_1 + W_2 y_2 + \ldots). \qquad (24)$$

149. Continuous Beams. Theorem of Three Moments.

The theorem in question refers to the case of a continuous beam, uniformly loaded, and resting on a number of supports at

Fig. 175.

the same level. It gives a relation between the values of the bending moment at any three consecutive points of support A, B, C.

Let $AB = a$, $BC = b$. We take the origin at B and the axis of x along BC. Then at points in the segment BC we find, by integration of Art. 145 (6),

$$M = EIy'' = \tfrac{1}{2}wx^2 + Ax + M_B, \qquad (1)$$

where the constant A is as yet undetermined. Hence

$$EIy' = \tfrac{1}{6}wx^3 + \tfrac{1}{2}Ax^2 + M_B x + EI\alpha, \qquad (2)$$

$$EIy = \tfrac{1}{24}wx^4 + \tfrac{1}{6}Ax^3 + \tfrac{1}{2}M_B x^2 + EI\alpha x, \qquad (3)$$

where α denotes the value of dy/dx at B. Putting $x = b$ we have, from (1) and (3),

$$M_C = \tfrac{1}{2}wb^2 + Ab + M_B, \qquad (4)$$

$$0 = \tfrac{1}{24}wb^3 + \tfrac{1}{6}Ab^2 + \tfrac{1}{2}M_B b + EI\alpha. \qquad (5)$$

Hence, eliminating A,

$$-6EI\alpha = bM_C + 2bM_B - \tfrac{1}{4}wb^3. \qquad (6)$$

Again, by taking the origin at B and the axis of x along BA, we should find

$$-6EI\beta = aM_A + 2aM_B - \tfrac{1}{4}wa^3, \quad\text{.............(7)}$$

the sign of M being unaltered when we reverse the direction of x. The quantities α, β both denote the gradient of the tangent at B, but since the directions of the axis of x are opposite in the two cases, we have $\beta = -\alpha$.

Hence, by addition,

$$aM_A + 2(a+b)M_B + bM_C = \tfrac{1}{4}w(a^3+b^3), \quad\text{.........(8)}$$

which is the theorem in question*.

Again, we find that the value of the shearing force $F(=-dM/dx)$ immediately to the right of B is, by (1) and (4),

$$F_{B+} = -A = \tfrac{1}{2}wb + \frac{M_B - M_C}{b} \quad\text{..............(9)}$$

Similarly, the value immediately to the left of B is

$$F_{B-} = -\tfrac{1}{2}wa - \frac{M_B - M_A}{a}, \quad\text{..............(10)}$$

the sign of F being reversed. The pressure on the support B is therefore

$$F_{B+} - F_{B-} = \tfrac{1}{2}w(a+b) + M_B\left(\frac{1}{a}+\frac{1}{b}\right) - \frac{M_A}{a} - \frac{M_C}{b}. \quad\text{...(11)}$$

It must be remembered that the preceding investigation is based on the supposition that the three supports are *exactly* at the same level. A slight deviation from this condition may seriously affect the results.

Ex. In the case of a beam resting on three supports A, B, C, of which two are at the ends, we have $M_A = 0$, $M_C = 0$, and therefore

$$M_B = \tfrac{1}{8}w(a^2 - ab + b^2), \quad\text{.........................(12)}$$

and the pressure on the middle support is

$$\frac{\tfrac{1}{8}w(a+b)(a^2+3ab+b^2)}{ab}. \quad\text{.......................(13)}$$

In the case of $a=b$, this agrees with the result of Art. 148, Ex. 3.

In the case of a beam resting on four supports A, B, C, D, of which A, D are at the ends, the formula (8) gives two linear relations between M_B and M_C. When these have been found, the pressures on B and C can be derived from (11). The pressures on the remaining supports A, D are then given by the ordinary rules of Statics.

* Due to B. P. E. Clapeyron (1857).

150. Combined Flexure and Extension.

We return to the investigation of Art. 146. We retain the hypothesis that the cross-sections remain plane in the strained condition; but the neutral line is no longer assumed to pass through the mean centre of the area. We will suppose for simplicity that the flexure and extension are parallel to a plane of symmetry.

The stresses across any section have now a resultant

$$\frac{E\omega h}{R}, \quad \text{......(1)}$$

where h is the distance of the mean centre from the neutral line, and R is the radius of curvature of the locus of the intersection of the neutral line with the plane of symmetry.

If η denote the distance of any point of the section from a line through the mean centre parallel to the neutral line, the extension at this point will be $(\eta + h)/R$, and the moment, with respect to the mean centre, of the stresses across the section will be

$$\frac{E}{R}\iint(\eta + h)\,\eta\,dx\,dy = \frac{E}{R}\iint \eta^2\,dx\,dy = \frac{E\kappa^2}{R}, \quad \text{......(2)}$$

exactly as in the case of pure flexure.

The distance from the mean centre of the line of action of the resultant is accordingly κ^2/h. In most applications this distance is known independently to be large compared with the dimensions of the cross-section, and accordingly with κ. The neutral line will then pass very nearly through the mean centre, and the distinction between R (as above defined) and the radius of curvature of the locus of the mean centre may be ignored.

The theory of a linear distribution of normal stress over an area has been treated without any restriction of symmetry in Arts. 94, 95, where we were concerned with centres of pressure in Hydrostatics. The more general form of the result just obtained is that the line of action of the resultant passes through the antipole of the neutral line with respect to the central ellipse of the area.

Such a distribution is assumed to hold in some other technical problems, as for instance in the case of a column excentrically loaded. Here it is usually important that the stress should have the same sign at all points of the area. For instance in a column of masonry the joints should not be subjected to *tension*. This limits the admissible

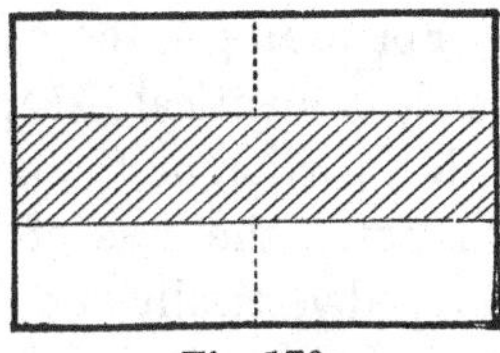

Fig. 176.

positions of the load, since the neutral line must lie outside the section. Thus for a rectangular section, whose sides are a, b, if the load be at a distance x from the centre, in a direction parallel to the sides b, the distance of the neutral line from the centre is κ^2/x, where $\kappa^2 = \frac{1}{12}b^2$. If this is to be greater than $\frac{1}{2}b$ we must have $x > \frac{1}{6}b$, i.e. the point of application of the load must lie within the *middle-third* of the depth of the section. The same principle applies to the voussoirs of an arch.

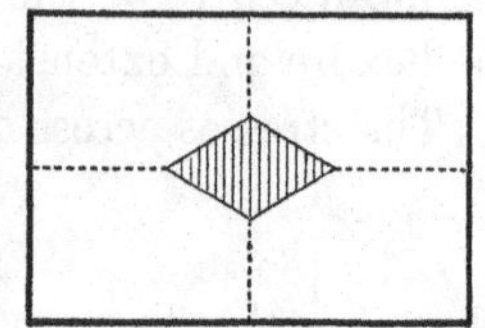
Fig. 177.

In any case the various positions of the point of application of the resultant, corresponding to positions of the neutral line which do not traverse the area, are included within a certain region, called the 'core' (Germ. *Kern*) of the area. In order that the stresses across the section may be all of the same sign, the line of action of the resultant must fall within the core. The boundary of the core is formed by the locus of the antipoles of all lines which touch the contour of the section. Thus for a circular section of radius a, the core is a concentric circle of radius $\frac{1}{4}a$. For a rectangular section it has the form shewn in the figure.

151. Finite Flexure of a Straight Rod.

We may extend the theory to the finite flexure of an originally straight elastic spring. We may consider the strained form as sufficiently represented by the 'central line,' i.e. the locus of the mean centres of the cross section.

Thus in the case of a spring bent by two equal and opposite thrusts $\pm W$ we have, taking moments, about any point of it, of the forces acting on either side,

$$\frac{E\omega\kappa^2}{R} = Wy, \quad \text{.........................(1)}$$

where y denotes distance from the line of thrust.

Hence if the spring be uniform we have

$$Ry = \text{const.}, \quad \text{.........................(2)}$$

exactly as in the case of the 'capillary curves' of Art. 127. The diagrams there given will serve to illustrate some of the forms which may be assumed.

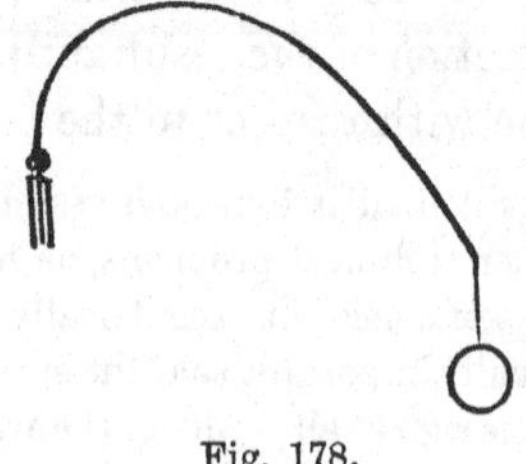
Fig. 178.

For example, the curve in Fig. 178, which is identical except as to position with a portion of the curve in Fig. 157, illustrates the case of a rod which is clamped vertically at one end and carries a weight attached to the other end. In Figs. 152, 155, we may suppose the forces $\pm W$ to be applied (along the dotted line) to two arms rigidly attached to the spring.

A case of some importance is that of a straight strut slightly bent by two opposing forces at the ends. This can be treated in a more elementary way as follows. Taking the axis of x in the line of the ends, we have $R^{-1} = y''$, approximately, where y is the deflection, and therefore

$$EIy'' = -Wy. \qquad (3)$$

The solution of this equation is

$$y = A \cos mx + B \sin mx, \qquad (4)$$

provided

$$m^2 = W/EI. \qquad (5)$$

If $y = 0$ for $x = 0$ and $x = l$, we have $A = 0$ and either

$$B = 0, \text{ or } \sin ml = 0. \qquad (6)$$

The latter alternative requires that

$$ml = s\pi, \qquad (7)$$

where s is an integer. Hence the slightly bent form is only possible if

$$W = s^2\pi^2 EI/l^2. \qquad (8)$$

The physical interpretation of this result is that if $W < \pi^2 EI/l^2$, the straight form is stable; i.e. the rod if accidentally bent will straighten itself again. But if W exceeds this value, the straight form is unstable, and the slightest disturbance will lead to a finite flexure. When W has the above value, exactly, the equilibrium is neutral. The higher values of s in (8) give values of W for which the equilibrium is neutral as regards particular types of deformation*.

The above interpretation is confirmed by a calculation of the energy in the bent state.

Suppose we have a vertical strut AB whose lower end A is fixed, whilst the upper end B carries a load W. It is assumed that B is constrained to lie in the same vertical with A, but that no *couple* acts there or at A. Imagine now that the rod is slightly bent into the form

$$y = \beta \sin mx, \qquad (9)$$

where $m = \pi/l$, if l denote the *altered* distance AB. If l_1 be the original (strained) length, we have

$$l_1 = \int_0^l \surd(1 + y'^2)\, dx = l\,(1 + \tfrac{1}{4} m^2 \beta^2), \qquad (10)$$

approximately. The *additional* extensions due to the flexure will be subject to the same relations as in Art. 145; and it is easily seen that the elastic energy of the bent rod is increased by the amount

$$\tfrac{1}{2} EI \int_0^l y''^2 dx = \tfrac{1}{4} EI m^4 \beta^2 l. \qquad (11)$$

* This theory is due to L. Euler.

On the other hand the potential energy of the load is diminished by

$$W(l_1 - l) = \tfrac{1}{4} W m^2 \beta^2 l. \qquad (12)$$

The nett loss of energy is therefore

$$\tfrac{1}{4} m^2 \beta^2 l \, (W - m^2 EI). \qquad (13)$$

If $W < \pi^2 EI/l^2$ this expression is negative, i.e. the system has *more* energy than when the rod is straight. But if on the other hand $W > \pi^2 EI/l^2$, the expression (13) is positive, and the bent form has less energy than the straight. The latter form is accordingly unstable.

152. Flexure of a Curved Bar.

We consider now the problem of flexure, parallel to the plane of symmetry, of a bar whose central line is, in the unstrained state, a plane curve, and whose section is symmetrical with respect to the plane of this curve.

The stress across any section may be resolved into a tension T tangential to the central line, a shearing force F normal to this line, and a bending moment M about a line through the mean centre perpendicular to the plane of flexure. Considering the equilibrium of a linear element δs which subtends an angle $\delta\psi$ at the centre of curvature, and resolving along the tangent and normal, we have, if there are no extraneous forces on the element,

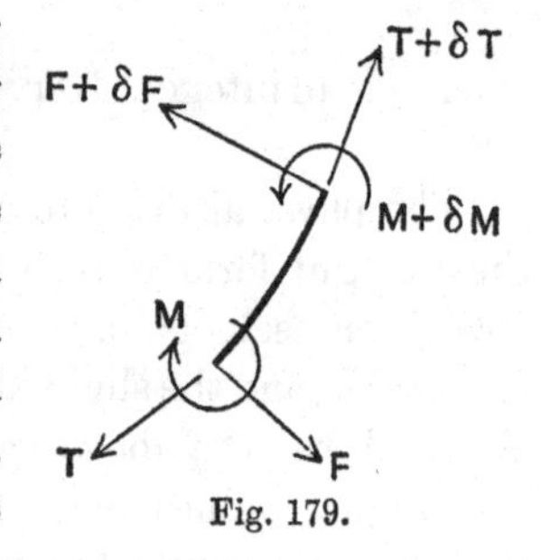

Fig. 179.

$$\delta T - F\delta\psi = 0, \quad \delta F + T\delta\psi = 0. \qquad (1)$$

Also, taking moments,

$$\delta M + F\delta s = 0. \qquad (2)$$

Hence

$$\frac{dT}{d\psi} = F, \quad \frac{dF}{d\psi} = -T, \quad \frac{dM}{ds} = -F. \qquad (3)$$

Hence

$$\frac{d^2 F}{d\psi^2} + F = 0, \qquad (4)$$

and therefore, over any portion of the bar which is free from external force,

$$F = A\cos\psi + B\sin\psi, \quad T = A\sin\psi - B\cos\psi, \quad (5)$$

where the constants A, B are arbitrary. These equations also follow at once from a consideration of the forces on a finite length of the bar.

To find the strains due to flexure, we denote by R_0 and R the radii of curvature, before and after the deformation, of the curve which is the locus of the intersection of the neutral line with the plane of symmetry, and by y the distance of any point P of the section from the neutral line. As a purely geometrical process we may imagine the bar to be bent from the unstrained condition until the central line becomes straight, and then bent back again into the strained condition, the position of the neutral line relative to the section being unaltered throughout. The extensions at P in the two stages of this process will be $-y/R_0$ and y/R, respectively, by Art. 145, so that the actual extension is $y/R - y/R_0$, and the corresponding tension

$$Ey\left(\frac{1}{R}-\frac{1}{R_0}\right). \quad \text{.......................(6)}$$

This differs from the result for a straight bar only in that it involves the change of curvature instead of the actual curvature. Hence the stresses across a section have a resultant

$$T = E\omega h\left(\frac{1}{R}-\frac{1}{R_0}\right), \quad \text{.....................(7)}$$

and their moment about a parallel to the neutral line through the mean centre of this section is

$$M = E\omega\kappa^2\left(\frac{1}{R}-\frac{1}{R_0}\right), \quad \text{.....................(8)}$$

where the symbols h and κ have the same meanings as in Art. 150.

In most cases of interest h is small compared with κ, for reasons similar to those indicated in Art. 150. The extension of the central line is then relatively negligible, and the symbols R and R_0 may be taken to refer to the curvature of this line. We further neglect, as in Art. 146, the geometrical effect of the shearing *strains* in the plane of the section, although the corresponding stresses have a resultant F which must be retained in the statical equations.

To examine more particularly the case of a *circular* bar, let u, v be the small displacements of a point on the central line along and at right angles to the original radius, so that the polar coordinates (referred to the centre) of this point are changed from (a, θ) to $(a + u, \theta + v/a)$. If ϕ be the angle which the tangent to the altered curve makes with the radius vector we have in the ordinary notation of the Calculus $\sin\phi = rd\theta/ds$, or in our present notation

$$\sin\phi = \frac{(a+u)\,d(\theta+v/a)}{ds} = \left(1+\frac{u}{a}\right)\left(1+\frac{v'}{a}\right)\frac{a\,d\theta}{ds}, \quad \text{...(9)}$$

where accents denote differentiation with respect to θ.

Again, $$\cos\phi = \frac{d(a+u)}{ds} = \frac{u'}{a}\frac{a\,d\theta}{ds}. \qquad\text{.............(10)}$$

Hence $$\left(\frac{ds}{a\,d\theta}\right)^2 = \left(1+\frac{u}{a}\right)^2\left(1+\frac{v'}{a}\right)^2 + \frac{u'^2}{a^2}. \qquad\text{...........(11)}$$

These formulæ are exact, but in the present application u and v are supposed to be small, and we further neglect the extension of the central line. Hence, putting $ds = a\,d\theta$, we have

$$u + v' = 0 \qquad\text{......................(12)}$$

to the first order. Also, from (10),

$$\phi = \tfrac{1}{2}\pi - \frac{u'}{a}. \qquad\text{........................(13)}$$

Hence if ψ denote the angle which the normal to the altered curve makes with the initial line of θ,

$$\psi = \theta + \frac{v}{a} + \phi - \tfrac{1}{2}\pi = \theta + \frac{v - u'}{a}. \qquad\text{...........(14)}$$

The altered curvature is therefore

$$\frac{1}{R} = \frac{d\psi}{a\,d\theta} = \frac{1}{a} + \frac{v' - u''}{a^2} = \frac{1}{a} - \frac{u'' + u}{a^2}, \qquad\text{.........(15)}$$

by (12). Hence, referring to (9), the differential equation to be satisfied by u is

$$\frac{EI}{a^2}(u'' + u) = -M, \qquad\text{....................(16)}$$

where $I = \omega\kappa^2$.

Since F and T, as well as M, are regarded as small quantities, whilst the difference between ψ and θ is of the first order, we may in the equations (5) substitute θ for ψ, and write

$$F = A\cos\theta + B\sin\theta,\quad T = A\sin\theta - B\cos\theta, \quad\text{...(17)}$$

$$M/a = -A\sin\theta + B\cos\theta + C. \qquad\text{...........(18)}$$

Substituting this value of M in (16), and integrating, we have

$$EIu/a^3 = -\tfrac{1}{2}A\theta\cos\theta - \tfrac{1}{2}B\theta\sin\theta - C + A'\cos\theta + B'\sin\theta, \text{...(19)}$$

and thence, by (12),

$$EIv/a^3 = \tfrac{1}{2}A(\theta\sin\theta + \cos\theta) - \tfrac{1}{2}B(\theta\cos\theta - \sin\theta) + C\theta$$
$$- A'\sin\theta + B'\cos\theta + C'. \text{......(20)}$$

A solution of this form holds for each segment into which the bar is divided by the points of application of the external forces. The constants in the solutions for adjacent segments are to be adjusted so that u, u', and v shall be continuous.

It is to be noticed that the solution includes, as it ought, an arbitrary displacement of the bar as a whole, without deformation.

Thus the terms in A' denote a translation parallel to the initial line, those in B' a translation at right angles to this line, whilst C' denotes a rotation about the centre. When the terminal and other conditions are of *force* only, these constants are undetermined, and may be omitted as irrelevant; but when *geometrical* conditions are imposed they must be retained.

Ex. 1. A finite bar extending (say) from $\theta = -\alpha$ to $\theta = \alpha$, is bent by two equal and opposite forces P at the ends.

We have obviously

$$M = Pa(\cos\theta - \cos\alpha), \quad \text{..............(21)}$$

and therefore

$$EIu/a^3 = P(\cos\alpha - \tfrac{1}{2}\theta\sin\theta),$$
$$EIv/a^3 = P(\tfrac{1}{2}\sin\theta - \tfrac{1}{2}\theta\cos\theta - \theta\cos\alpha), \quad \text{......(22)}$$

irrelevant terms being omitted.

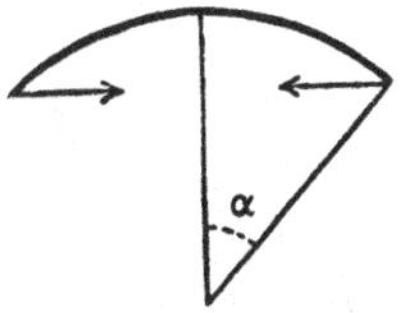

The length of the chord is diminished by

$$\frac{Pa^3}{EI}(\alpha + 2\alpha\cos^2\alpha - 3\sin\alpha\cos\alpha). \quad \text{.........(23)}$$

Fig. 180.

If the ring is almost complete ($\alpha = \pi$), this $= 3\pi Pa^3/EI$.

Ex. 2. A circular hoop is deformed by two equal and opposite forces P acting outwards (say) at the ends of a diameter.

We may, for definiteness, suppose the centre to be fixed, and the diameter in question fixed in direction. We take it as initial line of θ. On account of the symmetry we may confine our attention to either half of the hoop, say that extending from $\theta = 0$ to $\theta = \pi$. Considering the equilibrium of this half, we have obviously $F = \pm\frac{1}{2}P$, $T = 0$ at the ends. Hence, from (17) and (18),

$$F = \tfrac{1}{2}P\cos\theta, \quad T = \tfrac{1}{2}P\sin\theta, \quad M = -\tfrac{1}{2}Pa\sin\theta + M_0, \quad \text{......(24)}$$

where M_0 is the bending moment at the ends of the semicircle. The formulæ (19), (20), become

$$EIu/a^3 = -\tfrac{1}{4}P\theta\cos\theta - M_0/a + A'\cos\theta + B'\sin\theta, \quad \text{..................(25)}$$
$$EIv/a^3 = \tfrac{1}{4}P(\theta\sin\theta + \cos\theta) + M_0\theta/a - A'\sin\theta + B'\cos\theta + C'. \quad \text{...(26)}$$

Our assumptions require that the values of u for $\theta = 0$ and $\theta = \pi$ shall be equal, and also that $u' = 0$, $v = 0$ at these points. We find

$$A' = \tfrac{1}{8}\pi P, \quad B' = \tfrac{1}{4}P, \quad M_0 = Pa/\pi, \quad C' = -\tfrac{1}{2}P.$$

Hence, finally,

$$EIu/a^3 = \frac{1}{4}P\left(\sin\theta - \theta\cos\theta + \frac{1}{2}\pi\cos\theta - \frac{4}{\pi}\right), \quad \text{............(27)}$$

$$EIv/a^3 = \frac{1}{4}P\left(\theta\sin\theta + 2\cos\theta - \frac{1}{2}\pi\sin\theta + \frac{4\theta}{\pi} - 2\right). \quad \text{......(28)}$$

The diameter $\theta = 0$ is increased by

$$\frac{\pi^2 - 8}{4\pi}\frac{Pa^3}{EI} = \cdot 149\frac{Pa^3}{EI}, \quad \text{...........................(29)}$$

and the perpendicular diameter is diminished by

$$\frac{4 - \pi}{2\pi}\frac{Pa^3}{EI} = \cdot 136\frac{Pa^3}{EI}. \quad \text{...........................(30)}$$

When distributed forces act on the bar the equations (3) require modification. Taking for instance the case of a bar in a vertical plane, subject to gravity, and measuring ψ from the downward vertical, we must add terms $-w ds \sin\psi$ and $w ds \cos\psi$, respectively, to the tangential and normal components of force on the element. Thus

$$\frac{dT}{d\psi} = F + w\frac{ds}{d\psi}\sin\psi, \quad \frac{dF}{d\psi} = -T + w\frac{ds}{d\psi}\cos\psi. \quad \text{......(31)}$$

Hence for a circular bar

$$\frac{d^2F}{d\theta^2} + F = -2wa\sin\theta, \quad \text{.................(32)}$$

$$F = wa\theta\cos\theta + A\cos\theta + B\sin\theta, \quad \text{...........(33)}$$

$$M/a = -wa(\theta\sin\theta + \cos\theta) - A\sin\theta + B\cos\theta + C. \quad \text{...(34)}$$

Substituting this value of M in (16), and solving the equation, we find

$$EIu/a^3 = -\tfrac{1}{4}wa(\theta^2\cos\theta - 3\theta\sin\theta) - \tfrac{1}{2}A\theta\cos\theta$$
$$-\tfrac{1}{2}B\theta\sin\theta - C + A'\cos\theta + B'\sin\theta, \quad \text{...(35)}$$

$$EIv/a^3 = \tfrac{1}{4}wa(\theta^2\sin\theta + 5\theta\cos\theta - 5\sin\theta) + \tfrac{1}{2}A(\theta\sin\theta + \cos\theta)$$
$$-\tfrac{1}{2}B(\theta\cos\theta - \sin\theta) + C\theta - A'\sin\theta + B'\cos\theta + C'. \quad \text{...(36)}$$

Ex. 3. A hoop rests on a peg at its highest point, being deformed by its own weight only.

We may assume u to be an even, and v an odd function of θ, and therefore $A=0$, $B'=0$, $C'=0$. At the highest point ($\theta=\pi$) we put $u=0$, $u'=0$, $v=0$. This makes

$$B = \tfrac{1}{2}wa, \quad C = wa, \quad A' = (\tfrac{1}{4}\pi^2 - 1)wa.$$

Hence, putting $\theta=0$ in (35), we find for the increase in length of the vertical diameter

$$\frac{\pi^2-8}{8\pi}\frac{Wa^3}{EI}, \quad \text{..............................(37)}$$

where $W(=2\pi wa)$ is the total weight. The horizontal diameter is diminished by

$$\frac{4-\pi}{4\pi}\frac{Wa^3}{EI}. \quad \text{..............................(38)}$$

Comparing with Ex. 2 we see that these results are one-half the corresponding values when the weight of the hoop is concentrated at its lowest point.

153. Collapse of a Ring under Pressure.

We imagine a ring of radius a to be subject to a uniform normal pressure p per unit length, on the outside. The conditions of equilibrium of an elementary arc are then

$$\delta T - F\delta\psi = 0, \quad \delta F + T\delta\psi + p\delta s = 0, \quad \delta M + F\delta s = 0, \quad \text{...(1)}$$

in the notation of Fig. 179. Hence

$$F = -\frac{dM}{ds}, \quad T = -\frac{dF}{d\psi} - p\frac{ds}{d\psi}, \quad \frac{dT}{d\psi} = F. \quad \text{.....(2)}$$

We proceed to shew that for a certain value of p a slightly elliptic form of equilibrium is possible. If the ring be deformed into an ellipse of small excentricity e, without change of perimeter, the lengths of the semiaxes will be $a(1 \pm \frac{1}{4}e^2)$, and the radius of curvature

$$R = a(1 - \tfrac{3}{4}e^2 \cos 2\psi), \quad \text{..................(3)}$$

the zero of ψ being at an extremity of the major axis. The bending moment is therefore

$$M = \frac{3}{4}\frac{EI}{a} e^2 \cos 2\psi. \quad \text{......................(4)}$$

Hence, from (2), we must have, approximately,

$$F = \frac{3}{2}\frac{EI}{a^2} e^2 \sin 2\psi, \quad \text{......................(5)}$$

$$T = -pa + \left(\frac{3}{4}pa - \frac{3EI}{a^2}\right) e^2 \cos 2\psi. \quad \text{...........(6)}$$

The remaining equation of (2) is therefore satisfied, provided

$$p = \frac{3EI}{a^3}, \quad \text{..........................(7)}$$

which is independent of e.

Hence when p has this value the equilibrium force is indeterminate. The true interpretation of this result, as in the case of Euler's theory of struts (Art. 151), is that for smaller values of p the ring, if deformed and then left to itself, will recover its shape, whilst a greater value of p will cause it to collapse. This conclusion can be confirmed, as in the problem referred to, by calculations of the energy of deformation in the respective states.

For instance, if we assume

$$u = nA \sin n\theta, \quad v = A \cos n\theta, \quad \text{.......................(8)}$$

the energy of flexure is found to be

$$\tfrac{1}{2}\pi n^2 (n^2 - 1)^2 \frac{EIA^2}{a^3}. \quad \text{............................(9)}$$

The formulæ (8) satisfy the condition of inextensibility, to the first order. Proceeding to the second order, we find from the formula (11) of Art. 152 that the total increase of perimeter is

$$\tfrac{1}{2}\pi n^2 (n^2 - 2)\frac{A^2}{a}.$$

Since the ring is subject to a thrust pa (Art. 155), this gives a loss of compressional energy of amount

$$\tfrac{1}{2}\pi n^2 (n^2 - 2) pA^2. \quad \text{............................(10)}$$

The area enclosed by the ring is diminished by $\frac{1}{2}\pi n^2 A^2$. Hence the work done by the external pressure will exceed the gain of elastic energy if

$$p > (n^2-1)\frac{EI}{a^2}. \quad \text{.....................(11)}$$

The lowest critical pressure is given by $n=2$, in agreement with (7).*

The preceding investigation can be adapted to the collapse of a *tube* under external pressure by a modification of the elastic constant, as explained at the end of Art. 145. The critical value of the pressure (per unit area) is,

$$\frac{Eh^3}{4(1-\sigma^2)a^3}, \quad \text{.....................(12)}$$

where h is the thickness of the wall.

154. Torsion of a Bar of Circular Section.

The theory of the torsion of a wire or bar of *circular* section is very simple.

If the twist per unit length be θ, two cross-sections at a distance δz apart will be rotated relatively to one another through an angle $\theta\,\delta z$. Hence if PP' be a line drawn perpendicular to these sections, at a distance r from the axis, the displacement of P' relative to P will be $r\theta\,\delta z$ at right angles to PP' and to the axis. The deformation in the neighbourhood of P therefore consists of a shear of amount $r\theta\,\delta z/\delta z$, or $r\theta$. This implies a shearing stress $\mu r\theta$ tangential to the cross-section and at right angles to r. Hence, taking moments about the axis, and integrating over the section, we get a twisting couple

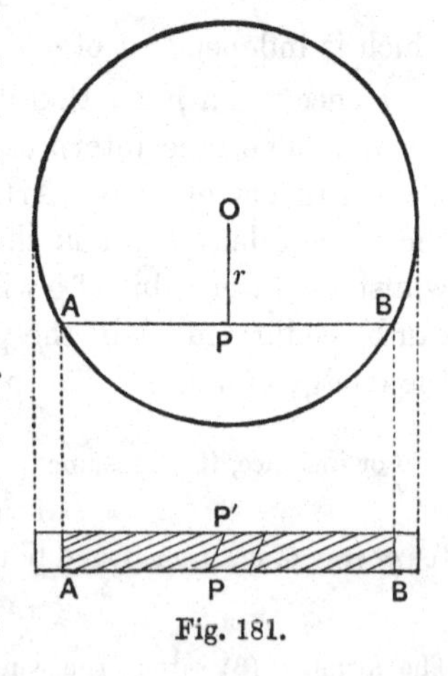

Fig. 181.

$$\int_0^a r\,.\,\mu r\theta\,.\,2\pi r\,dr = \tfrac{1}{2}\pi\mu a^4\theta, \quad \text{..............(1)}$$

where a is the radius. If we put

$$K = \tfrac{1}{2}\pi\mu a^4, \quad \text{.........................(2)}$$

* The investigation is due to Prof. G. H. Bryan.

this constant K, which measures the ratio of the couple to the twist per unit length, is called the 'modulus of torsion' of the particular rod.

From the value of K as found by experiment we can infer the value of the rigidity μ, so far as the material can be regarded as isotropic.

It may not be apparent at first sight why the limitation to the *circular* form of section has been introduced. If we imagine a bar of *any* uniform section (ω) to be twisted about an axis through the mean centres of the sections, without further deformation, we obtain a torsional couple $\mu\omega\kappa^2\theta$, where κ is the radius of gyration of the section with respect to the axis. But this couple would not of itself suffice to maintain the assumed state of strain. If we refer to Fig. 181, we see that on the principles of Arts. 135, 136 the strain implies tangential stress on longitudinal planes through the axis. Unless the section be circular, these planes will as a rule meet the surface of the rod obliquely, instead of at right angles, and the state of stress is therefore not consistent with an absence of external force on the surface. For equilibrium it would be necessary to apply certain longitudinal forces tangentially to the surface. These forces, it may be noticed, will have different signs in different parts of the circumference. If we imagine them to be at first applied along with the torsional couples, so as to maintain the special type of strain above postulated, and afterwards relaxed, the bar will yield further to the twisting couples, and there will be at the same time a warping of the sections, so that these are no longer accurately plane. The effective modulus of torsion will therefore be *less* than on the assumed hypothesis.

The action of an ordinary spiral spring, i.e. a wire of circular section coiled into a helix of small pitch, depends mainly on torsion*. Suppose that the coil hangs vertically and is stretched by a weight W hanging axially. The torsional couple on any section of the wire is then Wc, where c is the radius of the helix, and if θ be the twist of the wire we have

$$Wc = K\theta. \quad \text{..........................(3)}$$

To calculate the vertical displacement (z) of the lower end, due to the weight W, we may imagine the different elements δs of the wire to be twisted in succession. The vertical displacement due to the twist in any one element is $\theta\delta s \, . \, c$. There is also a horizontal displacement, but the horizontal components due to the different

* This seems to have been first pointed out in 1848 by James Thomson (1822–92), professor of engineering and mechanics at Glasgow 1873–89.

elements will on the whole neutralize one another. The total vertical displacement is therefore θlc, where l is the length of wire in the spiral. Hence

$$z = \theta lc = \frac{Wc^2 l}{K}. \quad \text{.......................(4)}$$

It should be added that unless the pitch be small there is an appreciable bending couple to be taken into account, in addition to the torsional couple above considered. The spring then tends to uncoil when stretched.

EXAMPLES. XXVI.

1. Compare the flexural rigidities of two rods of the same material, of square and circular section, respectively, the weight per unit length being the same. [π : 3.]

2. A bar whose section is a square 1 cm. in the side rests on two knife-edges a metre apart. A load of 1 kg. at the centre produces a deflection there of 1·15 cm. Find the value of Young's modulus for the material of the bar. [$2{\cdot}17 \times 10^8$.]

3. A light rod of length $2l$ rests symmetrically on two rigid supports at a distance $2a$ apart. If a load W be suspended from the centre, this point will sink through a space $\frac{1}{6} Wa^3/EI$, and the ends will rise through a space

$$\tfrac{1}{4} Wa^2 (l-a)/EI.$$

4. A light horizontal rod ABC is supported by smooth rings at A and B, and carries a weight W at C. Prove that the deflection of C is

$$\frac{1}{3}\frac{W}{EI}.AC.BC^2.$$

Also prove that the inclination to the horizontal at A is

$$\frac{1}{6}\frac{W}{EI}.AB.BC.$$

5. A bar of circular section (rad. $=a$) is made of material such that a length λ can hang vertically from one end without producing permanent extension. Prove that the greatest length which can rest horizontally on two supports at the ends, without permanent flexure, is $\sqrt{(2a\lambda)}$.

6. A uniform beam of length $2a$ rests on three supports, at the ends and the centre. What must be the depth of the middle support below the level of the other two, in order that the pressures on all three may be equal? [$\frac{7}{72} wa^4/EI$.]

7. A beam rests symmetrically on two supports at the same level; prove that the deflection at the centre will be up or down according as the distance between the supports is less or greater than ·523 of the total length.

8. One end of a horizontal beam is clamped, and the other is supported at the same level; prove that

$$EIy = \tfrac{1}{48} wx (l-x)^2 (l+2x),$$

where x denotes distance from the supported end.

Prove that the pressure on the supported end is $\frac{3}{8}wl$.

9. A weight is suspended from the free end B of a uniform cantilever AB. If the middle point of AB be supported so that it cannot droop, prove that the deflection at B due to the suspended weight is diminished in the ratio of 7 to 32.

10. If a uniform horizontal beam be clamped at both ends, the deflection at the centre is one-fifth of that of the same beam when merely supported at the ends.

11. A uniform beam rests on four equidistant supports at the same level, of which the two outer are at the ends. Shew that the pressures on the four supports are proportional to 4, 11, 11, 4.

12. Apply the formula $\quad \frac{1}{2} EI \int y''^2 \, dx$

to find the elastic energy of a uniform beam supported at the ends, and bent by its own weight. $[\frac{1}{240} w^2 l^5 / EI.]$

13. A cantilever has a uniform breadth, but its depth tapers uniformly to a point, from the fixed end. Prove that, the origin being at the free end, the form assumed when it is bent by its own weight is

$$y = \frac{w_0 l^2}{12 EI_0} (l-x)^2,$$

where the suffixes refer to the fixed end.

14. Prove that, in the preceding Ex., if the tapering had been in the breadth, but not in the depth, the result would have been

$$y = \frac{w_0}{72 EI_0} (l-x)^2 (3l^2 + 2lx + x^2).$$

15. Prove by partial integration that

$$\tfrac{1}{2} \int EIy''^2 \, dx = \tfrac{1}{2} [Fy + My'] + \tfrac{1}{2} \int wy \, dx,$$

where the integrations extend over the whole length of a beam, and the square brackets indicate that the difference of the values of the enclosed expression at the two ends is to be taken. (The beam is not to be assumed to be uniform.) Interpret this identity.

16. If a bar be subject both to extension and flexure, prove that the energy per unit length is

$$\tfrac{1}{2}\frac{EI}{R^2}+\tfrac{1}{2}E\omega\epsilon^2,$$

where ϵ is the extension of the axis.

17. A rod whose own weight may be neglected is clamped vertically at the lower end, and carries a load W at the upper end, which is free. Find the greatest value of W consistent with stability. [$\tfrac{1}{4}\pi^2 EI/l^2$.]

18. Calculate on Euler's theory (Art. 151) the greatest thrust which an iron bar 1 metre long, whose section is a square of 1 cm., can bear without bending ($E=2\times 10^9$). [160 kg.]

19. Prove that if both ends of a strut are clamped the greatest thrust which it can exert without bending is $4\pi^2 EI/l^2$.

20. Prove that the modulus of torsion of a circular tube whose inner and outer radii are a, b is to that of a solid rod of circular section, of the same material and weight per foot, in the ratio

$$(b^2+a^2)/(b^2-a^2).$$

21. A shaft whose section has a radius a is transmitting energy at the rate P, revolving steadily with angular velocity ω. Prove that the greatest shearing stress developed in the shaft is $2P/\pi\mu\omega a^3$.

Find the least diameter of the shaft in order that the stress may not exceed 4 tons per sq. in., when the shaft is working at 1600 horse-power, and making 25 revolutions per minute. [13·2 ins.]

22. A curved rod of any (plane) form is subject to a uniform normal pressure in its plane. Prove that the force-component of the stress at any point P is proportional to, and perpendicular to the radius vector OP drawn from a certain fixed origin O.

23. If equal and opposite couples $\pm M$ are applied to the ends of a circular bar, the form remains circular, but the radius is altered by Ma^2/EI.

24. A circular hoop of radius a and line-density ρ rotates freely about a diameter with angular velocity ω. Prove that this diameter is diminished, and the perpendicular diameter increased, by

$$\frac{\rho\omega^2 a^5}{3EI},$$

where E is supposed expressed in dynamical measure.

CHAPTER XVIII

STRESSES IN CYLINDRICAL AND SPHERICAL SHELLS

155. Stresses in Thin Shells.

Suppose we have a uniform thin *spherical* shell subject to an internal fluid pressure p_0. On account of the symmetry of the conditions, the total stress across a linear element drawn on the surface will be tangential to the surface and at right angles to the element. As in Art. 122, we infer that the intensity of the stress, per unit length of the element, is a constant; we denote it by T.

The principal axes of stress at any point in the substance of the shell will be respectively normal and tangential to the surface. If p_1, p_2, p_3 be the principal stresses, p_1 being that in the direction of the normal, we have $p_2 = p_3$, and T is the integral of p_2 or p_3 over the thickness.

The relation between p_0 and T is found by considering the forces acting on a hemisphere. As in Art. 124, the resultant fluid pressure $p_0 . \pi a^2$ is balanced by the tension on the rim, viz. $T . 2\pi a$, where a denotes the radius. Hence

$$T = \tfrac{1}{2} p_0 a. \quad \text{..........................(1)}$$

Take next the case of a *cylindrical* shell subject to a uniform internal pressure (p_0). The principal stresses in the tangent plane will be in the directions of the circular sections and the generating lines, respectively. We denote their amounts, when integrated over the thickness, by T_2, T_3. The former is sometimes called the 'hoop-tension.'

If the shell forms part of the surface of a closed vessel (e.g. a boiler), the longitudinal thrust on the ends will be $p_0 . \pi a^2$, where a is the radius, irrespective of the shape of the ends. Hence, considering the equilibrium of a portion cut off by a transverse plane, we have $p_0 . \pi a^2 = T_3 . 2\pi a$, or

$$T_3 = \tfrac{1}{2} p_0 a. \quad \text{..........................(2)}$$

Again, considering a portion bounded by a plane through the axis, and two transverse planes at unit distance apart, we have $p_0 . 2a = 2T_2$, or

$$T_2 = p_0 a. \qquad \text{..........................(3)}$$

The longitudinal tension is therefore half the transverse tension.

156. Thick Spherical Shell.

If we wish to take account of the thickness of the shell, we must have recourse to the elastic relations of Art. 138.

In the case of the sphere, if ξ be the radial displacement of a point at a distance r from the centre, the radial extension is $d\xi/dr$, by the same reasoning as in Art. 140. Also, since the points which originally lay on a circle of radius r in a diametral plane now lie on a circle of radius $r + \xi$, the transverse extension is ξ/r. We write therefore

$$\epsilon_1 = \frac{d\xi}{dr}, \quad \epsilon_2 = \epsilon_3 = \frac{\xi}{r}, \qquad \text{....................(1)}$$

and the formulæ (4) of Art. 138 give

$$E\frac{d\xi}{dr} = p_1 - 2\sigma p_2, \quad E\frac{\xi}{r} = (1-\sigma)p_2 - \sigma p_1. \qquad \text{......(2)}$$

Hence, eliminating ξ, we have

$$(1-\sigma)\frac{d}{dr}(rp_2) + 2\sigma p_2 = \sigma\frac{d}{dr}(rp_1) + p_1, \qquad \text{.........(3)}$$

as a necessary relation between the principal stresses, on the present hypothesis of symmetry.

We have next to introduce the statical condition. Consider a hemispherical stratum whose inner and outer radii are r and $r + \delta r$. The resultant of the normal stresses on its inner surface is a force $p_1 . \pi r^2$ inwards, at right angles to the plane of the edge. Against this there is a force

$$p_1 . \pi r^2 + \frac{d}{dr}(p_1 . \pi r^2)\,\delta r$$

outwards, due to the stresses on the outer face. Along the edge we have a force $p_2 . 2\pi r\,\delta r$ inwards. Hence

$$\frac{d}{dr}(r^2 p_1) = 2rp_2. \qquad \text{......................(4)}$$

If we eliminate p_2 between (3) and (4), we find after reduction, and division by $1-\sigma$,

$$r^2 \frac{d^2}{dr^2}(r^2 p_1) = 2r^2 p_1. \quad \text{.....................(5)}$$

If we assume, for trial, $r^2 p_1 = Ar^m$, we find that this equation is satisfied if $m = 2$ or $m = -1$. Since the equation is linear, solutions can be superposed, and we have

$$r^2 p_1 = Ar^2 + \frac{B}{r}.$$

Hence, and by (4),

$$p_1 = A + \frac{B}{r^3}, \quad p_2 = A - \frac{B}{2r^3}. \quad \text{...............(6)}$$

The constants A, B are determined by the conditions at the inner and outer surfaces. If $p_1 = -p_0$ at the inner surface ($r = b$), and $p_1 = 0$ at the outer surface ($r = a$), we find

$$A = \frac{b^3}{a^3 - b^3} p_0, \quad B = -\frac{a^3 b^3}{a^3 - b^3} p_0, \quad \text{...........(7)}$$

whence $$p_1 = -\frac{b^3(a^3 - r^3)}{r^3(a^3 - b^3)} p_0, \quad p_2 = \frac{b^3(a^3 + 2r^3)}{2r^3(a^3 - b^3)} p_0. \quad \text{......(8)}$$

Both p_1 and p_2 are greatest (in absolute value) at the inner surface. The value of p_2 there is diminished by increasing the outer radius (a), but not indefinitely, the lower limit being $\frac{1}{2}p_0$.

The dilatation Δ is given by the formula

$$\kappa\Delta = \tfrac{1}{3}(p_1 + 2p_2) = \frac{b^3}{a^3 - b^3} p_0, \quad \text{..............(9)}$$

and is accordingly independent of r.

If the thickness $a - b$ is small compared with the mean radius, we find

$$p_2(a - b) = \tfrac{1}{2} a p_0, \quad \text{.....................(10)}$$

in agreement with Art. 152 (1). Also if we write $r = b + x$, $a = b + h$, we have

$$p_1 = -\left(1 - \frac{x}{h}\right) p_0, \quad \text{.....................(11)}$$

approximately.

157. Thick Cylindrical Shell.

In the case of the cylinder, we assume that the deformation is symmetrical about an axis, and uniform along each generating line. We will also suppose that a cross-section remains plane after the deformation, so that the extension ϵ_3 in the direction of the length is constant.

If ξ be the radial displacement, the extensions in the direction of the radius vector, and at right angles to it, in a transverse plane, are

$$\epsilon_1 = \frac{d\xi}{dr}, \quad \epsilon_2 = \frac{\xi}{r}. \quad \text{.....................(1)}$$

If we eliminate p_3 between the equations (4) of Art. 138, we find

$$\left.\begin{aligned} E'\epsilon_1 &= (1-\sigma)\,p_1 - \sigma p_2 - \sigma E'\epsilon_3, \\ E'\epsilon_2 &= -\sigma p_1 + (1-\sigma)\,p_2 - \sigma E'\epsilon_3, \end{aligned}\right\} \quad \text{...........(2)}$$

where

$$E' = \frac{E}{1+\sigma}. \quad \text{........................(3)}$$

Substituting from (1), and eliminating ξ, we obtain

$$\frac{d}{dr}\{-\sigma r p_1 + (1-\sigma)\,r p_2\} = (1-\sigma)\,p_1 - \sigma p_2, \quad \text{......(4)}$$

as a necessary relation between the stresses.

Next, consider the equilibrium of an elementary shell whose internal and external radii are r and $r+\delta r$. A portion of this shell bounded by an axial plane, and two transverse planes at unit distance apart, is subject on the inner curved surface to normal stresses whose resultant is $p_1.2r$ inwards; and the resultant of the corresponding stresses on the outer face is

$$p_1.2r + \frac{d}{dr}(p_1.2r)\,\delta r$$

outwards. On the edges we have a pull $2p_2\delta r$. Hence

$$\frac{d}{dr}(rp_1) = p_2, \quad \text{......................(5)}$$

which is the statical equation.

Substituting in (4) we find

$$r\frac{d}{dr}\left(r\frac{d}{dr}\right)rp_1 - rp_1 = 0. \quad \text{.................(6)}$$

If we assume $rp_1 = Ar^m$, this equation is satisfied provided $m = 1$ or $m = -1$; and the general solution is therefore

$$rp_1 = Ar + \frac{B}{r}.$$

Hence $$p_1 = A + \frac{B}{r^2}, \quad p_2 = A - \frac{B}{r^2}, \quad \text{..............(7)}$$

whilst $$p_3 = \sigma(p_1 + p_2) + E\epsilon_3 = 2\sigma A + E\epsilon_3, \quad \text{...........(8)}$$

and is accordingly independent of r. Also

$$\kappa\Delta = \tfrac{1}{3}(p_1 + p_2 + p_3) = \tfrac{2}{3}(1+\sigma)A + \tfrac{1}{3}E\epsilon_3. \quad \text{......(9)}$$

The values of the constants A, B will depend on the conditions to be satisfied at the two cylindrical surfaces, whilst ϵ_3 will be determined by the conditions at the ends.

If there is a hydrostatic pressure p_0 on the inner face ($r = b$), whilst the outer face ($r = a$) is free from stress, we have

$$A = \frac{b^2}{a^2 - b^2}p_0, \quad B = -\frac{a^2b^2}{a^2 - b^2}p_0, \quad \text{...........(10)}$$

whence $$p_1 = -\frac{b^2(a^2 - r^2)}{r^2(a^2 - b^2)}p_0, \quad p_2 = \frac{b^2(a^2 + r^2)}{r^2(a^2 - b^2)}p_0. \quad \text{......(11)}$$

If there is no longitudinal stress, we have $p_3 = 0$, and

$$E\epsilon_3 = -2\sigma A = -\frac{2\sigma b^2}{a^2 - b^2}p_0. \quad \text{..............(12)}$$

If longitudinal extension (or contraction) is prevented, $\epsilon_3 = 0$, and

$$p_3 = 2\sigma A = \frac{2\sigma b^2}{a^2 - b^2}p_0. \quad \text{................(13)}$$

If longitudinal stress is produced by hydrostatic pressure on the ends of the cylinder, we have $p_3 . \pi(a^2 - b^2) = p_0 . \pi b^2$,

or $$p_3 = \frac{b^2}{a^2 - b^2}p_0, \quad \text{....................(14)}$$

whence $$E\epsilon_3 = \frac{(1 - 2\sigma)b^2}{a^2 - b^2}p_0, \quad \text{or} \quad \kappa\epsilon_3 = \frac{1}{3}\frac{b^2}{a^2 - b^2}p_0, \quad \text{...(15)}$$

by Art. 138 (8). This gives a means of determining κ directly by extensional measurements*.

158. Compound Cylindrical Shells. Initial Stress.

The formulæ (7) of Art. 157 may be applied to find the distribution of stress when one tube is shrunk over another, as in some processes of gun construction. There is then a discontinuity in the values of the constants A, B at the cylindrical surface of contact.

Thus if b be the inner radius of the inner tube, c that of the surface of contact, and a the outer radius of the outer tube, we assume for the inner tube

$$p_1 = A + \frac{B}{r^2}, \quad p_2 = A - \frac{B}{r^2}, \quad \text{..............(1)}$$

and for the outer tube

$$p_1 = C + \frac{D}{r^2}, \quad p_2 = C - \frac{D}{r^2}. \quad \text{..............(2)}$$

If there is no external or internal pressure, we have

$$A + \frac{B}{b^2} = 0, \quad C + \frac{D}{a^2} = 0, \quad \text{..................(3)}$$

whilst
$$A + \frac{B}{c^2} = C + \frac{D}{c^2}, = -P, \text{ say}, \quad \text{..............(4)}$$

since p_1 is necessarily continuous. Hence

$$A = -\frac{Pc^2}{c^2 - b^2}, \quad B = \frac{Pb^2c^2}{c^2 - b^2}, \quad \text{..............(5)}$$

$$C = \frac{Pc^2}{a^2 - c^2}, \quad D = -\frac{Pa^2c^2}{a^2 - c^2}. \quad \text{..............(6)}$$

When there is a hydrostatic pressure p_0 in the inner cavity, we must superpose on the stresses given by these formulæ the system (11) of Art. 157. Thus the hoop-tension in the substance of the inner tube will now be

$$p_2 = \frac{b^2(a^2 + r^2)}{r^2(a^2 - b^2)} p_0 - \frac{c^2(r^2 + b^2)}{r^2(c^2 - b^2)} P, \quad \text{...........(7)}$$

* A. Mallock, *Proc. R. S.*, 1904.

whilst for the outer tube

$$p_2 = \frac{b^2(a^2+r^2)}{r^2(a^2-b^2)}p_0 + \frac{c^2(r^2+a^2)}{r^2(a^2-c^2)}P. \quad \text{...........(8)}$$

The hoop-tension in the inner portion of the compound tube is thus diminished by the permanent stress depending on P.

159. Stresses in a Rotating Disk.

When a thin disk is subject only to forces and accelerations in its own plane, the stress p_3 in a direction normal to the plane may be assumed to vanish. In the case of a circular disk rotating in its own plane, the directions of the remaining principal axes will be along and perpendicular to the radius vector r; and we have by Art. 138 (4)

$$\left.\begin{aligned} E\epsilon_1 &= E\frac{d\xi}{dr} = p_1 - \sigma p_2, \\ E\epsilon_2 &= E\frac{\xi}{r} = p_2 - \sigma p_1, \end{aligned}\right\} \quad \text{....................(1)}$$

where ξ is the radial displacement. Hence, eliminating ξ,

$$\frac{d}{dr}(rp_2 - \sigma r p_1) = p_1 - \sigma p_2. \quad \text{................(2)}$$

If ρ be the density, and ω the angular velocity, the resultant of the centrifugal forces $(m\omega^2 r)$ on a semicircular strip $\rho.\pi r\delta r$, as given by the formula (2) of Art. 64, is

$$\rho.\pi r\delta r.\omega^2.\frac{2r}{\pi} = 2\rho\omega^2 r^2 \delta r,$$

in absolute units. The resultant stresses on the two circular edges are $p_1.2r$, inwards, and

$$p_1.2r + \frac{d}{dr}(p_1.2r)\,\delta r,$$

outwards, whilst the pull on the ends is $2p_2\delta r$*. Hence

$$\frac{d}{dr}(rp_1) - p_2 = -\rho\omega^2 r^2, \quad \text{................(3)}$$

the stresses p_1, p_2 being now, of course, supposed expressed in dynamical measure.

* A symbol for the thickness of the disk is omitted throughout. It would divide out in (3).

Eliminating p_2 between (2) and (3), we find

$$\left(r\frac{d}{dr}\right)^2(rp_1) - rp_1 = -(3+\sigma)\rho\omega^2 r^3, \quad \text{...........(4)}$$

whence

$$p_1 = A + \frac{B}{r^2} - \tfrac{1}{8}(3+\sigma)\rho\omega^2 r^2, \quad \text{.............(5)}$$

and, from (3),

$$p_2 = A - \frac{B}{r^2} - \tfrac{1}{8}(1+3\sigma)\rho\omega^2 r^2. \quad \text{...........(6)}$$

If the disk is complete to the centre, $B=0$, since the stresses must be finite. Also if the outer edge ($r=a$) is free,

$$A = \tfrac{1}{8}(3+\sigma)\rho\omega^2 a^2, \quad \text{..............................(7)}$$

whence

$$p_1 = \tfrac{1}{8}(3+\sigma)\rho\omega^2(a^2-r^2), \quad \text{...................(8)}$$

$$p_2 = \tfrac{1}{8}\rho\omega^2\{(3+\sigma)a^2 - (1+3\sigma)r^2\}. \quad \text{...........(9)}$$

It appears that both stresses are greatest at the centre, where they are equal, and that the maximum value depends on the circumferential velocity ωa. It may be noted also that p_2 is elsewhere greater than p_1.

If, on the other hand, the disk has a concentric aperture ($r=b$) which is free from external force, we find

$$A = \tfrac{1}{8}(3+\sigma)\rho\omega^2(a^2+b^2), \quad B = -\tfrac{1}{8}(3+\sigma)\rho\omega^2 a^2 b^2, \text{...(10)}$$

whence

$$p_1 = \tfrac{1}{8}(3+\sigma)\rho\omega^2 . \frac{(a^2-r^2)(r^2-b^2)}{r^2}, \quad \text{...................(11)}$$

$$p_2 = \tfrac{1}{8}(3+\sigma)\rho\omega^2\left(a^2+b^2+\frac{a^2b^2}{r^2}\right) - \tfrac{1}{8}(1+3\sigma)\rho\omega^2 r^2. \text{ ...(12)}$$

It appears that p_1 is a maximum when $r=\sqrt{(ab)}$, whilst p_2 is greatest at the inner boundary. When the radius b of this is small, the greatest value of p_2 is

$$\tfrac{1}{4}(3+\sigma)\rho\omega^2 a^2, \quad \text{.....................(13)}$$

which is double the value at the centre in the former case. For this reason a disk which is keyed on to a shaft is estimated to be only half as strong (in relation to centrifugal force) as if the shaft were in one piece with the disk.

160. Rotating Shaft.

The case of a rotating cylindrical shaft is somewhat different. We assume that there is no warping of the cross-sections except near the ends, if the length is large compared with the diameter. Hence ϵ_3 will be independent of r. Since the total longitudinal force across a section must vanish, we have

$$\int_0^a p_3 \,.\, 2\pi r\, dr = 0, \qquad \text{......................(1)}$$

the shaft being supposed solid, of radius a.

Since $$p_3 = \sigma(p_1 + p_2) + E\epsilon_3, \qquad \text{...................(2)}$$
we have in place of Art. 159 (1)

$$\left.\begin{aligned} E\frac{d\xi}{dr} &= (1-\sigma^2)\,p_1 - (\sigma+\sigma^2)\,p_2 - \sigma E\epsilon_3, \\ E\frac{\xi}{r} &= (1-\sigma^2)\,p_2 - (\sigma+\sigma^2)\,p_1 - \sigma E\epsilon_3. \end{aligned}\right\} \qquad \text{........(3)}$$

Eliminating ξ, and dividing by $1+\sigma$ we have

$$\frac{d}{dr}\{(1-\sigma)\,rp_2 - \sigma rp_1\} = (1-\sigma)\,p_1 - \sigma p_2. \qquad \text{........(4)}$$

The equation (3) of Art. 159 holds as before, and we find on elimination of p_2

$$\left(r\frac{d}{dr}\right)^2 (rp_1) - rp_1 = -\frac{3-2\sigma}{1-\sigma}\rho\omega^2 r^3. \qquad \text{...........(5)}$$

The solution which makes p_1 finite on the axis, and $=0$ at the surface, is

$$p_1 = \frac{3-2\sigma}{8(1-\sigma)}\rho\omega^2(a^2-r^2). \qquad \text{.................(6)}$$

Hence, from (3) of Art. 159,

$$p_2 = \frac{3-2\sigma}{8(1-\sigma)}\rho\omega^2 a^2 - \frac{1+2\sigma}{8(1-\sigma)}\rho\omega^2 r^2, \qquad \text{........(7)}$$

and from (2) above

$$p_3 = \frac{\sigma(3-2\sigma)}{4(1-\sigma)}\rho\omega^2 a^2 - \frac{\sigma}{2(1-\sigma)}\rho\omega^2 r^2 + E\epsilon_3. \qquad \text{......(8)}$$

The condition (1) then gives

$$E\epsilon_3 = -\tfrac{1}{2}\sigma\rho\omega^2 a^2, \qquad \text{.......................(9)}$$

whence $$p_3 = \frac{\sigma}{4(1-\sigma)}\rho\omega^2(a^2-2r^2). \qquad \text{..............(10)}$$

EXAMPLES. XXVII.

1. If a thick spherical shell is subject to hydrostatic pressure (p_0) on the outside only, the compression (i.e. the negative dilatation) is uniform throughout and equal to

$$\frac{a^3}{a^3-b^3}\cdot\frac{p_0}{\kappa},$$

where a, b are the external and internal radii.

2. Prove that a hydrostatic pressure p_0 in the interior of a spherical cavity of radius b in a large mass of metal produces an increase of the internal radius, of amount $\frac{1}{4}p_0 b/\mu$.

3. Prove that the elastic energy of a thick spherical shell subject to a hydrostatic pressure p_0 in the interior is

$$\frac{\pi b^3}{a^3-b^3}\left(\frac{a^3}{2\mu}+\frac{2b^3}{3\kappa}\right)p_0^2.$$

4. Prove that if a cylindrical tube, closed at both ends, be subject to an internal pressure p_0, the internal radius is increased by

$$\frac{b}{a^2-b^2}\left(\frac{a^2}{2\mu}+\frac{b^2}{3\kappa}\right)p_0,$$

and the external radius by $$\frac{ab^2}{a^2-b^2}\cdot\frac{2-\sigma}{E}p_0.$$

5. Find the greatest circumferential velocity of a steel disk of density 7·8, having a hole at the centre, if the hoop-tension is not to exceed 2×10^6 gm. per sq. cm. ($\sigma=\cdot3$). [175 metres per sec.]

6. Prove that the greatest tensile stress developed in a steel disk having a hole at the centre is about four-fifths of that in a thin steel ring rotating with the same circumferential velocity.

7. Prove that in the case of a hollow rotating shaft

$$p_1=\frac{3-2\sigma}{8(1-\sigma)}\rho\omega^2\frac{(a^2-r^2)(r^2-b^2)}{r^2},$$

$$p_2=\frac{3-2\sigma}{8(1-\sigma)}\rho\omega^2\left(a^2+b^2+\frac{a^2b^2}{r^2}\right)-\frac{1+2\sigma}{8(1-\sigma)}\rho\omega^2r^2,$$

$$E\epsilon_3=-\tfrac{1}{2}\sigma\rho\omega^2(a^2+b^2),$$

where a and b are the external and internal radii of the cross section.

EXAMPLES. XXVIII.

(Miscellaneous.)

1. A number of heavy particles are attached at equal intervals to a string, which hangs in the form of a regular polygon symmetrical about the vertical through the lowest particle w_0. If the weights of the successive particles on either side are $w_1, w_2, w_3, \ldots$, prove that

$$\frac{w_n}{w_0} = \frac{1+\cos a}{\cos a + \cos 2na},$$

where a is the external angle of the polygon.

Deduce the law of density of a chain in order that it may hang in the form of an arc of a circle.

2. Prove from the theory of the instantaneous centre that when the area of a convex quadrilateral of jointed rods is a maximum the quadrilateral is cyclic.

3. A rod is suspended by a string attached to the ends which passes over a smooth peg. The centre of gravity divides the length of the rod into two segments of lengths a, b. If θ be the angle which each portion of the string makes with the vertical, and ϕ the inclination of the rod to the vertical, prove that

$$\tan\theta = \frac{a-b}{a+b}\tan\phi.$$

If l be the length of the string, prove that

$$\sin^2\phi = \frac{(a+b)^2}{4ab}\left\{1 - \frac{(a-b)^2 l^2}{(a+b)^4}\right\}.$$

4. A hemisphere rests in the angle between a vertical wall and a horizontal plane, with its curved surface in contact with both. Its base is parallel to the horizontal line where these planes meet, and makes an angle θ with the horizontal. If μ be the coefficient of friction at each point of contact, the extreme value of θ is given by

$$\sin\theta = \frac{8\mu(1+\mu)}{3(1+\mu^2)}.$$

5. A pair of wheels of radius a are fixed to an axle, and are free to roll along the ground. A string wound round the axle (of radius b) leaves this on the under side at an inclination θ to the horizontal. Find the force and couple due to the reaction of the ground in order that the system may not move when the string is pulled with tension T.

If the reaction consists of a *force* only, in which direction will the wheels begin to roll?

6. A plane system of forces is equivalent to a couple M, and if the forces are turned through a right angle about the respective points of application they are equivalent to a couple N. Prove that if turned through a suitable angle θ they will be in equilibrium; and find θ. [$\tan\theta = -M/N$.]

7. A frame has the form of a regular pentagon $ABCDE$, with the diagonals AC, AD. Two equal and opposite forces P act outwards at B and E. Find by a diagram the stresses in the various members, and deduce numerically the stress in CD. [$\cdot 382\, P$.]

8. Find the mean centre of the volume included by the cylinder

$$x^2+y^2=2ax,$$

and the planes $z=0$, $z=mx$. [$\bar{x}=\frac{5}{4}a$, $\bar{z}=\frac{5}{8}ma$.]

9. OA, OB are two sides of a uniform plate in the form of a parallelogram, of lengths a, b, respectively, and include an angle α. Prove that the square of the radius of gyration about the diagonal through O is

$$\frac{a^2b^2\sin^2\alpha}{6(a^2+2ab\cos\alpha+b^2)}.$$

10. A solid hemisphere is pierced axially by a cylindrical hole of length h. Prove that the mean square of the distances of the elements of volume from the base is $\frac{1}{5}h^2$.

Also that the mean square of the distances of the elements of the curved surface from the base is $\frac{1}{3}h^2$.

11. A spherical segment of radius R rests on the top of a fixed sphere of radius R', and is free to roll but not to slide. The segment (of weight W) is rolled through a small angle θ; prove that the increase of potential energy is

$$\tfrac{1}{2}W\theta^2\left(\frac{RR'}{R+R'}-h\right),$$

where h is the initial height of the centre of gravity of the segment above the point of contact.

12. A uniform bar AB of weight W and length l hangs as in Fig. 109 (p. 137) by two equal crossed strings from two points C, D at the same level, such that $CD=AB$. Prove that if the bar be turned in a vertical plane through a small angle ψ from the horizontal position the increase of potential energy is

$$\frac{W(h^2-l^2)}{8h}\psi^2,$$

where h is the original depth of AB below CD.

13. A uniform bar of length $2b$ and weight W is suspended in a horizontal position from two points at the same level, by two strings of length l which make angles α, in opposite senses, with the vertical. If the bar be turned in a vertical plane through a small angle ψ, the strings remaining taut, prove that the increase in potential energy is

$$\frac{Wb(b+l\sin^3\alpha)}{2\sin^2\alpha\cos\alpha}.\psi^2.$$

14. A chain of variable density hangs in the form of a parabola whose axis is vertical. Prove that the tensions at any two points P, Q are to one another as TP to TQ, where T is the intersection of the tangents at P and Q.

15. A ship is at anchor in 5 fathoms in a tideway, and the drifting force on the ship is 1500 lbs. The anchor chain weighs 40 lbs. per fathom, and the anchor will not hold unless the pull on it is horizontal. Prove that the ship must have at least 20 fathoms of chain out.

16. A heterogeneous liquid fills an upright cylindrical vessel of sectional area A and height h, and ρ is the density at a height z above the base. Prove that if the contents were thoroughly mixed without change of volume of any portion the potential energy would be increased by

$$A\int_0^h \rho z\,dz - \tfrac{1}{2}Ah\int_0^h \rho\,dz.$$

17. A uniform log of rectangular section, of length l, floats partially immersed with two sides vertical, and carries a weight W at one end. Prove that the bending moment at a distance x from that end is

$$Wx\left(1-\frac{x}{l}\right)^2.$$

18. A perfectly flexible balloon contains a gas of total mass m. At the ground level it is at the same temperature as the surrounding air. Prove that it will exert the same lift at all heights if it remains at the same temperature as the air round it.

Prove that if the gas expands adiabatically, whilst the air is in convective equilibrium, the lift at height x will be less than at ground level by

$$mg\sigma\left\{1-\left(1-\frac{z}{H}\right)^{\frac{\gamma'-\gamma}{(\gamma-1)\gamma'}}\right\},$$

where σ is the ratio of the density of air to that of the gas under standard conditions, and H is the height of the atmosphere, and γ, γ' are the ratios of specific heats for the air and the gas, respectively.

19. A globule of liquid rotates about an axis with angular velocity ω, under no forces except the tension of the free surface. If the pressure at points of the axis is equal to that of the surrounding atmosphere, the equation of the meridian curve has the form

$$y^3 = c^3 \sin\psi,$$

where ψ is the inclination of the normal to the axis of rotation.

Prove that the curvature varies as y^2.

20. Prove that the form assumed by a horizontal beam is identical with that of a catenary whose line-density is such that the weight per unit length of the horizontal projection is proportional to the bending moment of the beam.

21. A uniform heavy beam of length $2a$ rests on two supports at the same level, at distances c from the centre. Prove that if

$$4\left(\frac{c}{a}\right)^3 - 12\left(\frac{c}{a}\right)^2 + 3 = 0$$

the centre and the ends will be at the same level.

Prove that this equation has only one relevant root, and find it approximately. [$c/a = \cdot 554$.]

22. A uniform beam of length l rests on two supports at the ends, at the same level, and carries a weight W at its centre. Prove that the strain-energy is

$$\frac{1}{EI}\left\{\frac{w^2l^5}{240}+\frac{5}{384}Wwl^4+\frac{W^2l^3}{96}\right\}.$$

Prove that this is also equal to

$$\frac{wl^5}{10240EI}+\frac{24EI\eta^2}{l^3},$$

where η is the depression at the centre.

23. The load per unit length on a horizontal beam supported at the ends is

$$\frac{6\,Wx\,(l-x)}{l^3},$$

where x is the distance from one end. Prove that the bending moment is

$$-\frac{Wx\,(l-x)}{2l^3}\{x\,(l-x)+l^2\}.$$

Draw the curves of shearing stress and bending moment.

24. Prove that in the case of a horizontal beam (not necessarily uniform) supported at the ends the inclination to the horizontal at a distance x from the end is

$$\int_0^x\frac{Mx}{EIl}\,dx-\int_x^l\frac{M\,(l-x)}{EIl}\,dx.$$

25. A light semicircular rod is clamped vertically at one end, and a weight W is suspended from the other end. Prove that the horizontal and vertical displacements of this end are

$$\frac{2\,Wa^3}{EI}\quad\text{and}\quad\frac{3\pi\,Wa^3}{2EI},$$

respectively.

26. A circular hoop is deformed by two equal and opposite forces P applied at the ends of a diameter, but is strengthened by a bar across the perpendicular diameter, initially unstrained, of extensibility λ. Prove that the thrust in this bar is

$$\frac{(8-2\pi)\,P}{\pi^2-8+4\pi\lambda EI/a^3}.$$

INDEX

[*The numbers refer to the pages.*]

CAMBRIDGE: PRINTED BY W. LEWIS. M.A., AT THE UNIVERSITY PRESS

Printed in the United States
By Bookmasters